Quantitative Investing

Lingjie Ma

Quantitative Investing

From Theory to Industry

Springer

Lingjie Ma
University of Illinois at Chicago
Chicago, IL, USA

ISBN 978-3-030-47201-6 ISBN 978-3-030-47202-3 (eBook)
https://doi.org/10.1007/978-3-030-47202-3

This Springer imprint is published by the registered company Springer Nature Switzerland AG
The registered company address is: Gewerbestrasse 11, 6330 Cham, Switzerland

To my wife, Jianping Zhuang

Preface

Is there a sound way to determine if the current US stock market is too hot? How can the relationship between the Chinese and US stock markets be evaluated through the use of data? Can one form a multi-factor stock selection strategy? What factors determine the price of oil and gold? Is the USD relatively weak compared to other currencies, if so what would be a proper strategy? What do most hedge/mutual funds do quantitatively to tackle the above issues? What are current innovative approaches? If one aspires to pursue a career in quantitative investments, what skill sets should he/she possess?

This book attempts to answer the questions addressed above based on a scientific approach. Before any more details are presented, I would like to emphasize that a successful quantitative investment strategy is not easy, but neither is it impossible. A successful quantitative investment strategy requires four pillars: a deep understanding of financial markets, investment theories, and econometric modeling as well as proficiency in programming that deals with real-world data sets.

A quantitative strategy is easy to fail if one of the four pillars is weak.

This book aims to provide readers a systematic approach on quantitative investing, step by step, from theory to industry. Each of the nine chapters begins with an investment topic, which is then followed by fundamental insights, explored by finance theories, analyzed by statistical models, and provided with quantitative results from real-world data. In each chapter, there is a section solely dedicated to industry insights, exposing readers to the real industry approach on specific issues. There is also an R programming section at the end of each chapter. The first seven chapters progress in order of increasing difficulty. The last two chapters introduce readers to the frontier approaches in finance: Chap. 8 focuses on the quantitative methodology side with an introduction to the distributional quantile regression, while Chap. 9 focuses on the investment side with an introduction to the quantamental approach which gained recognition after the recent financial crisis.

Throughout the book, within each chapter, we attempt to address real-world issues in investments, where typically, data sets are dirty, the conditions of quantitative methods are not valid, and the asset and portfolio performance do not coincide with finance theories. However, this does not mean finance theories are not beneficial, but rather, they provide a general framework and guidance; this does not mean we cannot use those quantitative methods, but rather they are much more useful when we are aware of the weaknesses and conditions and are more cautious on interpretation; this does not mean we should simply disregard the dirty data, but rather we need to gain a better understanding of the data and have a scientific analysis. Correspondingly and constantly, the book raises the following questions in various forms for readers to contemplate: How are financial theories applied to real-world investments? How does one build econometric models to analyze finance data for investment decisions? How does one tackle real-world data when they are very dirty containing outliers? How does one put all the pieces together implemented by a programming language (e.g., R)?

While the book will not provide all the answers, it will attempt to provide solutions for the following points: (i) Motivate readers on critical thinking: understanding fundamentals about financial markets and limits of finance theories. (ii) Teach readers to master the hard-core quantitative methodologies: not only know what they are and how to use them, but also know the assumptions and conditions. (iii) Foster the ability to tackle real-world industry finance data. This will involve big data analysis and programming to apply quant analysis to solve investment problems efficiently. (iv) Provide a hands-on approach with exposure to industry insights and learn to make sound investment decisions. (v) Introduce frontier approaches: quantile regression and quantamental investments.

I hope this book will fill the gap between the academia and the industry, providing a practical guidance on quantitative investing for both students and professionals.

I would like to take this opportunity to thank Gib Bassett who encouraged me on the book and my Ph.D. advisor Roger Koenker, who always helps and supports me on academic projects. I am extremely grateful to Laura Briskman, my editor for professional excellence, numerous suggestions, and great help. I would like to thank the companies I worked for, PanAgora Asset Management, Deutsche Asset Management, BMO Global Asset Management, and Northern Trust Global Investments. I thank my colleagues Larry Pohlman, Weidong Li, Brian Bruce, and Rosy Macedo, to name a few. I would like to express my appreciation to my current employer, UIC, for generous support and encouragement, from both faculty—Mike Mikhail, Lan Zhang, and Andriy Bodnaruk—and many graduate students in the MBA, MSF, and Ph.D. programs. I would like to specially thank Chung-Ming Kuan, Kevin Hallock, Zhijie Xiao, Bart Trescott, and Xuming He for encouragement and help. I thank graduate assistant Peter Royal and student worker Aurora Priego for proofreading the book.

And finally, my heartfelt appreciation to my family. Thank you, Ellen, Abby, and Rachel, for numerous dinner-table discussions about the book.

Chicago, IL, USA
December 28, 2019

Lingjie Ma

Contents

About the Author

Dr. Lingjie Ma has 15 years of experience developing global multi-asset investment strategies. He has worked as both a head of research and as a portfolio manager in the investment industry, overseeing full-spectrum investment process and business management. He joined the University of Illinois at Chicago in 2016 as a clinical associate professor in finance. Dr. Ma is a frequent public speaker on quantitative investing and quantamental strategies.

Chapter 1
Introduction

Abstract In this chapter, we introduce major concepts and layout of the book. The industrial practice of quantitative investing emerged in the 1960s developed rapidly in the 1990s and became on par with traditional fundamental investing in the 2000s. Naturally, mathematical and statistical modeling plays a critical role in quantitative investing. Major factors contributing to the development of quantitative strategies are the rise of modern investment theory, large-scale use of computers, greater availability of data sets, and development of programming languages. While building a successful quantitative strategy is not easy, it is achievable and requires a fundamental understanding of the market, strong knowledge of investment theory, mastery of quantitative methodologies, and proficiency in data analysis and programming. These constitute the four pillars of successful quantitative investment described in this book. In this chapter, we introduce quantitative investing and present a hall of fame for modern investment theory and quantitative methods. We then share industry insights with readers and briefly introduce R programming.

1.1 Can You Outperform the Market?

In general, the goal of a quantitative strategy is to outperform the market. Outperforming the market means doing better than the market average. It happens when your stock portfolio does better than the 7–10% annual average return of the US stock market over the last 50 years.[1]

Consider a long-only portfolio (that is buy and hold). We could employ a major index in the market as a benchmark, such as the S&P 500 index in the US stock market. In mathematical terms,

$$R_b = \sum w_i^b r_i, \qquad \sum w_i^b = 1, \quad w_i^b > 0,$$

where w_i^b is the benchmark weight and r_i is the return of security i. Similarly, for a portfolio with a group of selected stocks and weights, we have the portfolio return,

[1] We discuss the US stock market's performance in Chap. 2.

L. Ma, *Quantitative Investing*, https://doi.org/10.1007/978-3-030-47202-3_1

$$R_p = \sum w_i^p r_i, \qquad \sum w_i^p = 1, \quad w_i^p > 0,$$

where w_i^p is the security i's weight in the portfolio. If we can pick companies that tend to deliver higher than average returns, the portfolio will outperform the market, $R_p > R_b$.

$$\begin{aligned} R_p - R_b &= \sum w_p r_i - \sum w_b r_i \\ &= \sum (w_p - w_b) r_i \\ &= \sum_{r_i >= R_b} (w_p - w_b) r_i + \sum_{r_i < R_b} (w_p - w_b) r_i, \end{aligned}$$

where we drop the letter i for notational convenience. Thus, outperformance could be achieved in two ways: we could overweight the stocks with better than market returns or underweight those with less than market returns or both.

To explore this further, we examine fund performance across equity and fixed income asset classes in both US and international markets. Let us first take a look at the performance of US equity funds managed by professional investors in the retail space: the mutual funds. The S&P Dow Jones Indices (Liu et al. 2017) tracked outperforming funds that beat their benchmarks based on 3-year annualized returns, net of fees. The study then examined what percentage of "winning" funds continued to outperform during the subsequent 3 years from March 2000 through September 2016. The results are summarized in Table 1.1. During these years, about 25–30% of mutual funds beat the market (after subtracting management fees), and about 10–15% of these companies continued to outperform the market for 5 years, which is remarkable! This percentage is far from trivial.

If we assume the market returns follow a normal distribution, there should be the same number of funds on the left and right side of the mean return. That is, at most half can beat the market. So, 50% is a maximum in theory! The data clearly show that first, yes, there are funds that can beat the market; and second, it is very challenging to consistently outperform the market!

Table 1.1 The percentage of mutual funds that outperformed their benchmarks from March 2000 to September 2016

Fund category	Benchmark	Prior 3 Yr	1 Yr	2 Yr	3 Yr
Large-cap	S&P 500	30.83%	33.93%	13.62%	5.17%
Mid-cap	S&P MidCap 400	25.65%	30.39%	10.35%	3.24%
Small-cap	S&P SmallCap 600	30.58%	35.25%	13.30%	4.60%
International	S&P 700	23.89%	36.68%	15.56%	6.88%
Emerging markets	S&P/IFCI Composite	25.24%	38.39%	15.48%	5.22%

Source: S&P Dow Jones Indices

The statistics in Table 1.1 are for mutual funds that do not include institutional managers, which are usually regarded as the most sophisticated professional managers. We present below the institutional funds' performance relative to the market in both the US and international spaces across equity and fixed income asset classes. Table 1.2 lists the percentage of institutional funds that outperformed the benchmark (index) over 1-, 3-, 5-, and 10-year periods. The top panel is for the US stock market, and the bottom panel is for the international stock markets. We see that first, for the US equity funds, regardless of size (large-, small-, or mid-cap), the percentage of funds outperforming their benchmarks is about 45–60% for a 1-year period, 25–30% for a 3-year period, and 25–35% for 5-year and 10-year periods. Second, in the international equity markets, institutional managers do a much better job than mutual fund managers: the percentage of outperforming funds in emerging markets is about 30–55% over 3-, 5-, and 10-year periods and about 40–50% in developed markets. This is phenomenal!

For the fixed income asset class, Table 1.3 lists the percentage of institutional funds that outperformed their benchmarks over 1-, 3-, 5-, and 10-year periods in the US markets (top) and international markets (bottom). Within each panel, there are funds for corporate bonds, government bonds, and high-yield products. Consider the US fixed income market first. We see that for MBS, the percentage of outperforming funds is very high: about 90% for 1- and 3-year periods and 70–80% for 5- and 10-year periods. For other US fixed income products, the outperforming percentage is about 40–50% for 3-, 5-, and 10-year periods. In the international space, about 20–60% of fixed income funds outperform their respective benchmarks over 3-, 5-, and 10-year periods.

So again, we see the same picture from the institutional funds as for the mutual funds: it is possible to beat the market, but it is quite challenging.

Table 1.2 The percentage of institutional equity funds that outperformed their benchmarks

Fund category	Comparison index	1 Yr	3 Yr	5 Yr	10 Yr
U.S. equity funds					
All Domestic Funds	S&P Composite 1500	60.25	70.22	70.15	59.88
All Large-Cap Funds	S&P 500	55.72	69.20	69.41	62.88
All Mid-Cap Funds	S&P Mid Cap 400	34.77	71.39	61.31	77.01
All Small-Cap Funds	S&P Small Cap 600	44.90	74.15	74.26	72.92
International equity funds					
Emerging Market Funds	S&P/IFCI Composite	4.81	43.24	46.88	48.74
Global Funds	S&P Global 1200	41.91	58.12	54.31	59.79
International Funds	S&P International 700	55.88	51.14	44.15	57.71
International Small-Cap Funds	S&P Ex-U.S. Small Cap	35.80	27.78	25.00	32.69

Source: S&P Dow Jones Indices

Table 1.3 The percentage of institutional fixed income funds that outperformed their benchmarks

Fund category	Comparison index	1 Yr	3 Yr	5 Yr	10 Yr
U.S. FI funds					
Investment-Grade Funds	Barclays U.S. Credit	26.73	20.46	21.83	26.75
Government Funds	Barclays U.S. Government	60.61	58.10	60.00	66.09
High-Yield Funds	Barclays U.S. Corp. High Yield	56.33	67.94	65.10	77.33
MBS Funds	Barclays U.S. Aggregate MBS	9.04	11.11	21.76	30.08
Municipal Funds	S&P National Municipal Bond	64.39	62.80	59.45	63.16
Inflation-Linked Funds	Barclays U.S. TIPS	41.03	47.83	46.00	64.44
International FI funds					
Emerging Market USD Funds	Barclays EM USD Aggregate	16.13	61.40	69.39	50.00
Global Credit Funds	Barclays Global Aggregate Corp.	50.62	35.90	44.26	38.10
Global Government Funds	Barclays Global Treasuries	65.12	59.18	57.89	51.22
Global High-Yield Funds	Barclays Global High Yield	70.30	81.52	63.53	81.25

Source: S&P Dow Jones Indices

1.1.1 Why Is It So Challenging to Beat the Market?

Financial markets have become more and more efficient, which means that it is increasingly difficult to consistently outperform the market. Increasing efficiency arose from various sources, some major ones are greater access to information, technological advancement, and fewer investment barriers.

Greater Access to Information First, among institutions, everyone has the same access to the same information. Second, the gap in terms of information access between individual and institutional investors has become much narrower recently compared with 10 or 20 years ago. Third, the vendors who provide products of data, alpha, or risk models have made the utility of information more uniform for different investors. There are many highly competent people with very different skills who try to understand not only the market and companies but also the reactions of other strong players in the market. A new discovery of abnormal returns from competitive edge, the so-called alpha, becomes beta (index performance) fairly quickly due to technology and professional turnover across companies. In addition, public companies try to play the game with investors as well. In the 1950s, very few companies bought back shares. In the 2000s, more than 50% of companies in the S&P 500 index performed share buybacks, partially because professional investors view share buyback as a positive signal for stock returns.

Impacts of Technological Advancement Information spreads faster and more widely due to technological advancements. Internet has made information accessible simultaneously to investors. Computer's capacity to store and process large volumes of data has increased as well with greater speed and lower cost. Many financial service companies and trading platforms employ advanced technology, such as cloud computing to make information access global and efficient.

Fewer Barriers to Investment As for barriers to investment, here we refer to anything physically structural that prevents the market from being efficient and complete. The structural barriers to investment span legal, policy, and cultural aspects. Better regulation, increased market integration, strengthened public capacities, and improved access to capital markets all help to counter investment barriers and secure a better economic future. Relative to 10, 20, and 30 years ago, there are fewer investment barriers today. For example, financial filing, settlement, and public disclosure rules have eliminated barriers or made barriers less severe for investment.

Human Capital Flow and Professional Turnover Investment professionals are in a continuous flow from one company to another, vying for better positions with higher pay, better work environments, or promotion. Investment companies are also eager to learn key information about outperforming products offered by peer companies.

1.1.2 Conditions for Persistent Outperformance

Though challenging, there is still a significant (though small) percentage of portfolios that outperform the market. There must be systematic reasons. Mispricing and inefficiency exist in financial markets. Those with skillsets to identify these investment opportunities can potentially harvest higher returns than the market, thus achieving outperformance.

Sources of Inefficiency Inefficiencies do exist, which provides fundamental support for the possibility of outperformance. Where are these inefficiencies? How do they arise? Will they recur consistently? Sources of inefficiency include but are not limited to: Companies, Investors, Professionals, Regulation agents, and Non-economic structural barriers. More details will be discussed in Chaps. 4 and 5.

Identify and Capture Inefficiencies How can such inefficiencies be detected? How can investors realize them in an outperforming portfolio? As we discussed in the preface, there are many reasons for a quantitative strategy to fail, but a successful strategy requires the following pillars: Deep understanding of the financial market, strong knowledge of investment, disciplined investment process, strong execution capability. This will be stressed throughout the book.

Of course, Rome was not built in a day. Persistent outperformance requires continuous collective efforts:

- Efforts
  These include, but are not limited to, the hard work to dig into fundamental details, the mastery of quantitative methodologies, and continuous due diligence examination of companies.
- Collective
  Can you outperform an index as an individual? In general, the answer is no: as an individual, you cannot outperform the market consistently over time. What

about a strong team of professional investors? Even as a team, it is hard to beat the market consistently over time. This is why even highly skilled teams in the market employ several different strategies in case one does not work.

- Continuous
  The market changes all the time. Sometimes a change is temporary, and sometimes it is structural, meaning its effects will last longer. A very dynamic financial market requires investors to continue their collective efforts, not only to understand what is going on and adapt, but also foresee changes in the industry or recognize upcoming trends in the economy, etc.

Quantitative investing requires a solid background in quantitative methodologies and skills in data analysis with computer programming languages. We introduce quantitative investing in the following section.

1.2 What Is Quantitative Investing?

Quantitative investing employs quantitative methods and computers to analyze historical data to identify market inefficiencies and construct portfolios through an investment process. It is based on two principles: the law of large numbers and history repeats itself. Quantitative investment analysis traces its origins to the book *Security Analysis* (1934) by Benjamin Graham and David Dodd, in which the authors advocate detailed analysis of objective financial metrics of specific stocks.

Besides Graham, some influential academic scholars are also the early pioneers of quantitative investing. Financial market historians ascribe the beginning of quantitative investment strategies to the seminal work of portfolio theory by the economist Harry Markowitz in 1952. For the development of quantitative investing, other important figures from the academic side are William Sharpe for his work on risk models and Eugene Fama for return forecasting through multi-factor models. We present a brief history of quantitative investing in the following section.

1.2.1 History of Quantitative Investing

To help people invest with a disciplined approach, quantitative investing emerged in the 1960s. It advanced in the 1980s and gained momentum in the 1990s.[2] We divide the development of quantitative investing into four stages based on the following characteristics: (1) investment ideas and strategies; (2) data and technology; and (3) market environments and asset levels. We illustrate the history of quantitative

[2] I am very grateful to Dr. Ronald Kahn for the data and terms in his talk "Quant Investing: Past, Present, Future" in the 26th Annual Investment Seminar in London on September 10, 2012.

investing through quantitative equity strategies, which represent both the origination and the most developed area of quantitative investing.

1. Emergence: 1960s–1980s
   Quantitative investing began to emerge in the 1960s and was initially applied by a small set of industry practitioners. Pioneering thinkers such as Graham (1949), Graham and Dodd (1934), Markowitz (1952), and Sharpe (1964) developed fundamental concepts and theories in quantitative investing, including value investing, risk measurement, and portfolio construction. Meanwhile, industry pioneers such as Warren Buffett and Barr Rosenberg applied quantitative investing principles in money management. Index funds started during this period, risk models started to emerge, data vendors collected data (e.g. CRSP and Valueline) for both academia and industry, which enabled backtesting of investment ideas over a fairly long period of time. Technology, especially computers, was critical at this stage. PCs were introduced in the 1970s, enabling data storage and analysis by statistical languages such as S, SAS, and Matlab (1984).

   During this period, the industry started to see an influx of large amounts of money into model-driven products, with assets totaling about USD 3–4 billion by the end of the 1980s. Almost all quantitative equity portfolios at this time were long-only, using factors of value, profitability, etc. Quantitative strategies started to grow slowly but were still in their infancy. Indexing began in the 1970s. Quant investing started in the USA around that time, and used benchmarks (indices) for performance comparison. By the end of the 1980, quant investing had expanded from the USA to other developed economies, such as Japan, the UK, and Canada.
2. Development: 1990s
   In the 1990s, quantitative investing strategies gained recognition and large-scale industry development. In academia, asset pricing became an important topic. Researchers had been searching for factors based on pricing, such as the three-factor model by Fama and French (1992) and price momentum by Jegadeesh and Titman (1993). Research on fundamental ratios—particularly linking companies' financial filings to stock prices—expanded from finance to accounting faculties' research agendas, such as accruals-based earnings quality by Sloan (1996). Besides pricing and fundamental ratios, a new line of factors based on analysts' estimates appeared, such as earnings momentum, surprise, and diffusion. Practitioners in the quantitative industry started to use multi-factor models with various alpha sources to forecast returns, and a common investment process evolved. The risk management industry became large, with vendors of risk models like Barra and Northfield, which provide data, optimizers, and backtesting tools.

   Regarding data and technology, more data sets spanning longer periods became available. New data sets based on analysts' estimates were added to the industry, such as IBES and First Call. CompuStat also started to replace Valueline with better scope, accuracy, and scale. Many data sets on the US market went as far back as the 1970s. The rise of the Internet enabled more rapid dissemination of general scientific research.

During this period, quantitative investment advanced as a scientific subject. Wall Street hired many graduates in physics and mathematics. Quantitative equity products generally delivered consistent performance over this period through anti-crowd (fundamental products) bets. Money flowed into this new investing method, and quantitative assets grew to about $80–$100 billion, a significant portion of the entire money management business. Long-short strategies and high frequency firms grew over this period.

3. Rapid growth: 2000s

   Quantitative equity products are scalable because they are model- and process-driven. During this period, behavioral economics slowly gained acceptance, allowing academics to research potential market inefficiencies, such as herding and overreaction effects. Text mining methods and machine learning techniques also emerged. Given the momentum and solid performance of quant products (and hence the influx of investment and revenue), many similar products were launched very quickly by many companies. The number of quantitative shops increased dramatically. Meanwhile, academic and industry ideas—the so-called alpha factors—spread very quickly due to the accessibility of data, technological advances, and professional turnover. As a result, many new ideas were arbitraged away, and alpha quickly became beta.

   During this period, institutional money managers became the representative focus of quant equity products; they ran the most sophisticated quantitative strategies and were usually regarded as the thought and practice leaders in the field. Many new products were launched, including portable alpha, 130–30, market-neutral long-short, macro, and event-driven, many of which had leverages. Offering a whole suite and innovative products for investors, quant asset management companies are very competitive with traditional fundamental shops. Many pension funds, large and small, allocated significant amounts to quant products during this period. Overall, there was a dramatic influx of money into quantitative equity products, and assets under management reached about $1.5–2 trillion by the end of 2000.

   Regarding data and technology, granular industry-level data became available, such as SNL for financials. Data on international markets, both developed and emerging, also became available, with improving quality—broader coverage, better quality, and longer history. Technology made computation no longer a significant cost factor, and many quant shops established a separate IT team to deal with data and provide technology support (Fig. 1.1). Proprietary data, either from data vendors or self-conducted surveys, emerged during this period.

   Investment process became increasingly modular in terms of both process and team functions. A typical quant shop had teams for alpha and risk models, portfolio construction, rebalance and trading, and products and marketing.

   During this period, the equity markets had low volatility. Many quantitative factors worked well partially because crowded betting using common factors artificially inflated their efficacy. In short, quantitative investment strategies enjoyed a period of significant growth and perhaps overconfidence. This was the case until the subprime mortgage crisis, which sent an early warning to the

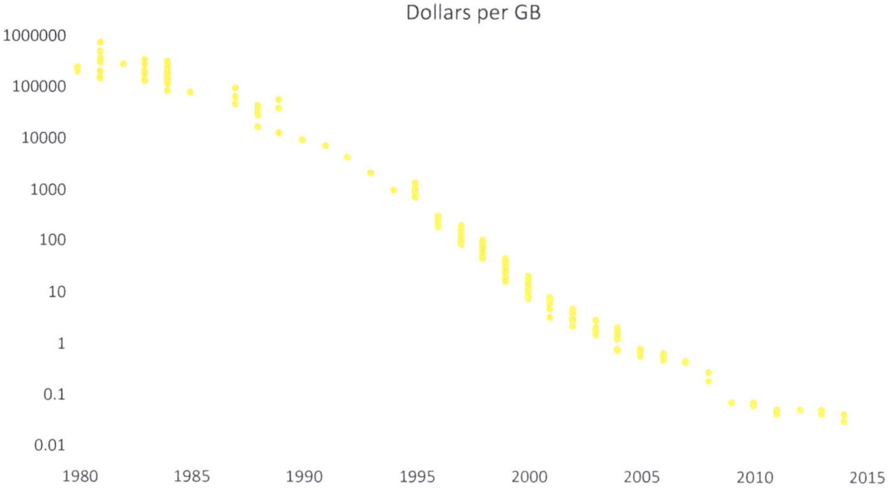

Fig. 1.1 Hard drive cost per gigabyte from 1983 to 2015. Data Source: Matt Komorowski

market about the pitfalls of quantitative strategies. There was a major downturn in quantitative investing during and after the financial crisis.

4. Reshaping: 2010s
During and after the financial crisis, the number of quantitative strategies decreased dramatically due to a 40% decline in market and mass withdrawals of money. Many people lost confidence in active quantitative portfolio products, which did not protect their wealth during the crisis. Assets under management of quant funds, measured by F13 filings, dropped by 75%!
One positive outcome of the crisis was that researchers and portfolio managers started to reconsider model based strategies. In light of overcoming the limits of quantitative approaches, there have since been three trends in quantitative investing:

- Learning from fundamental approaches.
A new investment approach—quantamental investing—emerged, which sought to combine quantitative and fundamental practices so as to add fundamental depth to the breadth of quantitative strategies. Interestingly, this was a mutual process: quants wanted fundamental inputs, but the fundamental portfolio managers also wanted to incorporate quant processes into their products. We discuss this new frontier approach in detail in Chap. 9.
- Quant becomes passive.
Smart beta products emerged during and after the financial crisis. Quant processes and optimizers are natural tools to provide such products. Smart beta products are characterized by low volatility or beta. The smartness comes from quality factors which are added to the portfolio construction process. These products are usually provided to investors in mutual funds at low cost

(as vehicles of exchange traded funds (ETF)). The percentage of smart beta quant products in mutual funds increased from under 5% to about 20% over the past 10 years. We will discuss this in detail in Chap. 7.
- New area of growth.
  Quant strategies found a new sweet spot after 2010. There has been a rapid growth of quant funds in the hedge fund space. During the last 10 years, quantitative hedge funds enjoyed 17% annual growth.

Recently, in terms of data and technology, big data and proprietary data have gained more favor, and machine learning and artificial intelligence have added value to alpha discovery, risk management, and trading cost and efficiency. These are the new frontiers of exploration and competition, though the results are yet to be seen. Perhaps there will be another cycle of quant strategies, but after each cycle, quant strategies will drive the market to become more efficient, and quant explorations will spread to other developed and emerging markets.

1.2.2 Quantitative Versus Fundamental Investing

Since the emergence of quantitative analysis, there has been an ongoing debate about the performance of fundamental and quantitative approaches. Each method has its advantages and disadvantages.

The fundamental approach analyzes the basics of business. When you buy a share, it is as if you were buying a piece of the company. To figure out the intrinsic value of the stock, you could start by looking at the company's financial information to see how much the whole company worth. The fundamental approach usually involves a screening or filtering process for stock selection. While the fundamental approach is transparent and easy to understand, it has limits, too. First, personal judgment is very important. A lot of fundamental information is "fuzzy" or "squishy," meaning it is often up to the personal interpretation for its investment significance. Secondly, even for a seasonal investor, the number of companies a person can cover is very limited.

To overcome the limits of the fundamental approach discussed above, the quantitative approach has emerged with the development of econometrics and computer technology. A quantitative approach is usually based on historical data, econometric models, and an investment process. Thus, comparing with a fundamental approach, it has several advantages, including greater breadth, more objectivity, less susceptibility to cognitive errors, and more effective implementation. However, we should recognize that a quantitative portfolio has limits too. For example, a lot of pricing information is hard to quantify.

We need to stress here that fundamental and quantitative investing are two distinct approaches but *not* mutually exclusive. Many investors are aware of this point, and some quant shops have started to adopt an approach that takes advantage of both methods. We further discuss this combined approach, quantamental investing, in the last chapter of this book.

In summary, some advantages that quantitative managers have relative to more traditional managers are a consistent and repeatable investment process, transparent risk control, and effective implementation. In the following section, we present a general process of quantitative investing in industry.

1.2.3 The Quantitative Investment Process

The quantitative investment process usually starts with a strategy designed for a particular investment universe. The strategy can come from either investors or clients. The most important components of a strategy are the target return and risk levels. The investment process is then designed to achieve specified target return and risk levels.

Strategy with return/risk targets

- Alpha model
- Portfolio construction with constraints and risk control
- Trading and rebalance
- Performance attribution

Strategy with Risk and Return Targets An investment strategy will define the main characteristics of the fund, such as long-only or long-short, target return and risk levels. A strategy is usually set up by clients or investors themselves. Other important information in a strategy includes the fee structure, such as management fees and/or performance fees. Some clients specify more details, such as overweight or underweight limits, number of stocks, industry preferences, specified risk profiles, turnover, and account management.

Alpha Model An alpha model serves to generate forecasts of future returns of the securities in an investment universe. A very common approach to build an alpha model is through a multi-factor framework, with each factor capturing certain information about the future returns of the assets in the selected investment universe. We discuss in detail how to build an alpha model in Chaps. 4–6.

Portfolio Construction Once we have the risk and returns (forecasts), we can construct a portfolio to maximize returns given the risk level defined in the strategy mandates, together with other constraints. Optimization is often achieved by an optimal objective function, with all targets and constraints expressed in mathematical terms. The constrained solution will produce the optimal set of portfolio weights for the securities in the investment universe. We discuss details of portfolio construction in Chap. 7.

Portfolio Rebalancing and Trading Once we have optimal portfolio weights, we can start to buy and sell securities to realize the optimal portfolio. Trading is part of portfolio rebalancing, as this is designed to adjust portfolio weights from the prior optimality to the updated optimality. A trading list is the securities with transaction amount equal to the difference between the optimal weight targets and existing security weights.

Performance Attribution Trading is not the final stage. We need to monitor the portfolio consistently over time. In particular, we need to understand and evaluate portfolio performance, including risk, returns, and constraints. Performance attribution helps us understand these profiles and provide feedback for refinements to the investment process. This will be addressed in Chap. 7.

1.2.4 Quantitative Investing: Information and Data

Based on the discussion above, we understand that quantitative investing relies on quantifiable information. For example, to build an alpha model in the investment process, we need to construct signals for each investment theme (factor in a model). A theme can comprise many signals, and each signal will measure that theme from different angles. In quantitative investing, the information for signals can be derived from three categories: financial reports or fundamental performance, pricing information or technical information, and macro-level information. Each category contains subcategories, which we present below.

- Financial reports
  - income statement
  - cashflow statement
  - balance sheet
- Pricing information
  - price
  - trading volume
  - dividend
- Macro-level information
  - regulations
  - industries
  - region/country

Regarding financial statements, all public companies in the USA are required to file quarterly reports to the SEC within 45 days of the fiscal quarter end, with large companies filing sooner than small companies. The three financial statements—balance sheet, income statement, and cashflow statement—have different functions and focus on different areas of the business performance of a public company. In terms of contents, income statements present change values, cashflow items present cumulative values, and the balance sheet presents stock values.

Pricing information mainly includes price, trading volume, and dividend, as the latter is related to total price (price plus dividend). Price is a continuous variable during trading hours. There are large volumes of information on pricing due to high frequency trading, especially for large companies. Trading volume can be measured by either shares or dollar values (number of shares multiplied by price). Dividends can be very important for dividend-issuing companies, particularly in certain industries, such as banking, utilities, and energy, where dividend issuance is a tradition, particularly in early years. Note that dividend policy is in the control of the board, not the CEO, and dividends have tax benefits in many countries.

While both financial reports and pricing information are at the company level, there is another set of information at the macro level. We can classify this type of information as regulation, industry, and region/country. Regulations applied to certain companies have special effects on the market. For example, regulation limiting investing activities for banks or a price ceiling for utility companies can result in off-balance-sheet activities and preset a profit margin immediately.

Also, given that different industries operate very differently, and each industry has its own special features, industry-level signals will be very different across industries. For example, when the price of oil goes up, this will impact the energy, utility, and airline industries more than other industries. Another example is interest rates, which are important for all companies, but banks will be more sensitive given that their core business is based on deposits and loans with interest differences.

At the country level, it is easy to understand that given legal, cultural, political, and economic differences between countries, the set of macro factors differs and, moreover, the same set of company-level information will have different effects on price movements. For example, momentum factors usually do not work well in Japan, as the Japanese economy has long been in stagnation. Another example is cash flow, which is very significant in the US market, but a general cash flow signal does not have much forecasting power for stock returns in the Japanese equity market because most Japanese companies are cash rich due to their family-controlled structures and interrelated board members.

In the following section, we briefly introduce our "Hall of Fame" for this book, a vivid illustration of modern investment theories from the last 100 years.

1.3 Hall of Fame: A Century of Modern Investment Theory

In this chapter, we explore important contributions to the development of modern investment theory. Based on four criteria—*impact, success, efficacy, and modernness*—we highlight several key figures for our hall of fame, spanning a century of modern investment theory and practice.

- Impact: degree of influence on the investment world in terms of both practice and theory development.

- Success: the degree to which a theory/hypothesis has successfully predicted performance in real-world scenarios.
- Efficacy: how effective a contributor has been in improving understanding of the financial markets and enhancing portfolio performance.
- Modernness: the contribution was made within the last 100 years. We choose this period because it begins after WWI, when modern economies emerged.

We compile a list of figures below with major contributions to quantitative investing.

- Benjamin Graham (1894–1976): value investing, margin of safety
- Warren Buffet (1930–): value investment and fundamental numbers
- Harry Markowitz (1927–): efficient frontier, mean-variance optimization
- William Sharpe (1934–): CAPM, Sharpe ratio
- Stephen Ross (1944–2017): APT model, binomial distribution
- Eugene Fama (1939–): MEH, 3-factor model with Kenneth French (1954–)

Note that this is just one version of Hall of Fame for quantitative investing, which are based on the relevance to the contents of this book. We introduce these figures one by one, with a focus on their major contributions to quantitative investing.

Investment Should Be a Science What is an investment? What is the difference between speculation and investment? Is there a sound scientific approach to investment? These fundamental questions needed solid answers at the beginning of the modern investment industry. To address this need, Benjamin Graham proposed foundations for scientific investment, Warren Buffett successfully practiced those principles through real money management. Both figures have made deep impacts on quantitative investing. While their work provided a great start for quantitative investing by demonstrating that it is possible to achieve outperformance relative to the market through deep understanding of financial markets, careful analysis of companies and their stock performance, adoption of a systematic approach (rather than personal feeling), to make quantitative investing a large-scale industry-wide standard practice requires further research and study.

Investment Is a Science To establish investment as a real science, or to establish a modern theory of investment, many questions needed to be researched and answered. First, since investment is risky, how can we define risk quantitatively? Supposing we know the risk and expected return, what is the best way to formulate a portfolio? Economist Harry Markowitz explored and answered these questions, paving the foundation for modern investment theory. Markowitz defines expected return for a security as the mean of its returns and risk as the standard deviation of its returns. Given the values of risk and expected return, how can we select securities to build a portfolio: the ones with the highest returns or the ones with the lowest risk? It turns out the solution should combine the two: given a risk level, select securities with the highest expected returns, or given a return level, select the ones with the lowest risk level. The combination of these selected securities thus forms a theoretically achievable optimum for a given financial market, the so-called efficient frontier.

In practice, how can we achieve the efficient frontier quantitatively given a set of securities in a certain market? Markowitz proposes the mean-variance approach: if we put the expected return and risk in an equation such that we maximize expected return given a certain level of risk, then we could solve this optimization problem and obtain an optimal portfolio.

Note that Markowitz defines risk and expected return, but he takes both values as given and does not define the relationship between risk and expected return. People have long guessed that there should be some relationship between return and risk. In general, return should be higher to compensate for higher risk, although the relationship may not be linear. Economist William Sharpe proposes a quantitative formula, the capital asset pricing model (CAPM), to capture the relationship between risk and return. The CAPM states that the expected return of a security is the risk-free return plus compensation for the over-market risk, where the over-market risk is measured by a term, β. So, β measures the sensitivity of a security's return to the market. A high value of β means higher volatility and hence higher expected returns according to the CAPM.

However, in reality, portfolio performance depends on future returns, and we do not know the market return or a security's return based on the CAPM. While risk can be measured by standard deviation or beta using historical returns, the performance of any investment depends on predicting future returns or price movements. Unfortunately, we cannot just use historical returns to assume future returns because financial markets change all the time. A scientific solution to forecasting future returns is key for modern investment theory and practical quantitative investing.

Quantitative Investing with Return Forecasting Is there a way to forecast "expected return?" What factors should we use to predict expected return? Several generations of economists, finance theorists, and industry practitioners have explored and answered these questions, with economists Stephen Ross and Eugene Fama among the most famous contributors. We should note that many others have made significant contributions to modern investment theory. However, given the concentration of this book, we focus on introducing basic concepts and theories proposed by Ross and Fama.

Note that the CAPM employs only one factor, the market return. As a further development of the CAPM, economist Stephen Ross proposed arbitrage pricing theory (APT) in 1976. APT assumes that markets sometimes misprice securities. Before the market eventually adjusts and securities return to fair values, arbitrageurs can take advantage of any deviations from fair market value. In the early years of APT, the following factors were explored and regarded as classical factors: gross domestic product (GDP) growth, inflation rate, gold price, the S&P 500 index return, and the risk-free rate. These factors clearly reflect the economic climate of the 1970s and 1980s, with high inflation, a good GDP growth rate, and gold as a hedging tool against inflation. In terms of quantitative investing, the contributions of APT are twofold: first, it states that the market can be inefficient, where securities are mispriced from the fair value; second, it provides a multi-factor framework to analyze systematic risk and security returns. The multi-factor framework of APT is used as the basis for many commercial risk systems used by quantitative investing. However, the limit of APT is that it focuses on systematic risks, while individual securities' prices are determined by many factors in addition to the macroeconomic situation, such as profitability and management quality.

While APT focuses on systematic macro-level factors, many studies have explored micro-level factors, particularly fundamental factors, such as value, size, and beta in the three-factor model of Fama and French (1992). The stock price of each individual company is related closely to its business performance, expected value, market sentiment, and many other factors. In addition to the FF three-factor model, there have been many other studies on factors at the company level, such as Jegadeesh and Titman (1993) on momentum, and Sloan (1996) on earnings quality. Industry practitioners have also made significant contributions, such as GARP (growth at a reasonable pace). These studies paved the foundation for security selection models because they address the ability to differentiate stocks with potentially different return forecasts given the macro situation. Most studies rely on quantitative analysis, and thus each contributes to quantitative investing.

1.4 Quantitative Methods

In the preceding section, we presented a broad overview of the development of modern investment theory, which is one pillar of quantitative investing. Another important pillar is the quantitative methods whose purpose is to apply econometrics to analyze data and formulate a portfolio based on investment theory and fundamental intuition.

We should stress here that while there are many advanced econometric methodologies for analyzing finance data, we should generally adhere to the more basic ones because more advanced methodologies require more assumptions for validity. Robustness of methodologies and models is especially important in quantitative investing, given that finance data are "polluted" by human beings, far from being "normal" or "natural" like data sets in other fields.

In this book, we begin our coverage of quantitative methods with univariate analysis, followed by bivariate and multi-factor models. Univariate and bivariate models help us understand each variable and the relationship between variables, which are very important preparations for multi-factor analysis. During our presentation of quantitative methods, we link the analysis with investment using real-world finance data.

Different quantitative investing strategies require different methodologies. For a stock selection strategy, most analytical work will focus on cross-sectional studies, whereas for commodities and currency, a significant portion should be devoted to time series analysis. Finally, we introduce a frontier quantitative methodology, quantile regression, which complements the classical approach with robustness and provides more information about tail behaviors. We present here a list of important figures for the areas mentioned above.

- William Sealy Gosset (1876–1937): Student's t-distribution, univariate analysis
- Karl Pearson (1857–1936): correlation coefficient, bivariate analysis
- Carl Gauss (1777–1855): ordinary least squares, multiple factor models
- Clive Granger (1934–2009): unit root and cointegration, time series
- Roger Koenker (1947–): quantile regression, frontier approach

Note that for multi-factor analysis, while we focus on the simplest linear form, we also examine nonlinear forms. Nonlinearity, if applied correctly, can enhance a model because it more closely approximates the real world and hence can further improve portfolio performance. We also present some simple nonparametric frameworks (Chap. 5) commonly used in quantitative investing.

Preparation: Univariate and Bivariate Models We begin our discussion of quantitative analysis with univariate analysis (Chap. 2), focusing on basic concepts of probability, the four moments, distribution, and hypothesis testing. We illustrate univariate analysis through an example using S&P 500 index returns.

For bivariate modeling, we use the examples of the US S&P 500 index and Chinese CSI 300 index—the former from the world's largest developed economy and the latter from world's largest emerging economy—to study the relationship between the two index return variables (Chap. 3).

Multi-Factor Analysis Through Least Squares Methods Studying univariate and bivariate models is necessary to prepare for exploring multi-factor models. In quantitative investing, models of either alpha or risk are mostly expressed by multi-factor models.

Quantitative investing usually involves forecasting returns of securities. For example, for a long/short equity portfolio, we would forecast returns for each stock and then buy (long) the ones with the highest (forecasted) returns and sell (short) the ones with the lowest (forecasted) returns. Since the price of a stock derives from many sources, we could use a factor to approximate each source and then put all factors together in an additive way, thus producing a linear multi-factor model.

We show (Chaps. 4 and 5) how to build a multi-factor model step by step and how to estimate parameters in a model. The most common traditional estimation method is least squares, for which we show the mathematical methodology, statistical properties, and assumptions for the validity of each of those properties. Furthermore, each statistical property has a different impact on quantitative investing. Therefore, we discuss the implications of each property for investment and the necessary conditions for each property in the context of quantitative investing. This is important for investing because we need to know which properties do not hold or barely hold for finance data and what conditions are necessary to make each property valid. When we apply a methodology to analyze data through a multi-factor model, we know what we are getting, and when we transfer this information into a portfolio, we know not only what is going on but also the conditions and assumptions behind the portfolio.

Time Series Analysis For special time series data and models, we cannot apply the least squares methodology directly because the data may have issues of unit root.

The unit root issue will cause spurious regression results. Unfortunately, a unit root exists in many data sets involving macroeconomic variables and commodities. We show through the example of the price of crude oil how to test for the presence of a unit root and what to do to estimate unit root cases. We also discuss important concepts, such as cointegration involving two or more commodities; and, if each commodity has a unit root, how to characterize the relationship between them, as in the example of the price of gold and the price of crude oil. We discuss details of time series analysis and its applications in quantitative investing in Chap. 6.

Portfolio Optimization To construct a portfolio with the information available, quantitative investing usually relies on an optimizer to derive a solution. For this, there is an objective utility function with constraints, some being "hard" and others "soft." Hard constraints cannot be violated, whereas soft constraints should be met if possible, but the utility function can tolerate their not being met to some extent. A general form of portfolio construction following modern portfolio theory is

$$\max_{W} \quad \mathcal{F}(\text{W, return, risk, trading cost})$$

$$s.t.\text{constraints on weights, industry, long or short, etc.}$$

with the purpose of achieving W, the optimal weights, that maximize the utility function $\mathcal{F}()$ while satisfying the constraints. Given the often sophisticated nature of the problem to be solved, sometimes there is no unique solution, and we can only find the "best" solution. Solving the optimal portfolio problem requires advanced math, such as linear programming, and a powerful computer for computation.

Fundamentally, we need to think carefully about how to formulate an optimization problem, perhaps imposing only minimal constraints. Constraints limit opportunity sets for higher levels of performance or values of a utility function. One common practice is to add constraints one by one, starting with the most important. Thus, we can observe the impacts of each constraint on a portfolio. We discuss details of portfolio construction including global and live portfolios in Chap. 7.

Frontier Approach: Quantile Regression While the traditional ordinary least squares method shows how factors impact the conditional mean of the response variable—usually the forward returns of securities in the investment context—it is not robust, nor does it tell us any information about other parts of the return distribution. We know that being robust and having information about tails is very important for portfolio strategies. For example, to control loss, we need to understand how factors impact the left tail of a return distribution, namely securities with the most negative returns. Quantile regression, introduced by Roger Koenker and Gib Bassett in 1978, provides a tool to tackle such issues. Through collaboration with Basset and others, Koenker has made original and most significant contributions to estimation and inference of quantile regression methods. He is also a major developer of computation algorithms and an empirical explorer of quantile regression.

During the last 20 years, quantile regression has become a mainstream econometric methodology and has been employed widely in many fields. In finance, the major reasons people use quantile regression are as follows:

1. Robustness: results do not change much with outliers or abnormal data.
2. Efficiency: more efficient for non-normal distributions with fat tails.
3. Tail behavior: full picture view of factor effects on security returns, especially at tails.

The applications of quantile regression in quantitative investing have steadily increased in the last decade. We present a detailed discussion about quantile regression and its applications in alpha modeling and portfolio construction in Chap. 8.

1.5 Industry Insights

Recall in the preface, we stressed four pillars of successful quantitative investing: fundamental understanding of the market, knowledge of investment theory, mastery of quantitative methods, and proficiency in data analysis with programming. A good fundamental understanding of the market requires years of experience, observation, exploration, and thoughts. This book shares with readers some industry approaches to quantitative investment strategies, where real-world issues were understood and analyzed by proper econometric methods with guidance from finance theory. We provide solutions by analyzing real-world data through R programming. We present industry insights on a series of topics, listed below in the order they appear in subsequent chapters of the book.

1. outliers
2. information asymmetry
3. multi-factor alpha modeling
4. alpha hunting and risk-adjusted factor diagnostics
5. using cointegration for pair trading
6. portfolio construction and backtesting
7. tail exploration for gold price forecasting and portfolio construction
8. quantamental approach for a Japanese stock selection strategy

Data Outliers Outliers in data are a serious issue in investment because it is usually those outliers that drive the performance of a portfolio. Therefore, the investment industry pays special attention to outliers, including outlier detection, determination whether outliers are genuine or errors, identification of outlier sources, and proper treatments. This is particularly true for quantitative investing, where high quality data is the foundation. Over the past 50 years, the quantitative investment industry has developed a systematic approach to deal with outliers. We get into details in Chap. 2.

Information Asymmetry With good data in hand, we are able to investigate a variable and its relationships with other variables, including security returns. It should be noted that there are many one-size-fits-all approaches, such as a correlation value to describe the relationship between the USA and Chinese stock markets. Usually, the correlation between these two stock markets is very low (around 10%), but we know that the two financial markets are somehow related given the fundamental links between the world's largest developed and emerging economies and events affecting both economies and stock markets. For example, when the US market is down, we do find that the relationship is much stronger—the USA affects China but not vice versa—and the relationship is much weaker when the US stock market is bullish. That is, the relationship between the USA and Chinese stock markets is not symmetrical! Why is this? Will this pattern hold in the future as well? How can we measure the asymmetry quantitatively? Can we build a trading strategy around these asymmetric effects? We explore these questions in detail in the industry insights section of Chap. 3 when we discuss bivariate models.

Multi-Factor Alpha Modeling Security prices and returns do not derive from one source. There are many factors that contribute to the movement of prices, and their impacts differ across different industries and countries. How can we build a multi-factor model for security returns? Moreover, is there a way to use multiple factors to forecast security returns? In Chap. 4, we show, using an example of a stock selection strategy, how factors are identified as either drivers or indicators of stock price movement, how to put them together to form a model, and how to estimate their impacts on security returns jointly. In particular, given that the quantitative investment industry has been developing for more than 50 years, many factors have become well known and widely used by most quantitative strategies such that the opportunity set has been explored away (from alpha to beta). As a result, people have been searching for new sources of returns (alpha), but how can we explore new sources and build a new factor in a "scientific" way? We address these questions in the industry insights section of Chap. 4.

Alpha Hunting and Risk-Adjusted Factor Diagnostics Existing factors explain less than 10% of returns in terms of forecasting. The remaining 90% is unknown. We specify a set of criteria for alpha hunting—namely new factors: intuitive, predictive, robust, add-value and executable (IPRAE). Following these guidelines, we show the industry approach, using both parametric and nonparametric methods, to conduct factor diagnostics, which include the factor distribution, factor efficacy, risk characteristics, and implementation cost. In terms of the forecasting power of a new factor, we incorporate risk into alpha—the risk-adjusted alpha is closer to the portfolio weight than the naked alpha. This is the solution to bridge the Grand Canyon between alpha and portfolio weights, as the latter is derived from alpha maximization with a given risk level.

Using Cointegration for Pair Trading We show how the cointegration concept can be used in a pair trading strategy and specify guidelines for entering and exiting conditions for pair trading strategies. We then demonstrate using an example of American Airlines and United Airlines.

Portfolio Construction and Backtesting We share with readers the industry approach to backtesting and portfolio construction with real-world constraints. We first describe the workflow with inputs, steps, and major points to consider. Using a stock selection strategy, we then show what a long-only and a long-short strategy look like and how they can be built with meaningful parameters and practical constraints. We also pinpoint the limitations of simulating portfolios and then introduce the procedure from a backtest portfolio to a dry run portfolio and from a dry run portfolio to a live portfolio.

Tail Exploration for Gold Price Forecasting and Portfolio Construction The traditional least squares estimation and mean-variance portfolio construction approaches rely on asset returns being normal distributed. However, returns of most asset classes are not normal but rather have fat and/or long tails. In practice, investors try to hold portfolios with long positions in the right tail and short (or no-hold) positions in the left tail of a return distribution. Therefore, tail behavior is very important for portfolio performance. We employ the quantile regression method to analyze tail behaviors and to forecast the price of gold in a multi-factor model framework. In addition to alpha, we also show how QR can be employed to construct portfolios with different percentiles of returns corresponding to requirements of different investment strategies. For example, a left tail (e.g., 5th percentile) portfolio is suitable for a pessimistic investor preferring loss aversion and wealth preservation, while a right tail (e.g., 90th percentile) portfolio is more appropriate for an optimistic investor who is eager for wealth growth.

Quantamental Approach for a Japanese Stock Selection Strategy The financial crisis in 2008 was a wake-up call for quantitative investing. A positive outcome is that many investors realized the shortcomings of quant strategies, such as relying too heavily on historical data/backtesting and lacking depth in company-level information and information about prospective events. To overcome these limits of quant investing, we introduce a frontier industry approach—quantamental investment—which combines fundamental and quant investing so as to have both depth and breadth in the portfolio. We describe a quantamental approach for a stock selection strategy in the Japanese market step by step, including fundamental analysis, data collection, company visits for forward information. We discuss how to the combine quant and fundamental elements, and portfolio construction using an investment process. We show that the quantamental portfolio does add value because it outperforms a pure quant portfolio with the same quantitative information.

1.6 Data Analysis Using R

"Big data" analysis has become a buzzword recently, but how can we analyze big data in finance? The investment industry has been a pioneer in big data analysis, driven by the desire for profits, the affordability of money management, advances in computer technology, and a wealth of skilled human capital. Programing languages have been very helpful for analyzing large data sets for quantitative investing. R is one of the most widely languages in the investment industry. In this book, we introduce R step by step at the end of each chapter, with the ultimate goal of enabling readers to write R code to conduct analytical explorations in quantitative investing.

Here we present a brief introduction to the R programming language, starting with installation.

1.6.1 R Installation

R is developed from the S language which was created by Rick Becker, John Chambers, and Allan Wilks at Bell Laboratories.[3]

R is a statistical package for data manipulation, calculation, and graphical display. Early versions of R were used mainly because of its flexibility and user-friendly interaction: users can write their own functions to deal with data, make calculations and graphs, and input and output results. However, there was one notable limitation of R in its early years: it could not handle large data sets, which was a very serious drawback before the early 2000s. This data capacity issue was eventually overcome, and R has become more and more popular since the mid-2000s. It has been used widely across academia, industry, government agencies, and beyond. In the finance industry, R is one of the most widely used software packages, especially in the quantitative investing world. One important reason for this is that investors can express ideas about alpha or risk management or data manipulation very easily through codes of functions in R and automate the whole process with powerful input and output capabilities. The following features contribute to the wide usage of R:

- Object-oriented: users have flexibility.
- Open-source: users have free and easy access.
- Rich packages (library): codes are developed with quality control and authors' responsibility.

We first go over installation and directory setup, then present basic features and commands when starting R. To begin the process, we can find the installation files and related information from the website of the Comprehensive R Archive Network (CRAN). Here are the installation steps:

---

[3]See the books by John Chambers and his co-authors.

1. Go to CRAN: www.r-project.org, click the "download R" link.
2. Select a mirror site and click the link.

To Install R on Mac (Check Operation Requirements)
Click on "Download R for (Mac) OS X."
Select a version under "Files."
Download and open the .pkg file, complete the installation.

To Install R on Windows
Click on "Download R for Windows."
Click on "Install R for the first time."
Click "Download R for Windows" and save the .exe file on your computer.
Run the execution and complete the installation.

Note that R can be run on both Windows and Mac operating systems in multiple languages.

1.6.2 R Basics

Now we can start to use R. When you open an R console, you will see a prompt where R expects input commands. The default prompt is ">". Before starting any action in the R console, one needs to know the work directory. The command *getwd()* will return the current work directory, and the command *setwd()* will set a new working directory for the session.

Set up work directory

```
> getwd()
[1] "/Users/lingjiema"
> setwd("/Users/lingjiema/Rdemo/")
```

The entities in R is called *objects*. An object can be a number, a vector, or a function. Each object is created and stored by an unique name during an R session. For example:

Object in R

```
> x=1:5
> x
[1] 1 2 3 4 5
```

R commands are case sensitive, that is, *A* and *a* are different objects. For execution, we can separate R commands by either a semi-colon (;) or a new line. Some basic commands for users include *help, list, rm, and q()*. During an R session, all objects can be stored for the future use. The data file is saved as *.RData* and the command lines are saved as *.Rhistory*, both are written to the working directory. This is very convenient as when we restart R later from the same working directory, both files are reloaded from the workspace.[4]

R has a built-in help facility. To get more information on any R command or function, such as *mean*, we can use the command *help()* or a question mark, *?*. When one does this, a new webpage will pop up with information about this command, usually accompanied by examples. The double question mark looks for packages containing the command. If you want to see all the objects in the current work directory, you can use *ls ()*. You can use *rm()* to remove an object.

Basic commands: help, ls, rm, q()

```
> help(mean)
> ?mean
> ??mean
> ls()
  [1] "a"              "aa"       "adjust.optimalPortfolio.weight"
  [4] "aim0817.practiceR"  "annual.perf"     "annual.performance"
  [7] "b"              "bb"       "bb0"
> a
[1] "IL" "IL" "MN" "MN" "ND" "ND"
> rm(a)
> q()
```

The last command, *q()*, will prompt the closing of the current R session. Before closing the session, you are asked if you want to "Save workspace image?" There are three options: Do not save, cancel, and save. Click on one to proceed.

Note that there is an online user group in each major city.

References

Fama, E., and K. French. 1992. "Cross-Section of Expected Stock Returns." *Journal of Finance* 47(2): 427–465.

Graham, B. 1949. *The Intelligent Investor*. New York: Harper & Row.

[4]R introduction by W.N. Venables, D.M. Smith and the R Core Team.

Graham, B., and D. Dodd. 1934. *Security Analysis*. New York: Whittlesey House, McGraw-Hill Book.
Jegadeesh, N., and S. Titman. 1993. "Returns to Buying Winners and Selling Losers: Implications for Stock Market Efficiency." *The Journal of Finance* 48(1): 65–91.
Liu, B., H. Preston, and A. Soe. 2017. "SPIVA Institutional Scorecard: How Much Do Fees Affect the Active Versus Passive Debate?" Research report, S&P Dow Jones Indices.
Markowitz, H.M. 1952. "Portfolio Selection." *The Journal of Finance* 7(1): 77–91.
Sharpe, W. 1964. "Capital Asset Prices: A Theory of Market Equilibrium." *The Journal of Finance* 19 (3): 25–442.
Sloan, R. 1996. "Do Stock Prices Fully Reflect Information in Accruals and Cash Flows About Future Earnings?" *The Accounting Review* 71(3): 289–315.

Chapter 2
Is the Current US Stock Market Overvalued? Univariate Analysis

Abstract In this chapter, we employ univariate analysis to evaluate the US stock market. Using the S&P 500 daily pricing data from 1950 to 2018, we illustrate the concepts of four moments, density and cumulative distribution functions, and hypothesis testing. We present two important figures: Benjamin Graham on investment and Student (William Sealy Gosset) on univariate analysis. We show nonnormality of asset returns and present an industry approach to outliers. In the last section, we introduce R and demonstrate simple calculations and plots.

2.1 The US Stock Market and the S&P 500: 1950–2018

In this section, we briefly trace the origins of the US stock market and the history of the S&P 500.

The first tradable share of a company was issued by the Dutch East India Company in 1602 and was traded on the Amsterdam Stock Exchange. In the USA, the New York Stock Exchange (NYSE) is the earliest exchange. It started in 1792 when 24 business people signed the Buttonwood Agreement on Wall Street. NYSE started with 5 securities, 4 of them were issued by the government. The Bank of New York was the first company traded on the NYSE. From those humble beginnings, the NYSE has grown dramatically and today has a list of 2282 companies with a total capitalization of nearly $24.2 trillion as of August 31, 2018.

When large numbers of stocks began trading on the NYSE, stock indices were created to measure the overall movement of the stock market. In 1860, Henry Varnum Poor founded a company, Poor's Publishing, that provided financial information and analysis. Poor's company introduced a "composite index" in 1923 to track the stock market performance. The Composite Index grew from a few stocks at the beginning to 90 stocks in 1926 and eventually expanded 500 in 1957. In 1941, Poor's Publishing merged with Standard Statistics, thus establishing Standard and Poor's Corporation. The Standard & Poor's 500, referred as the S&P 500, consists of 500 large companies whose common stock are traded in USA. The index components and weights of S&P 500 are based on the market capitalizations, with details determined by the S&P Dow Jones Indices. It is considered as a representation of the U.S. stock market.

L. Ma, *Quantitative Investing*, https://doi.org/10.1007/978-3-030-47202-3_2

Given its forward-looking character, the S&P 500 is often employed as a bellwether for the U.S. economy.[1]

Having briefly introduced the history of the US stock market, we present standard performance measurements for stock markets—return, risk, and annualization—in the following section.

2.1.1 Performance Measurement: Return, Risk, Annualization

We have daily price data for the S&P 500 index from 1950 to 2018. Using R scripts and daily closing prices, we present the index's price movements in Fig. 2.1. We see that the US stock market, represented by S&P 500, had an upward trend with dramatic ups and downs during this period. While visualization is helpful, is there a

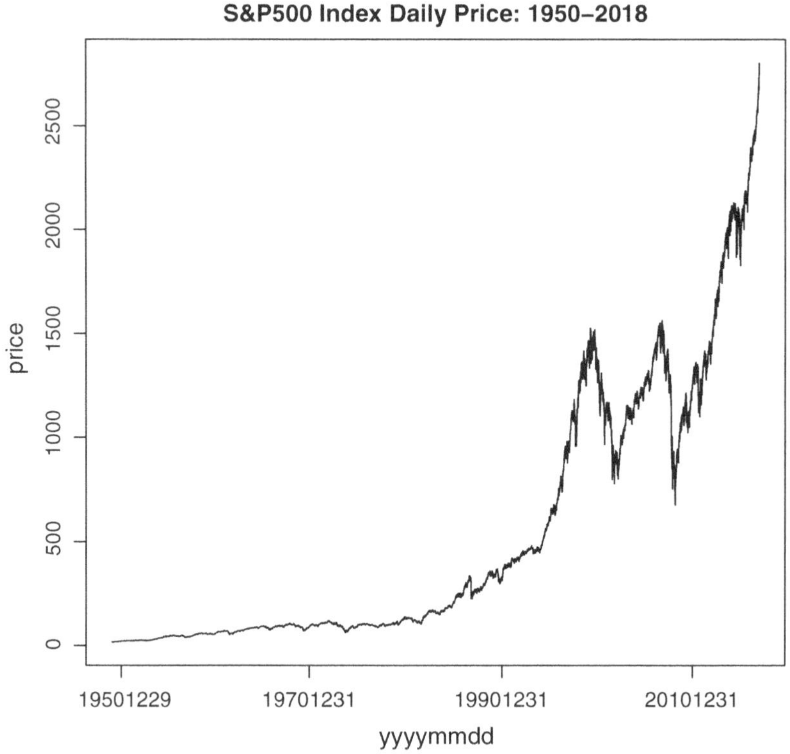

Fig. 2.1 The daily price movements of the S&P 500 index from 1950 to 2018

[1]The National Bureau of Economic Research has classified common stocks as a leading indicator of business cycles.

Table 2.1 S&P 500 daily price on Jan 4, 1950 and Jan 17, 2018

yyymmdd	Open	High	Low	Close	Daily.return
19500104	16.85	16.85	16.85	16.85	0.0114
20180117	2784.99	2807.04	2778.38	2802.56	0.00942

way to measure the market's performance quantitatively? A simple way is to examine return and risk: the former is simply the percentage price change, and the latter is usually measured by volatility (a.k.a. standard deviation). Table 2.1 shows the pricing data on the first and last day for this period. We can obtain the total return and risk for this whole period as follows:

$$\text{return} = 2802.56/16.85 - 1 = 165.324, \qquad \text{price volatility} = 629.6003.$$

An important measurement of asset performance is the Sharpe ratio, defined as annualized return divided by annualized risk. Annualization is employed to compare performance of different assets apples-to-apples. Below is the definition of annualized return and risk if the data are daily. Suppose there are T trading days for the asset return r_t at time t. Here we assume there are 220 business days per year for the US stock market.

$$\begin{aligned}\text{annualized return} &= [(1+r_1)\times(1+r_2)\times\ldots\times(1+r_T)]^{\frac{220}{T}}\\ \text{annualized volatility} &= sd(r_1, r_2, \ldots, r_T)\times\sqrt{220},\end{aligned}$$

where sd is for standard deviation. The Sharpe ratio is defined as

$$\text{Sharpe ratio} = \frac{\text{annualized asset return} - \text{risk-free return}}{\text{annualized risk}}.$$

Assuming the risk-free rate is zero, we calculate the S&P 500's performance from 1950 to 2018 as follows:

$$\begin{aligned}\text{annualized return} &= 165.324^{\frac{220}{17120}} - 1 = 6.78\%\\ \text{annualized volatility} &= 14.25\%\\ \text{Sharpe ratio} &= \frac{0.0678}{0.1425} = 0.48.\end{aligned}$$

In other words, the S&P 500 index has a 6.78% annual return on average over the period from 1950 to 2018, with a volatility of 14.25% indicating major fluctuations. The overall volatility-adjusted return—the Sharpe ratio—is 0.48.

We have given above a basic description of the performance of the US stock market over the last 70 years. To provide a better understanding of the equity market, we next discuss some fundamental issues in the public equity market, such as why do companies issue stocks.

2.1.2 Why Do Corporations Issue Stocks?

When a company issues equities to the public, it has changed its status from a private company to a public company. The stock market exists primarily because of the demand for capital raised from the public for business expansion, hence the reason for issuing shares in the first place and the trading of those shares, where the former is called the primary market and the latter is called the secondary market.

There are two types of share issuances: initial public offering (IPO) and seasoned equity offering (SEO). Typically, the issuing company, deal broker (usually an investment bank), and major institutional clients carry out an IPO, thus making it less open to the public. However, because of the marketing and price discount, an IPO usually generates abnormal trading volumes. The public chase the stock right after an IPO. There are many related aspects of an IPO, such as number of shares to be issued, the price and discount of each share, the shareowners' rights, etc., that are outside the scope of this book.

Different from an IPO, an SEO occurs only after a company has already gone public. Moreover, an SEO can go in either of two directions: the company can issue new shares or buy back existing shares. Usually, when companies issue new shares or repurchase existing shares, they go through a broker to inform the existing portfolio management companies, which may result in a larger number of attending clients than an IPO. Note that in earlier periods, an SEO barely got involved with any share repurchasing, but during the last 30 years, it has become a trendy corporate action, and over 80% of S&P 500 companies now conduct share buybacks.

In this book, we mainly focus on the secondary equity market. An important question to address at this moment is who the participants in this market are. This will provide a foundation for our discussion on market efficiency and alpha modeling in subsequent chapters.

- Company: issuer and carrier of shares
- Market maker: liquidity provider
- Investors: make profits from price changes
- Professionals: fee-motivated
- Government agents: regulation

Note that each group has its own function, but they are all connected, and as a whole they form the stock market. From the price movements in Fig. 2.1, we see that stock prices moved up and down, so there should be huge opportunities to make profits in the stock market if we can buy low and sell high at the right time. In the history of investment in the stock markets, people have made money and people have lost money. Is there a scientific way to invest in the stock market?

Having introduced basic elements of the equity market, we introduce Benjamin Graham, an important figure in investment, in the following section.

2.2 Investment Is a Science: Benjamin Graham (1894–1976)

Benjamin Graham (Fig. 2.2) is widely known as the "father of value investing."[2] He published two books that paved the foundation for scientific investing: *Security Analysis* (1934) with David Dodd and *The Intelligent Investor* (1949).

Graham's major contribution was to approach investment as a science: it is different from speculation, and there is a scientific approach to investing in financial markets. In particular, he proposed value investing: what is it and how to do it in a systematic way. He coined many fundamental terms that professional investors all over the world still use today, such as margin of safety, investor psychology, fundamental analysis, concentrated diversification, buy-and-hold investing, activist investing, and contrarian mindsets. All of these have had deep fundamental impacts on investment. Rule-based value investment and his scientific approach made Graham not only the father of value investing, but also a pioneer of quantitative investing.

2.2.1 Investment Versus Speculation

Investment should be a science; speculation is not.

> An investment operation is one which, upon thorough analysis, promises safety of principal and an adequate return. Operations not meeting these requirements are speculative.
> — Benjamin Graham

Benjamin Graham proposed definitions for investment and speculation in his book *Security Analysis* (with Dodd) in 1934, in the midst of the Great Depression. Graham pinpointed the difference between investment and speculation, which was

Fig. 2.2 Benjamin Graham, 1894–1976

[2]Photo source: Equim43, http://mejorbroker.org.

an insightful and timely critique. The principles and insights Graham laid out back then are still valid today, and these concepts will hold as long as there are price fluctuations in the financial market.

We present the price movements of the S&P 500 index from 1918 to 1949 in Fig. 2.3. We see that price movements during this period were highly volatile, especially around the Great Depression (1929–1933). In the 1920s, the US stock market had a 20% average increase with shares trading volume doubled to 5 million per day at the end of this period. One contributing factor for this boom was that investors can borrow money from a broker, up to 80–90% of the stock price, "on margin." To buy 1 million dollar stocks, an investor needs only to put down 10,000–20,000 dollars. If the stock price went up, both the broker and the investor became

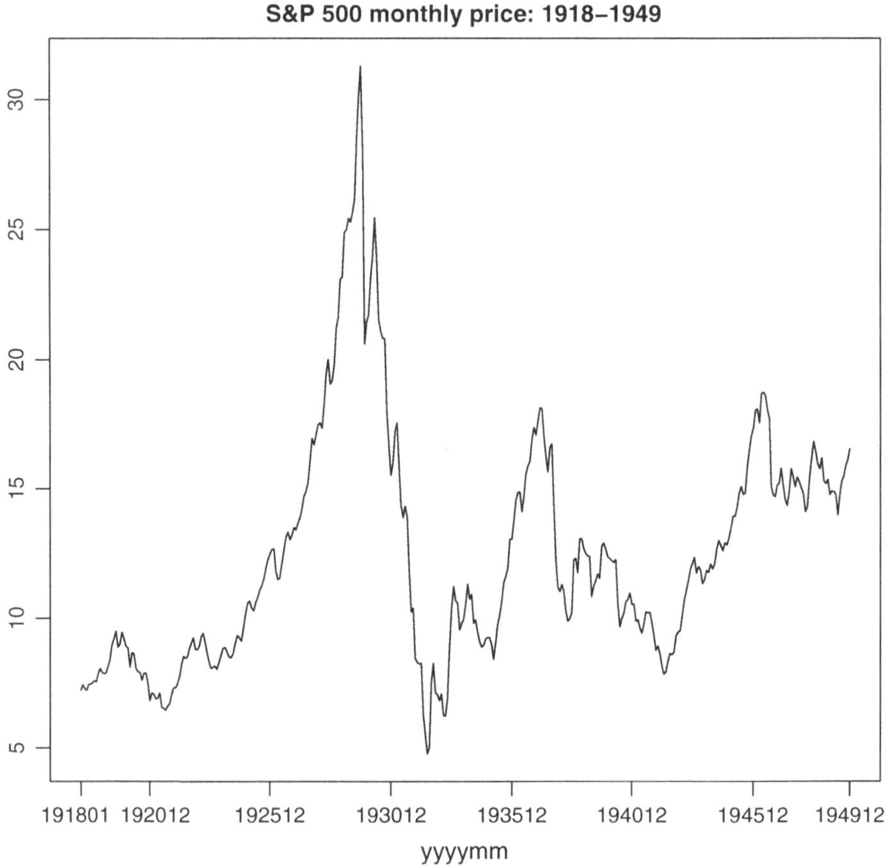

Fig. 2.3 S&P 500 index price movements, 1918–1949

millionaires. This rampant speculation led to erroneously high stock prices.[3] With the mindset of a wealth creation vehicle, the U.S. stock market underwent rapid expansion in the 1920s, reaching its peak in August 1929. By then, real economy had declined and unemployment started to rise, resulting in an excessive overvalue of the stock market. Among the other causes of the eventual market collapse were low wages. The situation got worse when large bank loans started to be illiquid with the proliferation of debt and a struggling agricultural sector. This eventually led to the stock market crash on October 29, 1929 (Black Tuesday). During this period, the speculative investors could not make their margin calls, prompting a massive sell-off which pushed the stock market further down.

After the Great Depression, the US stock market continued to have ups and downs but of a much smaller magnitude. The overall stock market did not perform well over this period from 1918 to 1949. The annualized return was 3% and annualized risk was 20%, resulting in a very low Sharpe ratio of 0.13.

S&P 500 performance: 1918–1949

```
> dim(sp5.graham)
[1] 384   4
> sp5.graham[c(1,384),]
  sp5.avgPrice sp5.mth.chg sp5.mth.ret yyyymm
         7.21        0.41        6.03  191801
        16.54        0.43        2.67  194912

> # annualized risk
> sd(sp5.graham$sp5.mth.ret)*sqrt(12)
[1] 20.62379

># annualized return, 6.8 is the price of 191712
> (16.54/6.8)^(1/32)-1
[1] 0.02816622

> #Sharpe ratio
> 0.0282/0.2062
[1] 0.1367604
```

So what is the underlying difference between a speculator and an investor? According to Graham, the major difference is the approach to investment: "an

[3] This all sounded familiar in the 2000s when house buyers could purchase a home with 0% down, and house prices increased by about 20% every year before the financial crisis. History does repeat itself!

investor attempts to profit in the long term by purchasing stocks at a substantial discount to their *intrinsic value*, while a speculator attempts to profit by predicting the short-term fluctuations of a stock."

Examples of speculation or even smart speculation include buying a stock simply because you believe it will rise or because it has dropped precipitously, or being more likely to buy a stock when the market is up and sell a stock when the market is down. Clearly, this emotional urge to trade has nothing to do with investing or attempting to ascertain the underlying value of a common stock.

The next question is how to determine the intrinsic value of a security. This can be done with a good understanding of the business, company, and market and a thorough analysis to arrive at a proper business value. By comparing the intrinsic value with the market price, we know if the security is undervalued or overvalued and roughly by how much. We get into more details in the following section.

To harvest the potential returns from undervalued stocks, timing is an important factor. An investor waits patiently for an expected return or the price to approach the intrinsic value and then closes the position. The difference for return realization between an investor and a speculator lies in the defensive nature of the investor or, more specifically, the probability that the transaction will not result in a loss of principal. In other words, the distinction is largely a function of risk.

2.2.2 Value Investing

While the preceding section differentiates investment from speculation conceptually, how should we actually carry out an investment? According to Graham, the core of investment is value investing, that is, to find the intrinsic value of a security. In their book, *Security Analysis*, Graham and Dodd provide groundwork for value investing: purchasing undervalued securities and harvesting returns with the safe of margin. About 15 years later, Graham expanded on the idea of value investing in another book. In 1949, he published the seminal book *The Intelligent Investor*, which is regarded by many professionals the bible of value investing. In the 1970s, the idea of value investing was applied to quantitative investment.

Benjamin Graham proposed the following formula for intrinsic value discovery of a security:

$$V = EPS \times (8.5 + 2g)$$

where V = intrinsic value,

EPS = trailing 12-month EPS of the company,

8.5 = P/E ratio of a zero-growth stock,

g = long-term growth rate of the company,

where EPS is earnings per share and P/E is price to earnings. This formula was revised later to incorporate other factors such as bond yield.

The central concept of value investing is to avoid loss of capital as its first rule. Graham advocated a simple formula to investigate companies with low P/E and P/B (price to book) ratios. One should also analyze reports of financial statements and footnotes to understand whether companies have serious issues that are potentially unnoticed by the market. Since accounting items and estimates of growth rate will always have errors, to make value investing practical, we need to have enough space for these errors, hence the term *margin of safety*, which Graham discussed at length in both his books. In simple terms, capital preservation is the first goal and wealth growth is the second goal of value investing.

$$\text{Margin of Safety} = \text{market price} - \text{intrinsic value}.$$

The concept of the margin of safety and its application are organic parts of value investing. However, forecasting the future earnings of a company is never easy. Graham suggested that a company's growth rate g could be estimated by analyzing a company's assets, core business products, and financial performance. Moreover, evaluation of a company's future can be more of an art than a science. Two different investors who analyze exactly the same valuation data on a company can arrive at very different decisions.

To be on the safe side, the higher the uncertainty, the larger the margin of safety is required. To determine exactly how much a margin of safety should be for a value investment depends on many factors, and this should always be analyzed case-by-case. In general, these factors include, but are not limited to, macro-level (country) policy and events, micro-level (company) performance dynamics, an industry's fundamental outlook, and the investor's risk tolerance and expected return. When an investor has high confidence on the accuracy of her estimate of the underlying value of the stock, she can set a thin margin, while if the estimation of intrinsic value is for a starting company in a new industry, the margin of safety should be set higher to compensate for the uncertainties. On the other side of the equation, if an investor expects a higher return from a security, then he or she should select securities with a much higher margin of safety, all else being equal. Usually, a decent return should be higher than the index return for the same asset class. For example, one could use the S&P 500 index returns as a benchmark in the US stock market.

If a stock has a current price of \$10, and the calculated intrinsic value of that stock is \$15, then the margin of safety is about \$5, or using the ratio definition, 50%. If the expected return is 50%, there is no room for safety. However, if the expected return is 20%, then the investor has enough room to feel safe, as she can close the position when the price increases to \$12 if something happens down the road (for example, the real intrinsic value is \$13).

Since Graham proposed the original definition for intrinsic value, many people have used the formula in practice that the alpha has been gradually explored away from this method. However, the principle—if you know the underlying value of something, you can save a lot of money when you buy it with a bargain—continues to be valid. In the context of quantitative investing, these are reflected in factors and formulated in a model, and the intrinsic values of securities are then estimated using an econometric methodology. As long as there is a larger proportion of securities with estimated intrinsic values close to their true values, a quant strategy can yield profits based on the law of large numbers, rather than for an individual security.

A fundamental question we need to address is why some stocks are undervalued. The short answer is that the stock market is fairly efficient (in developed countries) in the long term but can be very inefficient in the short term. Efficiency means that market prices reflect information correctly and completely. We get into details about this in Chap. 4. The realization of value investment requires patience and time. This premise is reflected in Warren Buffett's summation (1987 Annual Letter) of Graham's philosophy: "In the short run, the market is a voting machine, but in the long run, it is a weighing machine."

In the following section, we apply value metrics, including the P/E ratio as explored by Graham, to evaluate the US stock market.

2.3 How Can We Evaluate the Current US Stock Market?

In this section, we discuss traditional methods to evaluate the US stock market. To begin, we calculate the annual performance of the S&P 500 index by using return, risk, and the Sharpe ratio. We then discuss the business cycle and P/E ratio for the purpose of market evaluation over time.

2.3.1 *Annual Performance*

By applying the annualization formula for daily data on the S&P 500, we can calculate annual returns and volatility from 1950 to 2018. The results are presented in Table 2.2 and Fig. 2.4. Notice that the annual returns range widely from 40% to −40%. Over 74 years, the S&P 500 index has 57 years with positive returns and 17 years with negative returns.

2.3.2 *Historical Perspective: Business Cycles*

From a long-term perspective, the stock market is an indicator of economic health, so financial market movements and business cycles should generally follow the same pattern. Of course, there will always be periods with significant gaps between the two. For example, the stock market may be a few months ahead of business performance, or the stock market may react quickly to incentives, such as an immediate money supply or tax cut, while it usually takes time for these policies to impact business.

Table 2.2 S&P 500 annual returns (%), volatility (%), and Sharpe ratios from 1950 to 2018

Year	Return	Vola	Sharpe	Year	Return	Vola	Sharpe	Year	Return	Vola	Sharpe
1950	22.63	13.95	1.62	1973	-17.37	14.79	-1.17	1996	20.26	11.01	1.84
1951	16.35	10.07	1.62	1974	-29.72	20.39	-1.46	1997	31	16.94	1.83
1952	11.77	7.38	1.59	1975	31.56	14.42	2.19	1998	26.67	18.96	1.41
1953	-6.62	8.94	-0.74	1976	19.14	10.38	1.84	1999	19.53	16.89	1.16
1954	45.02	8.79	5.12	1977	-11.51	8.48	-1.36	2000	-10.14	20.76	-0.49
1955	26.4	14.34	1.84	1978	1.05	11.77	0.09	2001	-13.04	20.14	-0.65
1956	2.62	11.62	0.23	1979	12.31	10.13	1.21	2002	-23.37	24.32	-0.96
1957	-14.31	12.14	-1.18	1980	25.77	15.38	1.67	2003	26.38	15.95	1.65
1958	38.06	8.39	4.53	1981	-9.73	12.57	-0.77	2004	8.99	10.37	0.87
1959	8.49	8.78	0.97	1982	14.76	17.06	0.87	2005	2.99	9.61	0.31
1960	-2.97	9.71	-0.31	1983	17.27	12.45	1.39	2006	13.61	9.37	1.45
1961	23.13	9.45	2.45	1984	1.4	11.91	0.12	2007	3.53	14.94	0.24
1962	-11.81	15.45	-0.76	1985	26.33	9.5	2.77	2008	-38.49	38.28	-1.01
1963	18.88	8.06	2.34	1986	14.62	13.73	1.06	2009	23.46	25.49	0.92
1964	12.97	4.92	2.64	1987	2.03	30.03	0.07	2010	12.78	16.87	0.76
1965	9.06	6.48	1.4	1988	12.4	15.97	0.78	2011	0	21.75	0
1966	-13.09	11.02	-1.19	1989	27.25	12.2	2.23	2012	13.4	11.93	1.12
1967	20.09	7.84	2.56	1990	-6.56	14.9	-0.44	2013	29.6	10.34	2.86
1968	7.67	8.77	0.87	1991	26.3	13.36	1.97	2014	11.39	10.62	1.07
1969	-11.36	9.38	-1.21	1992	4.46	9.05	0.49	2015	-0.73	14.48	-0.05
1970	0.1	14.16	0.01	1993	7.06	8.04	0.88	2016	9.53	12.24	0.78
1971	10.78	9.54	1.13	1994	-1.54	9.2	-0.17	2017	19.42	6.25	3.11
1972	15.63	7.44	2.1	1995	34.1	7.29	4.68	2018	4.82	6.19	0.78

Following industry convention, we use the NBER definition of US business cycles. We collect data on US GDP and unemployment rate as well as S&P 500 index price from 1900 to 2018. We transform the quarterly GDP data into a growth rate and make the data on a monthly basis (remains the same for consecutive 3 months). We also add a 2-month lag to GDP and unemployment rate to reflect the information as it was available during the period under consideration. The S&P 500 returns are calculated on a 12-month basis before the peak and trough dates. We present the data in the plots of Fig. 2.5. Since the numbers tend to be much larger in the early years, they are displayed as two periods, 1900–1953 and 1954–2018. To make the numbers comparable, we standardize each variable by subtracting the mean and dividing by the standard deviation. The unemployment rate is displayed as negative for visualization convenience.

We see that during the first period from 1900 to 1953, the US economy and stock market performance were almost perfectly synchronized. The relationship still held but became weaker during the second period from 1954 to 2018. There was less volatility in GDP growth rate and unemployment rate in recent years.

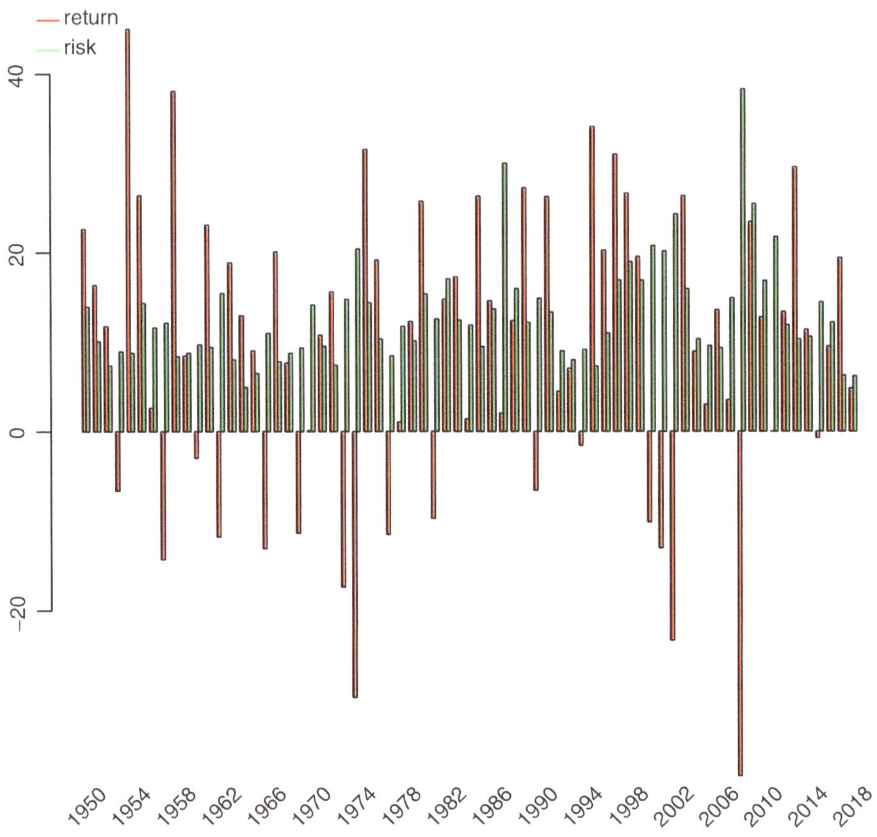

Fig. 2.4 The annual returns (%) and volatility (%) of the S&P 500 index from 1950 to 2018

For the most recent period, we see that while the US GDP growth rate and unemployment rate are in good condition, the stock market is overwhelmingly bullish, compared with other business cycles. From a business cycle perspective, the current stock market may be overpriced.

2.3.3 *Company Valuation: P/E*

One conventional measure of market soundness is the market-level price-to-earnings (P/E) ratio. Intuitively, there should be a fundamental price level for the earnings per share at an overall economic level: the price supported by earnings or the price investors are willing to pay for that earnings level. If the P/E ratio is too high, this may indicate that the market is too hot because earnings are overpaid.

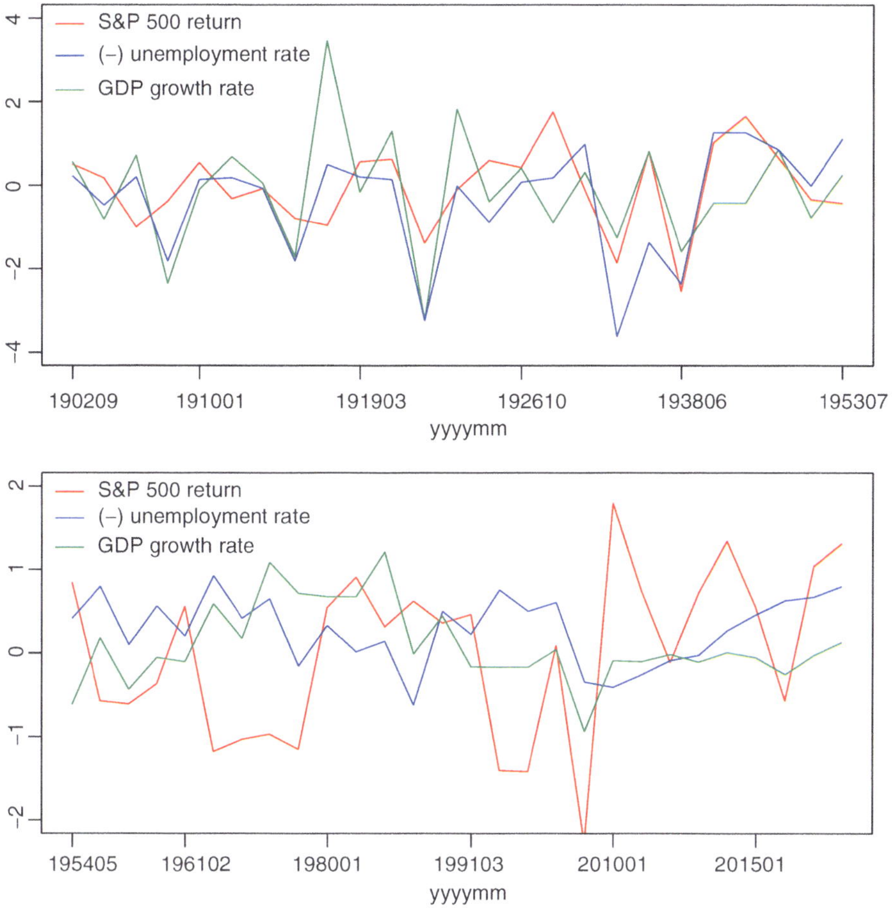

Fig. 2.5 US business cycles, unemployment rate, GDP growth rate, and S&P 500 returns, 1900–1953 (top plot) and 1954–2018 (bottom plot)

We present in Fig. 2.6 the S&P 500 index-level P/E ratio from 1900 to 2019 on a monthly basis. The price-to-earnings ratio is calculated based on trailing 12-month as-reported earnings.[4] During the period from 1990 to 2019, extremely high values of P/E occurred during the financial crisis in 2008 and 2009 because of extremely low earnings (see Table 2.3).

After removing extreme P/E values, we plot P/E ratios and the average (the horizontal line) in Fig. 2.6. We also list summary statistics for the whole and subperiods in Table 2.4. First, we see that the P/E ratio stayed within the same range, from 5 to 25, over a long period from 1900 to 1997. The first time it exceeded

[4]Source: Robert Shiller and his book *Irrational Exuberance* for historical S&P 500 P/E ratios.

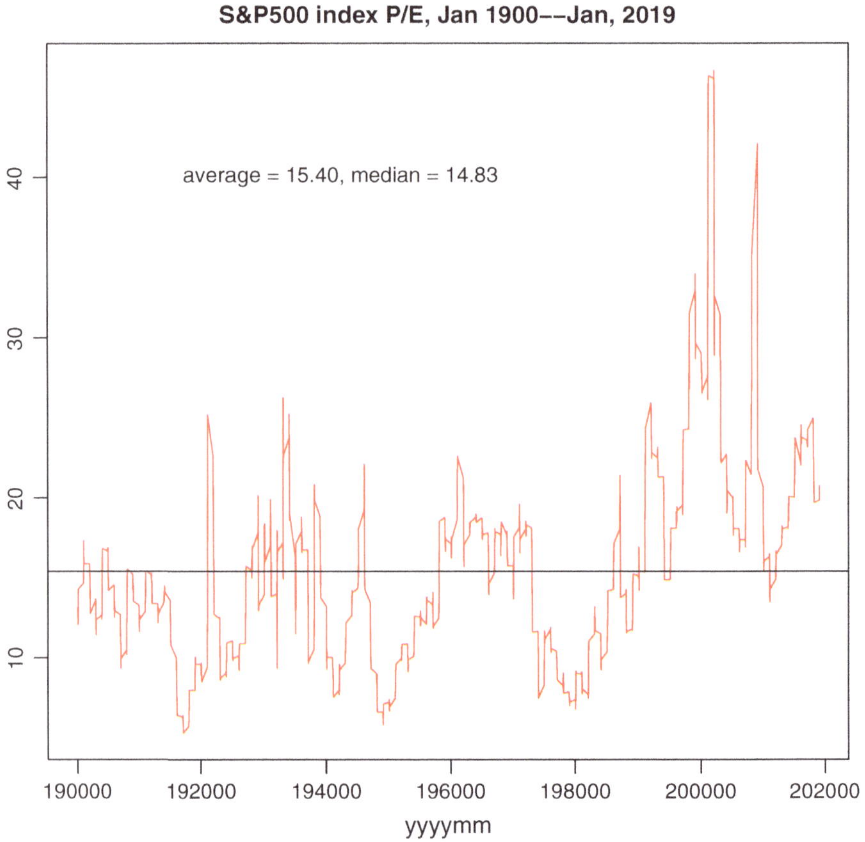

Fig. 2.6 The P/E ratio of the S&P 500 index from January 1900 to January 2019

Table 2.3 S&P 500 extreme P/E values

yyyymm	P/E	yyyymm	P/E
200812	58.98	200901	70.91
200902	84.46	200903	110.37
200904	119.85	200905	123.73
200906	123.32	200907	101.87
200908	92.95	200909	83.3

30 was in Nov 1998 (P/E=30.25), and it reached 40 in Nov 2001. The ratio fluctuated wildly during the recent financial crisis in 2008 and 2009. The P/E ratio has stayed around 20–25 during the last 4 years, with the most recent value being 20.76 on January 31, 2019.

Therefore, judging from the P/E ratio's long-run average and trend, the current US stock market seems overvalued.

Table 2.4 Summary statistics for the S&P 500 index P/E ratio from Jan 1900 to 2019

yyyy	Min	25th	Median	Mean	75th	Max
1900–2019	5.31	10.99	14.83	15.40	18.22	46.71
1900–1950	5.31	9.99	13.00	13.06	15.28	26.27
1951–2000	6.68	10.88	15.29	15.14	18.20	34.00
2001–2019	13.50	17.88	21.10	22.50	24.80	46.71

2.4 Univariate Analysis: The Four Moments, Density, and CDF

In the previous section, we discussed the US stock market's performance using S&P 500 data. In particular, we discussed returns, volatility, and other factors. However, we know that a single number is not enough to describe the market given its dynamics over time. A natural question is how to describe a stock market quantitatively. In this section, we introduce univariate analysis and present basic terms such as random variable, the four moments, and distribution.

2.4.1 *Random Variable: Stock Price Movement*

We know that the stock price for a public company moves up and down and is determined by many factors.

Definition 2.1 In probability and statistics, a random variable describes event outcomes that change due to chance or randomness.

A random variable is expressed as X, with numerical values for the outcome of random events. Random variables can be classified into two types: discrete and continuous. For a discrete random variable, its probability is measured by the count of outcomes, $P(X = \{x1, x2\})$, such as the probability of having snow in Chicago on Dec 31, 2019. For a continuous random variable, its probability is defined by a range: if x is in real values and $x \in (-\infty, +\infty)$, then the probability for x to have values between x_1 and x_2 can be expressed as $P(x_1 < X < x_2)$. $P(X < x)$ indicates all the values from negative infinity to x. The price of a stock is a continuous variable.

For a random variable, we want to know the chances of outcomes. For example, for a stock, we would like to know the probability of its price moving up or down.

Definition 2.2 The probability of an event is the measure of the chance that the event will occur, between 0 and 1. Prob = likelihood of an event/total possible outcomes.

The following are apparent: Prob = 0, impossible; Prob = 1, certain; SUM (prob) = 1.

2.4.2 The Four Moments

How can we describe the outcomes of a random variable? Probability measures the chances of an outcome to occur. Once we have all the information about past outcomes, how can we summarize those outcomes? In statistics, those outcomes can be described by four moments: mean, standard deviation, skewness, and kurtosis. We introduce them one by one below. Before getting into definitions, we present a dialogue about stock market performance to help the reader understand the four moments' relevance to the stock market.

Suppose we have a sharp grandma who understands the stock market very well but has no formal training in statistics. Grandma asks about stock market performance, and we offer two types of answers: answer A uses ordinary language, while answer B uses statistical terms from the four moments.

Dialogue about the Four Moments: How is the stock market doing today?

- *Mean*

  Grandma: Is today's market roughly good or bad?
  A: Well, on average, it is up.
  B: The mean is positive.

- *Standard Deviation*

  Grandma: Are there any ups and downs?
  A: It is pretty volatile.
  B: The standard deviation is very high.

- *Skewness*

  Grandma: Are there more ups than downs?
  A: Yup, a good day, many more ups than downs.
  B: It is skewed to the right.

- *Kurtosis*

  Grandma: Are there big movements today?
  A: Yes, the price dropped to an extremely low level then bounced back.
  B: There is a fat tail on price movement today, that is, kurtosis is high.

Having illustrated the link between the statistical terms and stock markets, we now present mathematical expressions of the four moments.

Mean: average

$$\mu = \frac{\sum_{i=1}^{n} x_i}{n}$$

Standard deviation: variation

$$\sigma^2 = \frac{\sum_{i=1}^{n}(x_i - \mu)^2}{n}$$

Skewness: symmetry

$$\gamma = \frac{m_3}{\sigma^3} = \frac{\frac{1}{n}\sum_{i=1}^{n}(x_i - \mu)^3}{\sigma^3}$$

Skewness measures the lack of symmetry in data distribution. A symmetrical distribution will have a skewness value equal to zero. There are two types of skewness: positive and negative. Positive values imply a long right tail, while negative values imply a long left tail.

Kurtosis: fat tail

$$k = \frac{m_4}{\sigma^4} = \frac{\frac{1}{n}\sum_{i=1}^{n}(x_i - \mu)^4}{\sigma^4}$$

Kurtosis measures a distribution's peakedness. In terms of security returns, high kurtosis of a return distribution implies more probabilities for the occurrence of extreme returns.

Applying the formulas above to S&P 500 index daily returns data, we use R statistical language to compute the values of the four moments. Our data set for the S&P 500 is called *sp*5.

Four moments of S&P 500 index return

```
> sp5[1:2,]
  yyyymmdd  open  high   low close  return
1 19500104 16.85 16.85 16.85 16.85 0.01140
2 19500105 16.93 16.93 16.93 16.93 0.00475
> mean(sp5$return)
[1] 0.0003457465
> sd(sp5$return)
[1] 0.009606054
> library(moments)
> skewness(sp5$return)
[1] -0.6449608
> kurtosis(sp5$return)
[1] 24.0427
```

We see that during the period from 1950 to 2018, the average daily return of the S&P 500 index is 0.03% or 3.5 basis points (called bps in the industry), with the volatility of 0.96% indicating huge ups and downs on a daily basis. The negative skewness value of −0.64 indicates that the daily return distribution is not symmetrical but skewed to the left of the mean, while the very high kurtosis value of 24 implies a very fat tail, due to many extreme days in the history of the US stock market.

To have a better sense of the skewness and kurtosis, we plot the density of the monthly returns of the S&P 500 before and after the two biggest financial crises in the history of the US financial market, 1929 and 2008 (Fig. 2.7). We see that before each financial crisis (the left panel), the distribution of returns is left skewed (due to many more months of positive returns than negative returns). However, after the 1929 financial crisis, the skewness changed direction to the right, indicating big downturns in the market during the after-crisis period. While after the 2008

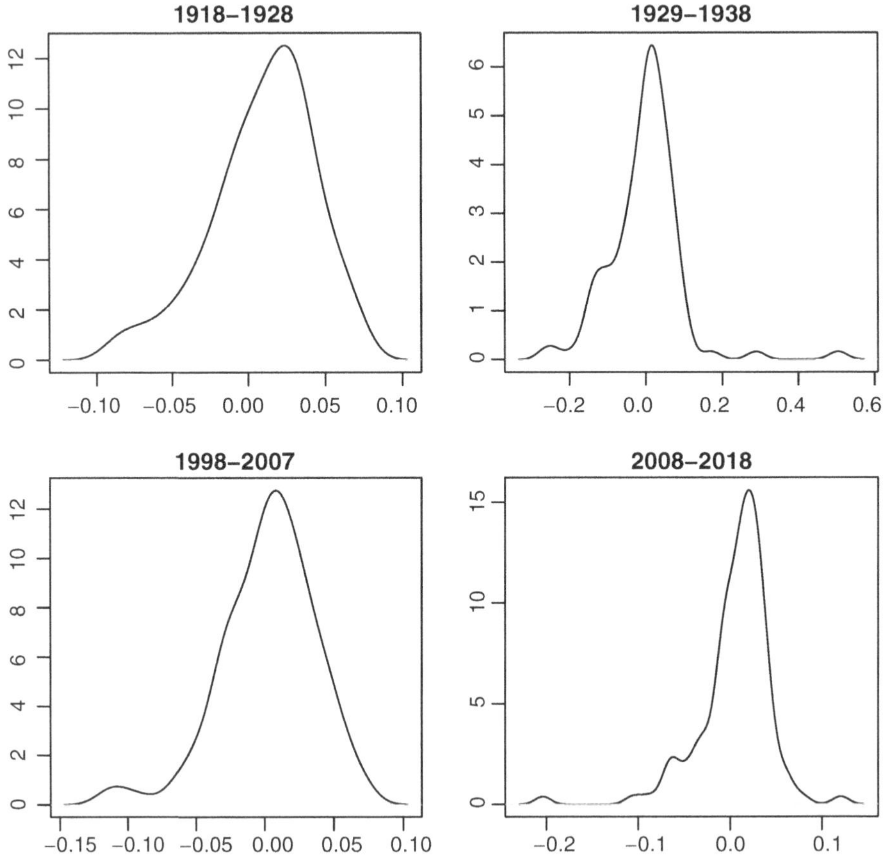

Fig. 2.7 Skewness and kurtosis visualization: the density plots of 10-year monthly returns of the S&P 500 before and after the stock market crises of 1929 and 2008

Table 2.5 Values of skewness and kurtosis for S&P 500 monthly returns before and after the financial crises of 1929 and 2008

yyyymm	Skewness	Kurtosis	Mean	Median	Min	Max
1918–1928	−0.66	3.27	0.009578	0.015200	−0.0896	0.070400
1929–1938	1.17	10.71	−0.00267	0.00720	−0.26470	0.50300
1998–2007	−0.77	4.28	15.29	15.14	18.20	34.00
2008–2018	−1.62	9.93	21.10	22.50	24.80	46.71

financial crisis, the skewness became even more negative (to the left), partially due to the implementation of effective government policies dealing with this financial crisis with quantitative easiness. The market then recovered very quickly and stayed bullish for a long period.[5] Regarding the kurtosis, we see that all four plots have long tails, especially after the crisis. The values for skewness and kurtosis, together with other summary statistics of monthly returns, are listed in Table 2.5.

2.4.3 Density Function and CDF

We calculated the four moments for S&P 500 index returns in the previous section. However, are the four moments enough? For some standard distributions (such as the normal distribution, which we discuss later), the four moments are enough to describe the probability of outcomes for that random variable. But in the real world, empirical data do not follow standard forms of distribution. Consequently, the four moments are not enough if we want a more vivid picture of outcomes. In this section, we discuss distribution and two important related statistical terms: density and cumulative distribution function (CDF).

The distribution of a random variable shows how outcomes are distributed according to their probabilities. There are three special distributions: uniform, normal, and t. We give a short description of each below.

Uniform Distribution Values are distributed with equal likelihood. In other words, probabilities spread with the same chance within a given range.

Normal Distribution Occurs "normally" in experiments and nature, also known as a Gaussian distribution. We discuss the properties of a normal distribution in the following section.

t-Distribution It is also called Student's t-distribution and is close to normal when sample size is large. It is often used for hypothesis testing for population parameters with a small sample size and/or unknown population variance.

[5]The government bought bad assets from companies by pumping new money into the economy and market.

The shape of a distribution can be described by its density function. A density function is defined as follows:

Definition 2.3 Density function: is a function that defines the probabilities of all the outcomes of a random variable. It is expressed in a mathematical form, $f(x)$. Typically, the probability density function is viewed as the shape of the distribution.

Since the density function represents probabilities of outcomes of a random variable, its values will be from zero to one, and the sum is equal to one, $\sum_i f(x_i) = 1$.

$$\text{Discrete variable: } f(x) = P(X = x), \quad \text{Continuous variable: } f(x) = P(X = x).$$

If we have x-y axes and plot the values of X on the x-axis and the probability of those values, $f(x)$, on the y-axis, we get the shape of the distribution function. If we add up all densities in the order of the X values, we get the cumulative distribution function (CDF).

Definition 2.4 Cumulative distribution function (CDF): Derived by adding up all densities starting from the lowest value to the highest value of a random variable, X. A CDF can be expressed as $F(x) = P(X \leq x)$.

Based on the definition above, the CDF for discrete and continuous random variables can be written as

$$F(X) = \sum_{i=1}^{N} f(x_i)$$

$$F(X) = \int_{-\inf}^{x} f(t)dt.$$

We can also derive expected values from the density function. For example, for a continuous variable, we obtain the expected mean and expected variance as follows:

$$\mu = E(X) = \int x f(x)dx \tag{2.1}$$

$$\sigma^2 = Var(X) = E(X^2) - \mu^2 = \int x^2 f(x)dx - \mu^2. \tag{2.2}$$

Thus, we see the advantage of density and CDF functions: we can express ideas in a mathematical form. This is very important for quantitative investing.

We generate a sample of 1000 observations for a uniform, a normal, and a t-distribution and present the shapes of their density and CDF functions in Fig. 2.8. We see that for a uniform distribution, its density plot is close to a horizontal line and its CDF plot is a straight 45° line, while *t*- and *normal* distributions are similar with the t-distribution having fat tails on both sides.

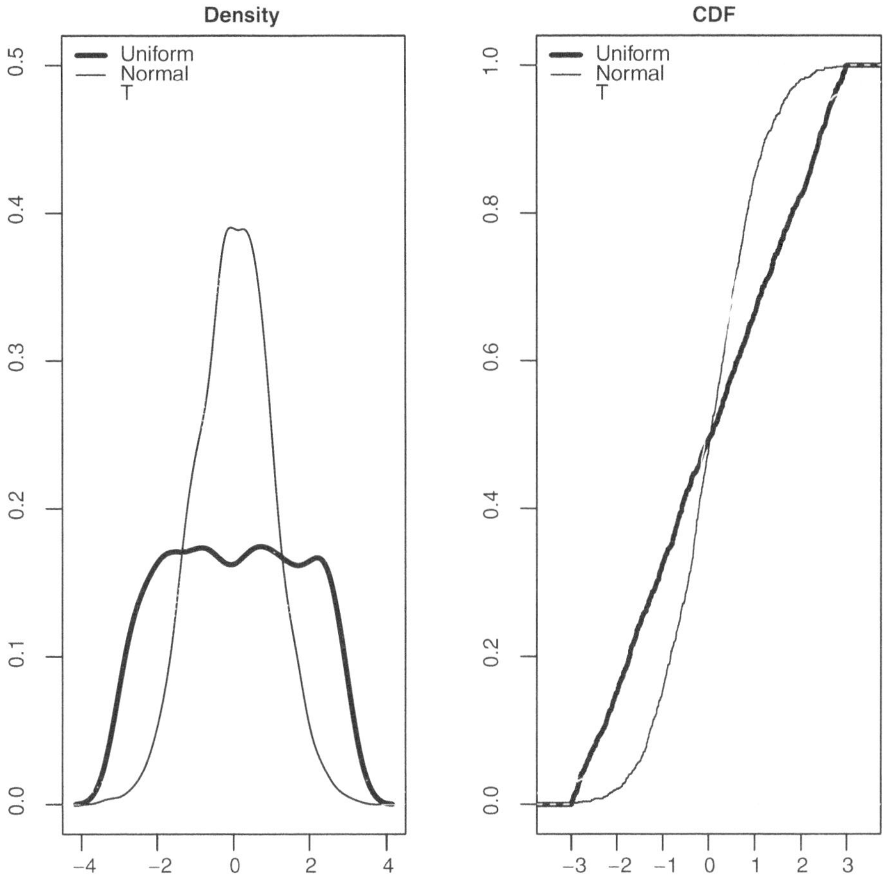

Fig. 2.8 The density and CDF plots of a uniform, a normal, and a t-distribution

2.4.4 Uniform Distribution and Normal Distribution

Among all distributions, there are two special ones: the uniform and normal distributions, which have been studied and applied extensively in real-world analytics. In this subsection, we focus on these two distributions, discussing their density functions, expected values, and special features.

2.4.4.1 Uniform Distribution

Recall that a uniform distribution has outcomes spread with the same likelihood within a range. The density function of a uniform distribution can be expressed as

$$f(x) = \frac{1}{b-a}, \ a \le x \le b; \qquad f(x) = 0, \ \text{otherwise}.$$

Using (2.1), we can obtain the expected mean and expected variance as follows:

$$E(X) = \frac{a+b}{2} \qquad V(X) = \frac{(b-a)^2}{12}.$$

For example, if we think a stock is equally likely to take a price from \$10 to \$15, then the expected average price would be 12.5, and the variance would be $25/12 = 2.08$.

2.4.4.2 Normal Distribution

Most outcomes do not occur with equal likelihood. Rather, they cluster around an average value with fewer extreme values. The normal distribution represents this typical scenario and is the most studied among the bell-curve distributions. The name "normal" comes from the fact that this distribution occurs normally in nature and many experiments.[6]

A random variable X follows a normal distribution:

$$X \sim N(\mu, \sigma^2)$$

with the density function

$$f(x|\mu, \sigma^2) = \frac{1}{\sqrt{2\pi\sigma^2}} e^{-\frac{(x-\mu)^2}{2\sigma^2}}.$$

We see that a normal distribution can be characterized fully by just the first two moments, mean and standard deviation, where the first moment is called location and the second moment is called scale. If we move the location by $-\mu$ and divide the scale by σ, we obtain a standardized normal distribution. Moreover, all normal distributions can be standardized with mean zero and variation one, with the standard normal density function

$$f(x|0, 1) = \frac{1}{\sqrt{2\pi}} e^{-\frac{x^2}{2}}.$$

The CDF of a normal distribution can be obtained by the integration

$$F(x) = \frac{1}{\sqrt{2\pi}} \int_{-\infty}^{x} e^{-\frac{t^2}{2}} dt.$$

[6]Normal distribution is also called Gaussian because it was discovered by Carl Friedrich Gauss.

Since a normal distribution depends only on the first two moments, the third and fourth moments must be fixed. Because a normal distribution is symmetric, its skewness value is zero. Since the fourth moment of a normal distribution is $3\sigma^4$, thus by definition, its kurtosis is 3. We will see later in this chapter that $\mu + / - 3\sigma$ covers about 99% of all outcomes.

The normal distribution produces a symmetric bell curve without a fat tail. Many natural phenomena follow a normal distribution, which has many elegant properties. Here we list the main ones:

- Symmetric: around the mean.
- Stable: the addition of two normal distributions is still normal.
- Scalable: standardization of a normal distribution is still normal, $(X - u)/\sigma \sim N(0, 1)$.

The normal distribution is employed very often due to its useful properties and the central limit theorem. Major usages are

- Sample testing
- Central limit theorem: a distribution converge to normal as sample size increases
- Models, assuming normal distribution of the error term

In finance, we need to think twice—financial events have long/fat tails, and they usually do NOT follow a normal distribution! We present evidence of this in the industry insights section.

2.5 Univariate Analysis: Hypothesis Testing

We have discussed the financial markets, stock price movements as a random variable, and the basic quantitative analysis of a random variable, including the four moments, distributions characterized by a density function and CDF. In the real world of investments, investors make judgments about the status of a stock market—is it very hot now or in a trough?—then take appropriate actions. What is a scientific way to make a sound judgment about a financial market? We have talked about the business cycle, economic fundamentals, and corporate earnings, but all of these metrics need to be tested.

In this section, we introduce hypothesis testing and illustrate with the S&P 500 data how to carry out an empirical study. Before getting into the details of hypothesis test, we first need to introduce some basic concepts.

Population and Sample A population consists of *all* elements from a set of data. A sample includes *one or more observations* drawn from the population. If the sample is random and large enough, we can collect the information derived from the sample to make the best guess about the population. For example, we cannot taste all the strawberries in an orchard, but we could taste a few or many to tell whether the strawberries in that orchard are good or bad.

Population mean: the true average value of the population, which we do not know. However, we can use a sample to approximate the population, thus estimating population statistics from the sample.

Sample mean: an estimate from a sample for the real ("theoretical") population mean.

Estimation Error from Samples Every estimate deserves an error. Because the sample mean may change with different samples, the variation of sample means is defined by standard error (SE). Suppose the sample size is n, and the sample standard deviation is s, we have SE:

$$SE_{\bar{x}} = \frac{s}{\sqrt{n}}.$$

Hypothesis: a hypothesis is a proposed conclusion made on the basis of evidence as a starting point for further investigation. Hypothesis testing is the use of statistics to estimate the probability whether a given hypothesis is true and to what degree. Using the observed data from a sample, hypothesis testing is basically an assumption that we make about the population parameter.

Confidence interval: an interval with bounds from both sides, in which the estimate of a parameter lies with an associated probability. In statistics, we construct a confidence interval from the statistics of the observed data. The interval may or may not contain the true value of an unknown population parameter.

Next, we discuss Student's t-test, two types of error, and one-sided and two-sided tests. We then conduct an empirical study on the current status of the US stock market.

2.5.1 Student's t-Test

Once we have a real-world data set, how do we carry out a hypothesis test? First, we need to know the distribution of the sample, then compute a confidence interval to make a judgment whether to reject or accept a proposition.

A long time ago, there was a smart young man named William Sealy Gosset (Fig. 2.9) who worked at a brewing company in London. He thought hard about how to control the quality of beer using a scientific approach. He found that if a sample comprises observations from a normal distribution, it will follow a t-distribution. Moreover, if the sample is large enough, say, greater than 30, the t-distribution approaches to a normal distribution. Gosset published the idea in a journal using the pen name "Student," hence the birth of the term Student's t-distribution, or simply t-distribution (Student 1908). This paved the foundation for hypothesis testing by using the information from a sample to estimate the population.

In rigorous quantitative terms, if a sample consists of observations from a normal distribution that are independently and identically distributed (i.i.d.) with population mean μ, then the following sample statistical value follows a t-distribution:

Fig. 2.9 William Sealy Gosset (1876–1937), known as "Student" for developing t-statistics

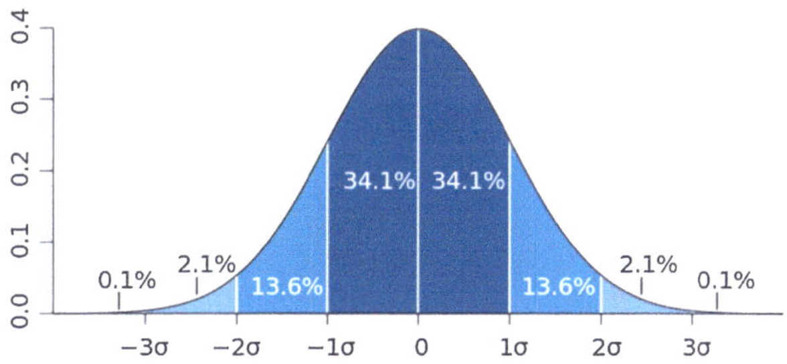

Fig. 2.10 t-Distribution and probability coverage

$$\mathcal{T} = \frac{\bar{x} - \mu}{s/\sqrt{n}} \sim \text{Student's t-distribution.}$$

One immediate application of Student's t-distribution (Fig. 2.10) is to construct confidence intervals for a population value (the true value), which is critical for hypothesis testing. Suppose we have a sample value, T, with a 95% probability of lying in the range from –A to A:

$$P(-A < T < A) = 0.95$$

$$P(-A < \frac{\bar{X}_n - \mu}{S_n/\sqrt{n}} < A) = 0.95$$

$$P(\bar{X}_n - A\frac{S_n}{\sqrt{n}} < \mu < \bar{X}_n + A\frac{S_n}{\sqrt{n}}) = 0.95.$$

We know that for a random variable following a t-distribution, 95% of values fall within 1.96 standard deviation from the mean, hence $A = 1.96$. Thus, the 95% confidence interval for the population mean μ would be

$$\text{upper 95\% limit} = \bar{x} + 1.96 \text{ S.E.}$$

$$\text{lower 95\% limit} = \bar{x} - 1.96 \text{ S.E.},$$

where S.E. is the standard error of the sample.

2.5.2 *Hypothesis Testing, Type I Error, and Type II Error*

With knowledge of t-statistics and confidence intervals in hand, we can test a hypothesis by using sample values to estimate a population value.

The form of a hypothesis consists of two parts, a proposal and its opposite, where the former is called the "null" (denoted by H_0) and the latter is called the "alternative" (denoted by H_1 or H_a). The null and alternative hypotheses cover all possible outcomes of a random variable, so the sum of the probabilities of the null and alternative hypotheses is equal to one. In other words, the only possible outcomes are that the null is true or the alternative is true. In statistical terms:

$$Null - -H_0, \qquad Alternative - -H_1.$$

An example is

$$H_0 : \text{wolf is not coming}, \qquad H_1 : \text{wolf is coming}$$

or in finance,

$$H_0 : \text{The S\&P 500 is overvalued}, \qquad H_1 : \text{The S\&P 500 is not overvalued.}$$

2.5.2.1 Type I Error and Type II Error

When testing a hypothesis, it is possible to make two types of errors: (1) rejecting the null when it is true or (2) accepting the null when it is false.

Type I: the error of rejecting H_0 when it is actually true.

Example: The truth is that the Big Bad Wolf is not coming, but we trusted the boy who cried wolf. Thus, we made a type I error.

$$H_0 : \text{No BBW coming;} \qquad H_1 : \text{BBW is coming.}$$

Type II: the error of accepting H_0 when it is actually false.

Example: The truth is that you are sick, but the doctor says you are perfectly fine. Thus, the doctor accepted the null that is false.

$$H_0 : \text{You are not sick;} \qquad H_1 : \text{You are sick.}$$

Many statistical theories revolve around the minimization of either one or both errors. However, it should be addressed that first, a complete elimination is impossible; second, minimizing type I error increases the chances of type II error and vice versa.

In the context of investment, we see both type I and type II errors almost everywhere. For example, in quantitative investing, people carry out backtests for a factor using historical data. A type I error is the mistake of accepting a hypothesis when it is not true: the backtest suggests that a factor resulted in high investment returns during a given period, but this could also be due to chance or explained by something else instead. This is one of the major drawbacks of quantitative backtesting. However, some factors do work in general, such as value and momentum. Completely ignoring those two factors in a quantitative strategy may result in a type II error.

2.5.2.2 Hypothesis Testing

We specify the following three steps for a hypothesis testing:

Step 1: State H_0 and H_a with a Significance Value Based on what you expect, state the null and alternative hypotheses (e.g., H_0: wolf is coming; H_1: wolf is not coming). To test a hypothesis, we need to set a critical probability, also known as significance level, to minimize type one error. In simple terms, the significance value is the probability of making an incorrect decision. In general, type one error is considered the more "grievous" kind of error to make. A typical significance value is from 0.01 to 0.10, corresponding to a confidence level of 90% or 99%.

Step 2: Calculate a Test Statistic Based on Sample Data and Construct a Rejection Area Based on the distribution (e.g., t-distribution) of the test-statistic values, we get a critical value corresponding to a significance level. This critical value is used as the minimum value for rejecting the null. Use the critical value to build a rejection region for the null.

Step3: Draw a Conclusion About H_0 and Make a Business Decision If the test value falls within the rejection area, we reject H_0 with the confidence level of (1-significance probability). Nowadays, most statistical packages provide a p-value, the probability of a test-statistic value greater than what you observe. In simple words, H_0 is rejected if a p-value is greater than the significance level.

Note that we need to be cautious on its business implications if a hypothesis testing is involved. We discuss this in detail in the next section.

2.5.3 Is the Current US Stock Market Overvalued?

To investigate whether the US stock market is overvalued, we presented in previous sections the plots of price and P/E ratio for the S&P 500 index over a long period from 1900 to 2019. We see based on both price and P/E ratio that the current market is at a historical high, indicating that the US stock market is probably hot.

However, visualization can be vague. We now conduct univariate analysis to explore the answer. Consider the price first. The 4-moment calculation and percentile comparison show that the current S&P 500 daily price is over 2-standard deviation and at the 95th percentile of the price distribution, indicating that the current US stock market may be overvalued. Now consider the P/E ratio. A t-test for the sample (see the results generated by the R scripts below) has the t-value equal to 0 and 95% confidence interval of (14.21, 14.69), a fairly narrow band for the population mean.

Test of PE ratio

```
> ## R command for t-test is t.test
> t.test(SP500$PE)
One Sample t-test

data:  SP500$PE
t = 0, df = 1336, p-value = 1
alternative hypothesis: true mean is not equal to 14.45
95 percent confidence interval:
 14.20519 14.68652
sample estimates:
mean of x
 14.44586
```

With the confidence in the mean accuracy, we calculate the standard deviation (sigma) of the current P/E ratio. To ensure robustness, we employ the average of the mean and median of the most recent 6-month P/E ratios to represent the current P/E value. We also calculate the quantile of the current P/E value in the distribution of all P/E values from 1900 to 2019. Table 2.6 presents the results in detail. The current P/E ratio is at 1.42 standard deviations from the mean and at the 91st percentile of the distribution. This implies that the US stock market is approaching the stage of being overvalued in the context of the long history measured by the P/E ratio.

For hypothesis testing, results based on historical data tell only one part of the story. We have to be very cautious when interpreting the test results, being mindful especially of the assumptions and conditions required for the test results to be valid. Note that Student's t-test is based on the assumption that observations are i.i.d. from a normal distribution. However, we know that, first, the S&P 500 index price data do not

Table 2.6 The summary statistics about P/E ratios

Summary name	P/E summary	Quantile	Quantile P/E
Mean	14.4458564	0.85	18.92
Std	4.4857108	0.90	20.68
Current P/E	20.81	0.95	22.50
(x-mean)/std	1.42	0.975	23.57
Quantile	0.91	0.99	24.13

follow a normal distribution and deviate very far from a t-distribution, and second, the data—daily prices in our test—are not i.i.d. at all: today's price has a strong relationship with yesterday's price, and every day new events can cause the index price to follow a very different distribution, so the data are neither independently nor identically distributed.

The limitations may make these tests seem somewhat useless. However, they at least give us some implications. We should interpret the test results with great caution: they may be significant, nonsignificant, or in a gray area. For example, in our test of S&P 500 index price, we can conclude that the current level does reach a historical high based on the historical data, but whether it will persist or not depends on many factors.

We now apply the "value investing" principles of Benjamin Graham. Recall Graham's original formula for intrinsic value, and let us apply this formula at the market level. During the 1920s, the US GDP growth rate was about 4.8%, so we have

$$V/E = 8.5 + 2 * 5 = 18.5.$$

Considering the margin of safety, value investing in the equity market requires a P/E ratio much lower than 18.5. Modifying the formula for the most recent years, 2015–2018, using the GDP growth rate of 4%, we calculate V/E and P/E ratios corresponding to different levels of safety margins and expected returns:

$$V/E = 5 + 2 * 4 = 13, \quad V/E = 10 + 2 * 4 = 18, \quad V/E = 15 + 2 * 4 = 23,$$

$$V/P - 1 = \text{expected return} + \text{margin of safety}$$
$$P/E = \frac{V/E}{1 + r + s}.$$

We present the values in Table 2.7, which shows that the investable P/E ranges from 15.33 to 17.69. Based on Graham's principle, there is no investable opportunity for long-term buy-hold investment in the current US stock market, where the current P/E is around 21, well overvalued.

When we make business decisions, a simple test is not good enough. We need to examine multiple factors in depth and context, including the economic situation,

Table 2.7 Calculation of investable P/E based on the values of V/E, expected return, and margin of safety

V/E	Expected return	Margin of safety	Investable P/E
23	10	20	17.69
23	10	30	16.43
23	15	20	17.03
23	15	30	15.86
23	20	20	16.43
23	20	30	15.33

political stability, and global business cycles. For example, in the previous sections, we measured the US stock market from two different angles based on business cycles and P/E value. As of today, we can see that all these measures indicate that the US stock market is overvalued at a historical high. We need to investigate further, for example, will the current level persist or go even higher? If we think the market has peaked and will start to decline, what is a proper investment strategy? We defer the answers to these questions for now. Hopefully, after a few more chapters, our readers will have a better understanding and be able to reach their own conclusions.

2.6 Industry Insights: Distribution, Outliers, and Treatment

We have discussed the US stock market with a focus on the S&P 500 index and univariate analysis in the previous sections. In this section, we share with readers some industry approaches, starting with the analysis of asset returns.

2.6.1 Asset Returns and Their Distribution

We mentioned earlier in this chapter that finance data, especially asset returns data, barely follow a normal distribution. Asset returns can be extremely low or high, thus yielding long tails in their distribution. Moreover, given different market situations, the price moves in different directions, thus causing serious skewness. We illustrate these aspects here using S&P 500 index data.[7]

In Fig. 2.11, we present density plots of S&P 500 index returns from 1918 to 1949 (top panel) and 1950 to 2018 (bottom panel). For each time period, the left graph has two density plots: the solid line represents the empirical density of S&P 500 returns, while the dashed line is the theoretical normal distribution with the same mean and standard deviation as the returns data. Therefore, if the S&P 500 returns follow a normal distribution, the two plots should be identical. However, we see that the empirical density plots deviate significantly from the normal density plots. The

[7]More evidence is provided for public equity markets in other countries and other asset classes.

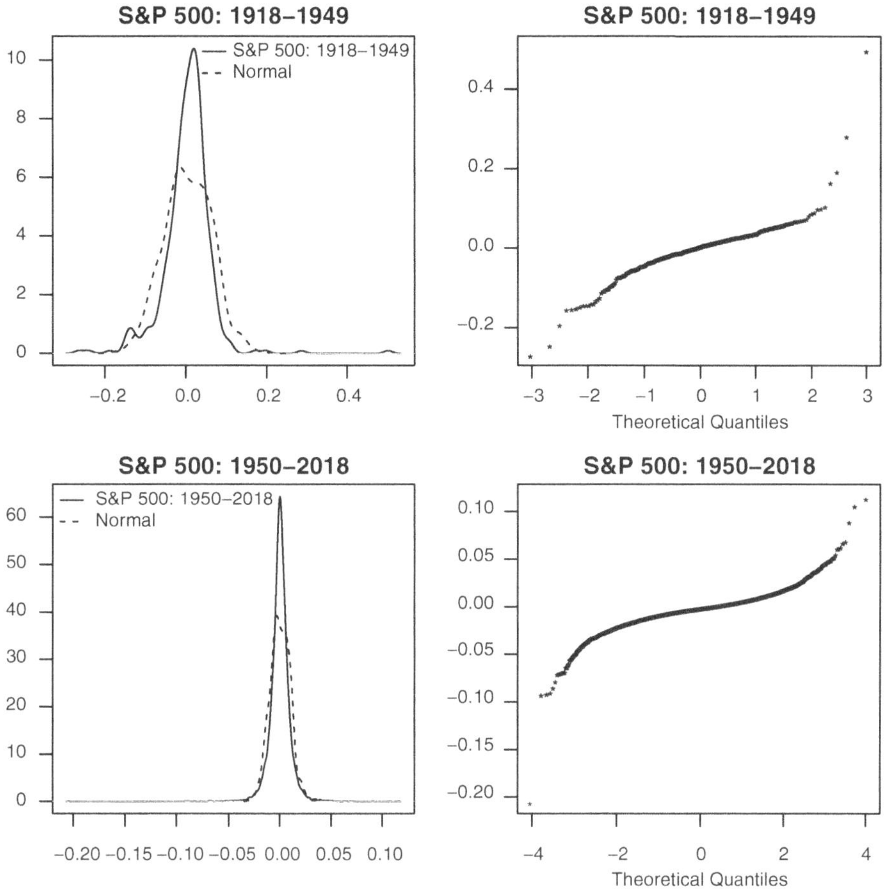

Fig. 2.11 Density (left) and normality (right) plots of S&P 500 index returns from 1918–1949 (top panel) and 1950–2018 (bottom panel)

empirical plot for 1918–1949 has a very fat long *right* tail because the market had many days with dramatic increases, while 1950–2018 has a very fat long *left* tail because the market had many days of extreme lows. This is further confirmed by the graphs on the right, where the returns are plotted against a theoretical standard normal distribution by percentiles. If the returns follow a normal distribution, the plots should be straight lines.

Since the US stock returns do not follow a normal distribution, the four moments are not enough to capture the characteristics of the distribution. In this case, we can complement the four moments with percentiles or quantiles to summarize a returns distribution.

In fact, it is well known in the industry that asset returns are not normal, so what should investors do with this information? Regarding portfolio performance impacts,

Table 2.8 The frequency outside the bound of 1–5 sigma per year for the S&P 500 daily returns and a normal distribution

Sigma	Return bound	Freq/year	Normal prob	Normal freq/year
1	−0.93, 0.99	53.31	0.6827	79.33
2	−1.89, 1.96	11.82	0.9545	11.38
3	−2.85, 2.92	3.53	0.9973	0.68
4	−3.81, 3.88	1.52	0.9999	0.01
5	−4.78, 4.84	0.67	1	0

Fama and French state excellently: "investors should expect extreme returns, with negative and positive returns having equal probabilities to every fair price." Second, investors should adopt a systematic approach to deal with fat tails in the investing process, because when extreme returns happen, they are usually very large, resulting in catastrophic effects. Table 2.8 presents the frequency of returns outside of the 1–5 sigma range for the S&P 500 daily returns and a normal distribution. Using daily returns data of the S&P 500 from 1950 to 2018, we first calculate the mean and sigma (standard deviation), then compute the bound for 1–5 sigma deviations from the mean return, and finally we get the frequency per year by using the number of daily returns outside of the bound. Now, assuming daily returns for this period follow a normal distribution, we compute the probability of returns outside of the bounds and hence the frequency per year for returns occurring outside of the bound, assuming 250 trading days per calendar year. It is expected but still shocking to see that in the real world, about 3.5 days per year will have returns lower than −2.85% or higher than 2.92% (outside of the 3-sigma range), while it is expected to happen only 0.68 times per year if returns follow a normal distribution. For the probability of events outside of the 4-sigma range, there are about 1.5 days per year in which the S&P 500 price dropped more than 3.81% or increased more than 3.88%, while it would take 100 years to have such a day if returns followed a normal distribution. Extreme events do happen in the real world, especially for asset returns—they do not follow a normal distribution, so investors need to be prepared for long and/or fat tails.

In quantitative investing, non-normal distributions require investors to pay special attention to outliers, modeling, and portfolio construction. We describe in detail the outlier issue and treatment in the following section. On the modeling side, a good practice for the alpha generation process is to separate the outlying returns out. A good forecast for a few outliers and a bad forecast for massive returns can still result in a highly fitted model, but this will cause investments to fail if those extreme events do not happen in the future, especially if the industry portfolio follows monthly or annual performance ranking against index or peers. But we can prepare for this through extreme event investment and/or risk management. Regarding portfolio construction, the nonnormality of return distributions imposes significant limits on portfolio optimization relying on mean and variance. We analyze this further in Chaps. 8 and 9.

Table 2.9 A long/short portfolio based on P/E

Company name	Price	Earnings	P/E
ABC	16	0.1	160
XYZ	25	1000	0.025

2.6.2 *Outliers: A Systematic Approach to Detection and Treatment*

In the real world, data is not "normally" distributed, and outliers may be present. Unfortunately, outliers usually determine the outcomes of investments. For example, if you were to form a long-short portfolio based on the P/E ratio of 10 companies, the company with the highest P/E would be in the long position and the lowest would be in the short position (Table 2.9).

Given the significant role outliers play in quantitative investment, knowing how to deal with outliers is very important. We list below a three-stage approach used widely in the investment industry.

$$Detection \Rightarrow Judgment \Rightarrow Treatment$$

- Detection
  - Plot, visualization
  - Sigma, percentile
  - Output: flag
- Judgment
  - Correct or incorrect?
  - Structural or temporary?
- Treatment
  - Truncation or winsorization?
  - Further exploration such as distribution forcement

This will be discussed in detail in the context of factor treatment for alpha modeling in Chap. 4.

2.7 Introduction to R: Importing Data, Simple Calculations, and Plots

In this section, we introduce R with a focus on importing data from an external source as well as making simple calculations and plots. In particular, applying what we have learned about the univariate model, the reader should be able to achieve the following goals:

- Download R: Windows or Mac, 32-bit or 64-bit
- Import data: read.csv or read.table
- Data summary: summary(data.set$variable.name)
- Univariate model: calculate the four moments (mean, sd, skewness, kurtosis)
- Simple plots: scatter, time series, hist, density, normality
- Simple test: t-test
- Calculate asset performance: annualized return and risk, Sharpe ratio

2.7.1 *Importing Data and Simple Calculations*

One important feature of R is that users can easily import data to the work directory. Three elements are required: the data format, data location, and data name.[8] The command in R is *read.table*, which imports a table-format file into R session and creates a data frame. Here we use a simplified version, *read.csv*, to import the S&P 500 daily prices data set in *csv* format. Note that the symbol "#" is used for comments in R.

Import data from an external source

```
##read.csv(file, header = TRUE, sep = ",", quote = "\"",
##      dec = ".",  fill = TRUE, comment.char = "", ...)
> sp5.path="/.../teach.2019spring/R/data/"
> sp5.name="SP500.indexDailyPrice.csv"
> sp5.full=paste(sp5.path, sp5.name, sep="")
> sp5.full
[1] "/.../teach.2019spring/R/data/SP500.indexDailyPrices.csv"
> sp5=read.csv(file=sp5.full, sep=",", header=T)
> dim(sp5)
[1] 17435     6
> sp5[1:2, ]
  yyyymmdd  open  high   low close  return
1 19500104 16.85 16.85 16.85 16.85 0.01140
2 19500105 16.93 16.93 16.93 16.93 0.00475
```

[8] This is analogous to a mail carrier delivering mail to an address.

In the *read.csv* parameters, “sep” specifies the character used to separate columns, and “header” is for the column names. The command *dim* specifies the dimensions of a data frame. In R, a data frame has rows and columns, and the contents of each column should be in the same format: either numbers or characters. However, data type can differ between columns. This feature of R is very important for quantitative investing, as investors can have a column with stock identifiers, such as company name or cusip, and another column with prices, which is very convenient.[9] In addition to data frames, there are other forms of data:

- scalar
  can be a letter or character, e.g., x=5 or x=“a.”
- vector
  can be either numbers or characters but not both, e.g., $y = c(1, 2, 3, 4, 5)$ or y=c(“a,” “b,” “c,” “d,” “e”).
- matrix
  can contain either numbers or characters but not both, e.g., z=matrix(1:10, nrow=2, ncol=5). An element is expressed by the row and column numbers. For example, the second row and third column would be z[2, 3].
- data frame
  w=data.frame(stock=c("IBM","APPL","BOA"), price=c(10,20,30)).
- list
  multiple dimensions, list.example=list(2,3,4).

If a scalar is numeric, the R commands for addition, subtraction, multiplication, and division are +, -, *, *and* / , the power command is ^, and the log function is *log(x, base)*, with the default being the natural log without a base value.

Simple calculations: scalar

```
> ## quantitative results of working hard
> ## by doing just a little bit every day
> x=1.01
> y=0.99
> x^365/y^365
[1] 1480.66
> x^365-y^365
[1] 37.75792
```

For a vector, R requires the format *vector name = c(element1, element2)*, where “c” is mandatory and elements are separated by commas. Special cases include when the values are generated from functions and continuous numbers, such as $x = 1:5$,

[9] A cusip has 9 digits. It provides a unique identification for all stocks and registered bonds in the USA and Canada.

which is the same as $x = c(1, 2, 3, 4, 5)$. If two vectors are numeric, then the chosen calculation, such as addition or multiplication, will be applied to each element.

Simple calculations: vector

```
> ## mix of a uniform and normal distribution
> vector.uniform=runif(100)
> vector.norm=rnorm(100)
> summary(vector.uniform+vector.norm)
   Min. 1st Qu.  Median    Mean 3rd Qu.    Max.
-1.9870 -0.2199  0.4780  0.4095  1.2110  2.3610
```

A matrix can contain vectors as either rows or columns. The R command for matrix multiplication is %*%. An element can be obtained by specifying the row and column numbers, such as matrixName[2, 7]. To get the entire second row, use matrixName[2,], which by default does not specify column numbers.

Simple calculations: matrix

```
> mm1=matrix(1:10,nrow=5,ncol=2)
> mm2=matrix(11:16,nrow=2,ncol=3)
> mm1
     [,1] [,2]
[1,]    1    6
[2,]    2    7
[3,]    3    8
[4,]    4    9
[5,]    5   10
> mm2
     [,1] [,2] [,3]
[1,]   11   13   15
[2,]   12   14   16
> mm1%*%mm2
     [,1] [,2] [,3]
[1,]   83   97  111
[2,]  106  124  142
[3,]  129  151  173
[4,]  152  178  204
[5,]  175  205  235

> eigen(matrix(1:9,nrow=3))
```

```
$values
[1]  1.611684e+01 -1.116844e+00 -5.700691e-16

$vectors
           [,1]       [,2]       [,3]
[1,] -0.4645473 -0.8829060  0.4082483
[2,] -0.5707955 -0.2395204 -0.8164966
[3,] -0.6770438  0.4038651  0.4082483
```

Different from a matrix, a data frame can have mixed columns of numbers and characters. The special character *$* is used to call a specific column.

Simple calculations: data frame

```
> sp5[1:2,]
  yyyymmdd  open  high   low close  return
1 19500104 16.85 16.85 16.85 16.85 0.01140
2 19500105 16.93 16.93 16.93 16.93 0.00475
> summary(sp5$return)
      Min.    1st Qu.     Median       Mean    3rd Qu.       Max.
-0.2047000 -0.0040300 0.0004700 0.0003457 0.0049600 0.1158000
> dim(sp5)
[1] 17120     6
> cumprod(sp5$return+1)[17120]^(220/17120)-1
[1] 0.06807084
> sd(sp5$return)*sqrt(220)
[1] 0.1424808
> SharpeRatio=0.068/0.1425
> SharpeRatio
[1] 0.477193

>## Find days where daily returns dropped by more than 20%!
> bad.days=which(sp5$return < -0.20)
> sp5[bad.days,]
      yyyymmdd    open    high     low   close   return
9497  19871019  282.70  282.70  224.83  224.84 -0.20467
```

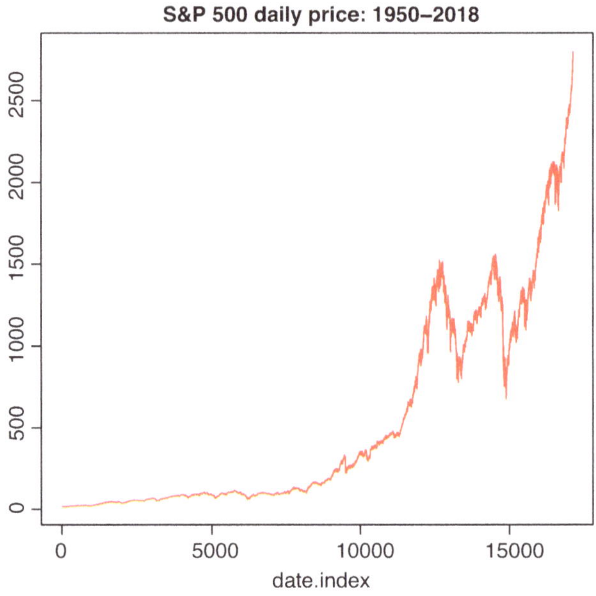

Fig. 2.12 Example 1: plot of closing prices for S&P 500 daily returns

2.7.2 Simple Plots

In the previous subsection, we presented basic R commands for simple calculations. Next, we demonstrate how to use R commands to draw graphs. R is very powerful in this respect.

One of the most important R commands is *plot*, a generic function for plotting R objects. Other useful R commands related to plots are *hist, barplot*, and *boxplot*, which produce graphs of histograms, bar charts, and box-cox statistics, respectively. Here we use the data set of S&P 500 returns for illustration purposes. There are three main categories that distinguish plots of functions: generic, distribution, and comparison functions.

Graph: Generic—plot, point The most common plotting function in R programming is the plot() function. In R, plot() is a generic function to make graphs. There are many parameters used in the plot() function, including *x, y, type, xlab, ylab, main, col,* etc. In R, a plot can be exported to an external folder as a PDF file, using the R command *pdf*. The result is shown in Fig. 2.12.

Graphs: plot

```
# options for colors, labels, and titles
> pdf("/.../book/quantInvesting/chapter1/RplotEx1.pdf")
> plot(sp5$close,col="tomato",xlab="date.index",ylab="",
  main="S&P 500 daily price: 1950-2018",type="l")
> graphics.off()
```

Graph: Distribution—hist, density, boxplot There are many ways to visualize the distribution of a variable. In R, some commonly used commands are *hist*, *plot(density())*, and *boxplot*. We present example plots in Fig. 2.13. A histogram, also

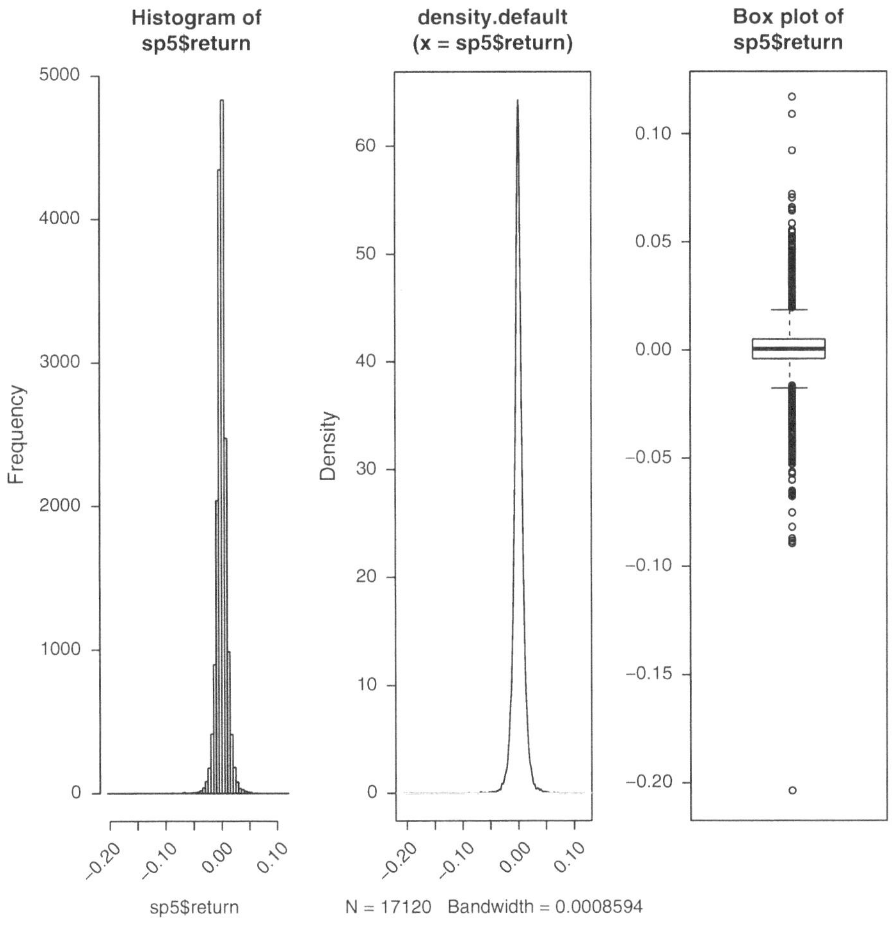

Fig. 2.13 Example 2: plot of the distribution of S&P 500 daily returns

called a frequency plot, breaks down data into classes or "bins." The number of bins is defined by hist(x, breaks), but there are many other parameters in the command. While a histogram is a categorical description of the data distribution, plot(density()) produces a continuous distribution. A box plot summarizes data with five statistic values: minimum, 25th percentile, median, 75th percentile, and maximum.

Graphs: hist, plot(density), boxplot

```
>## produce a 3x1 panel of plots using par(mfrow=c(1,3))
> pdf("/.../book/quantInvesting/chapter1/RplotEx2.pdf")
> par(mfrow=c(1,3))
> hist(sp5$return,breaks=50)
> plot(density(sp5$return))
> boxplot(sp5$return,main="Box plot of sp5$return")
> graphics.off()
```

Graph: Comparison—barplot, pie, heatmap Sometimes, we need to visualize the composition of or differences between data points. For such instances, one can use the R commands *barplot, pie, or heatmap*. Using the S&P 500 data, we show the pie and bar plots in Fig. 2.14. The meaning of bar and pie charts is straightforward. A heatmap is commonly used to look for hotspots in two dimensions.

Graphs: barplot, pie, and heatmap

```
> pdf("/.../book/quantInvesting/chapter1/RplotEx3.pdf")
> par(mfrow=c(2,1))
> pie(sp5.sector[,6],labels=sp5.sector[,1],col=rainbow(10),cex=0.5)
> barplot(t(xx),beside=T,names.arg=sp5.sector[,1],cex.names=0.3)
> graphics.off()
```

Keywords, Problems, and Family Project

Part I: Keywords

Four moments, distribution, density and CDF, normal/t/uniform distribution, hypothesis test, type 1 and 2 errors

S&P 500, P/E ratio, business cycles, annualized return/risk, Sharpe ratio
Process of dealing with outliers, long/fat tails of asset returns
R commands: import data, plots, scalar, list, matrix, data frame, t-test

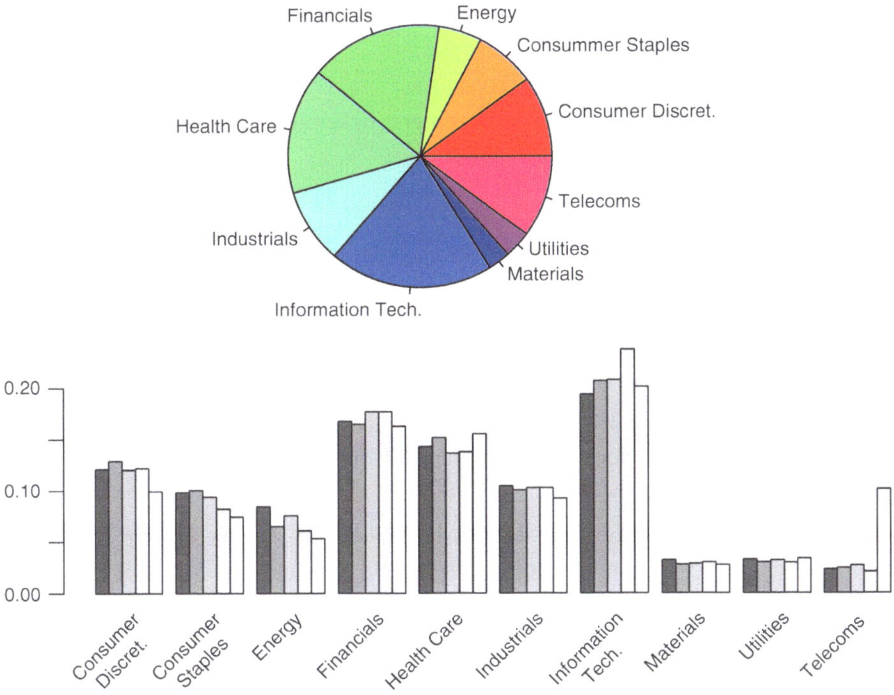

Fig. 2.14 Example 3: pie chart and bar chart of sector weights for returns of the S&P 500

Part II: Problems

Problem 2.1 Understand the US stock market by using R to analyze the S&P 500 daily price data from 1950 to 2019.

(1) Calculate annualized return, risk, and Sharpe ratio for the entire period.
(2) Calculate annual performance for each calendar year.
(3) Plot the price movement of the index.
(4) Calculate four moments of the daily returns.
(5) Plot the density of daily returns and judge if it follows a normal distribution.
(6) Identify extreme returns and explain what happened.

  (i) Extreme positive returns? When did they happen and explain why.
  (ii) Extreme negative returns? When did they happen and explain why.

Problem 2.2 Test if the current US stock market is overvalued.

(1) Set up a hypothesis test and carry out a t-test using R.
(2) Interpret the results and investigate whether the current market is overvalued.
(3) Validate the distribution assumption for the test.
(4) What are the business implications for quantitative investing?

Problem 2.3 Pick a stock in the S&P 500 index, conduct fundamental analysis of the company.

(1) Company origination and development
(2) The IPO date, price, and market identifiers (ticker and cusip)
(3) Lines of business and core business
(4) Management team
(5) Peers or competitors in the industry
(6) Company performance during the last 3 years
    (i) business performance: total revenue, net income, total asset, cashflow
    (ii) EPS, DPS (if the company issues dividends)
(7) Stock market performance since public
    (i) market performance: return, volatility, and Sharpe ratio
    (ii) compare the performance with the S&P 500 index and its peer companies

Part III: Family Project

Problem 2.4 Try to apply what you learned from this chapter to help with your family asset. Talk with your grandparents/parents or other family members, pick a financial/real asset owned by your family, write a 3–5 page report.

(1) Why did your family have that asset? What is the area you could help?
(2) Understand that asset and make a proposal for help.
(3) A plan to get data and conduct quant analysis in R.
(4) From a family member (owner of that asset): a paragraph of comments about your conversation and plan.

References

Graham, B. 1949. *The Intelligent Investor*. New York: Harper & Row.
Graham, B., and D. Dodd. 1934. *Security Analysis*. New York: Whittlesey House, McGraw-Hill Book.
Student. 1908. "The Probable Error of a Mean." *Biometrika* 6(1): 1–25.

Chapter 3
What Is the Relationship Between the Chinese and US Stock Markets? Bivariate Analysis

Abstract In this chapter, we explore the relationship between two random variables using the example of the Chinese and US stock markets. How does the performance of the S&P 500 impact the Chinese stock market and vice versa? Are the impacts the same when the US market is bullish and bearish? How long do the impacts last? Is there a scientific way to measure the impacts and formulate an investment strategy? We answer all of these questions in this chapter. First, we introduce the Chinese stock market with a focus on its origins, development, and special features. We then introduce bivariate analysis, the concepts of correlation and rank correlation, and how to apply correlation to measure the relationship between the S&P 500 and CSI 300. We present industry approaches, such as spillover effects and information decay, and demonstrate how to explore the asymmetry of the relationship between the two stock markets. On the programming side, we show how to write a function in R and introduce various methods for loops.

3.1 Introduction to the Chinese Stock Market

Over the last 30 years, the biggest change in the world economy has been the emergence and development of the Chinese economy, which is clearly reflected in global financial markets. Table 3.1 and Fig. 3.1 show the top ten stock markets by capitalization in 2008 and 2018. We see that China became the second largest stock market in 2008.[1] The market weight of the Chinese stock market globally increased from 10% in 2008 to approximately 14% in 2018.

In this section, we first describe three segments of global financial markets—developed, emerging, and frontier—then introduce the Chinese stock market within the context of global stock markets and China's recent economic development. In particular, we review the origins, development, and special features of the Chinese stock market.

[1]This includes China's mainland market and the Hong Kong stock market.

L. Ma, *Quantitative Investing*, https://doi.org/10.1007/978-3-030-47202-3_3

Table 3.1 The top ten stock markets by percentage of global stock market capitalization in 2008 and 2018

Year	USA	Japan	China (mainland)	Hong Kong	UK	France	Germany	Canada	India	Switzerland
2008	34.22	9.87	5.08	5.48	6.49	4.47	3.03	3.27	1.81	2.54
2018	39.81	7.76	7.54	6.75	4.41	3.14	2.83	2.74	2.72	2.11

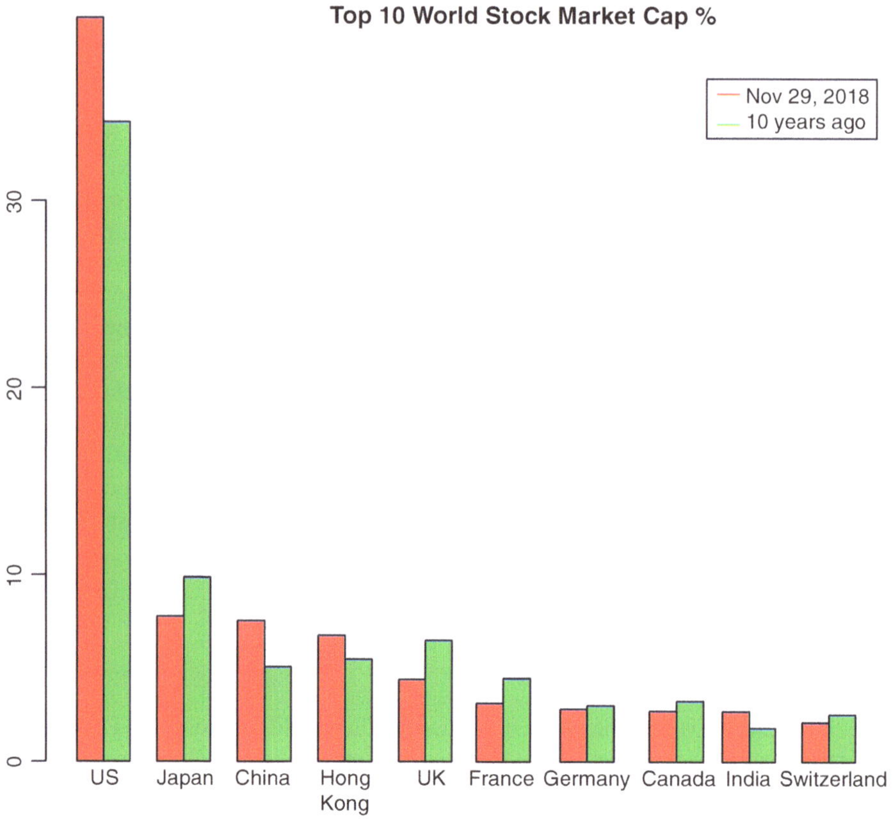

Fig. 3.1 The top ten stock markets by percentage of global stock market capitalization in 2008 and 2018

3.1.1 Global Equity Markets: Developed, Emerging, and Frontier

In the financial industry, global stock markets are divided into three segments—developed, emerging, and frontier—based on the economic development and liquidity of the financial markets. This is not only for conceptual convenience, but also for practical investment purposes, as members of each group share common features, making it easy to establish a benchmark and portfolio based on those common characteristics. Institutional benchmark index providers have done a great job of categorizing financial markets in a timely and dynamic manner.[2]

Below we summarize each market segment briefly so readers have an idea of what constitutes each group and how country weights vary over time within each group.

3.1.1.1 Developed Stock Markets

Developed markets (DM) comprise markets in developed economies in North America (NA), some western European countries (WEU), and countries in the Asia Pacific region (AP).

- NA: the USA, Canada
- WEU: most west EU countries and the UK
- AP: Australia, New Zealand, Japan, Singapore, Hong Kong, Taiwan, and South Korea

The plots in Fig. 3.2 present the number of stocks and weight distribution across countries in the DM from 1992 to 2012. It should be mentioned that the USA has long been the largest among developed markets, representing about 40% of the total weight; this is to be expected given the relative size of the US economy. One noteworthy feature is that the weight share of Japan was about 40% in the early 1990s but dropped gradually to less than 10% in the 2000s due to Japan's economic stagnation. There are several other interesting facts about countries in this group. For example, Canada has more than half of the large capitalization companies cross listed in the US stock market. The Canadian stock market is quite unusual given the country's economic relationship with the largest developed economy, the USA, and the largest emerging economy, China. It should be stressed here that, from a practical investment perspective, most institutional portfolio managers have treated South Korea and Taiwan as emerging markets even though they are formally classified as developed markets because in reality both markets have issues of liquidity, settlement, and currency exchange limits. These special features of the DM stock markets need to be considered when we construct quantitative investment portfolios.

3.1.1.2 Emerging Stock Markets

The emerging stock markets (EM) are made up of stocks in the emerging economies, including countries in the BRICS block, some countries in Europe, Southeast Asia, Mexico, and South America.

- BRICS: Brazil, Russia, India, China, South Africa
- Eastern EU countries
- AP: Malaysia, Indonesia, Thailand

[2]For example, the MSCI World index represents the developed stock markets, and the MSCI Emerging index represents the emerging stock markets.

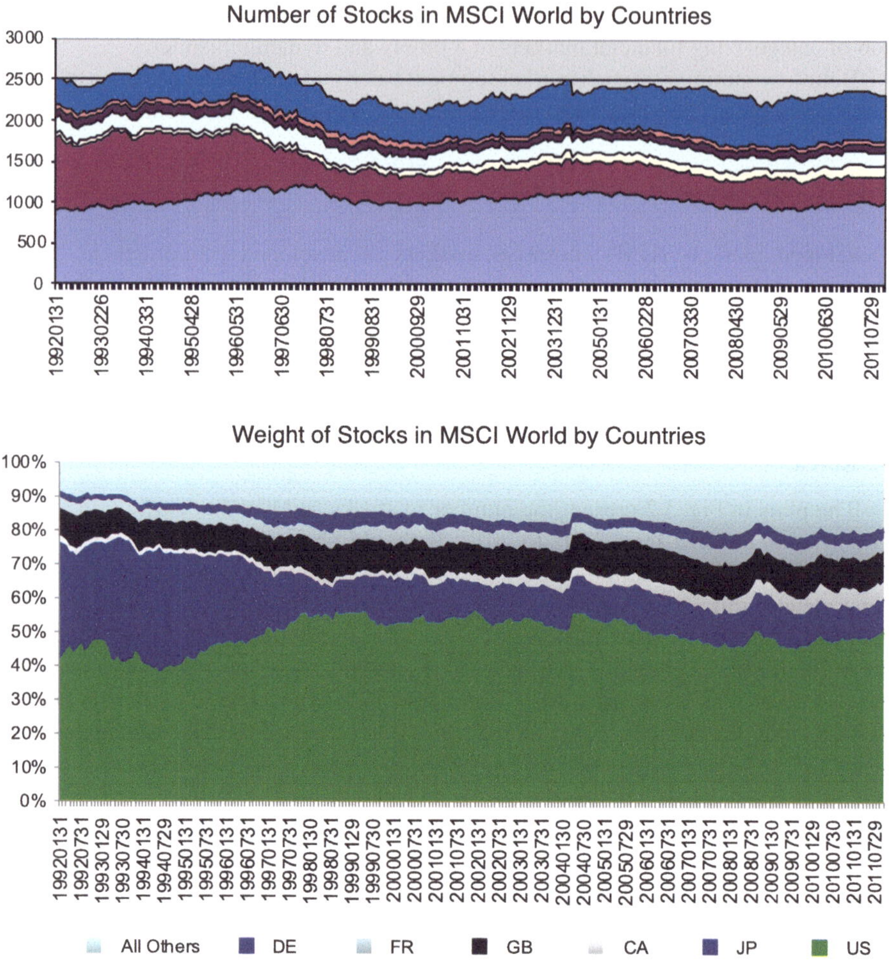

Fig. 3.2 The number of stocks (top plot) and weights of market cap (bottom plot) for stocks in the MSCI World index by country from 1992 to 2012

One prominent feature of EM is the significant weight changes of countries over time. For example, from 1995 to 2018, South Africa exhibited a downward trend, and China exhibited an upward trend, the former dropping from 40% to less than 10% and the latter increasing from less than 5% to about 40%. Currently, the market capitalization weight of Chinese stocks among the emerging markets is about the same as that of US stocks among the developed markets.

Over the past 20 years, the total market capitalization of EM has been about 5–15% of that of DM, with the percentage increasing over time. In terms of market performance, we present annualized performance in Table 3.2, where the MSCI Emerging and MSCI World indices are used for the EM and DM equity markets,

Table 3.2 Index returns (%) as of Nov 29, 2019

	1 Yr	3 Yr	5 Yr	10 Yr	Since 2000
MSCI EM	7.28	9.01	3.12	3.33	8.77
MSCI World	14.53	12.35	7.75	9.34	5.38

Data source: MSCI report

respectively. We see that over the last 20 years, EM has outperformed DM by 3.4% annually. However, EM has underperformed DM in recent years.

3.1.1.3 Frontier Stock Markets

The frontier stock markets refer to developing countries where the stock markets are in the infancy stage. One of the biggest risks in the FM comes from social and political instability, which are associated with the restrictions on foreign investments such as share ownership, currency exchange, and trade settlement. All of these create huge uncertainty in the FM market. In recent years, there has been an increasing trend in the investments in the FM, albeit with many limits, such as illiquidity, restrictions imposed on foreign investors, and extremely high volatility.

We present below the criteria of a major FM index, FSTE's frontier index[3]:

- A formal stock market regulatory authority
- No significant restrictions on repatriation of capital
- A rare occurrence of failed trades
- T + 5 or better (clearing and settlement)
- A timely trade reporting process

where T is the trading date, $T + 5$ means the settlement date is within 5 days after the trading date.

Having described the big picture of the global stock markets, we focus on the Chinese stock market in the rest of this section, including its origins, development, and special features.

3.1.2 Emergence and Development of the Chinese Stock Market

The Chinese stock market started in Shanghai in the late 1860s, when China was forced to open to foreign trade. The Shanghai Sharebrokers Association, formed in 1891, is the first stock exchange in China. Since 1920s, Shanghai had become the trade and economic center in the Far East region. After the merging of two exchanges

[3]FTSE Russell report "Frontier Markets: Accessing the next frontier," 2018.

in 1929, the official name "Shanghai Stock Exchange" (SSE) started. Financial products traded in the exchange include stocks, bonds, and futures. However, after Japanese troops invaded Shanghai on December 8, 1941, the SSE stopped its business. After the Japanese were defeated, the SSE resumed its business in 1946. Three years later in 1949, the People's Republic of China emerged and the SSE halted its operations.

After the cultural revolution in 1976, China adopted an economic policy open to the West (developed countries). To start a capital market, like Western countries, the Shanghai municipal government instituted a temporary policy on public shares issuing in July 1984. Immediately, a local company, Feiyue Musical Company, jumped at the opportunity and issued 10,000 shares of stock to the public with a face value of CNY50 on November 18, 1984. This was the first public stock in China after the stock market closed in the 1950s. About 2 years later, on September 26, 1986, the Shanghai branch of the Chinese Industry and Commercial Bank opened a new business called JingAn Stock Business, which established a trading desk for the public to trade stocks. This was the first stock exchange in China after the 1950s.

Four years later, on Nov 20, 1990, the Shanghai Stock Exchange was incorporated. It opened for business the same year on December 19. The Shanghai Comprehensive Index (popularly known as the Hu Index) was launched on July 15, 1991. The base (100) for the Hu index is the closing price of Dec 19, 1990.[4] On July 3, 1991, another stock exchange, the Shenzhen Stock Exchange, opened for business. Shenzhen developed rapidly from a small village into a modern city because it borders Hong Kong, where there are well-established stock exchange and fairly efficient financial market.[5] The trading systems of both stock exchanges were computer-based. From then on, the Chinese stock market was established and became a significant part of the capital markets.

There were only 8 stocks trading at these stock exchanges in the beginning, the number then increased to 15 in 1992. To avoid extreme volatility, price control rules such as a daily 1% price ceiling were adopted. On May 21, 1992, a new trading rule—adopting $T + 0$ and canceling the price movement ceiling—was established. This triggered a surge in the stock market: the prices of all 15 stocks skyrocketed, and the Hu index rose from 617 at the previous close to 1266 in a single day! The consequences of this have been very serious and long-lasting for both issuing companies and investors. Even today, many people think they can get rich overnight by taking a private company public and putting money into the stock markets. Note that the dramatic price change was caused by changes in rules and policies. This pattern has become a standard model for price movements in the Chinese stock market ever since: the central government makes a rule or policy change, the stock market reacts with dramatic price movements over a very short period, followed by

[4] Hu is the short name of Shanghai. In China, each major city and province have a short name based on culture and history.

[5] Recall from the previous section that the HK stock market is part of the DM.

a long period of reversal. This cycle has continued but at different levels, nowadays with more mature policies, more public companies, and a much bigger stock market.

The Chinese stock market includes three types of shares: A, B, and H. A-shares are domestic shares denominated in local currency (CNY). B-shares, designed for foreign investors, are denominated in foreign currency. Both A and B shares are traded on the Shanghai and Shenzhen stock exchanges. H-shares, denominated in Hong Kong dollars, are traded on the Stock Exchange of Hong Kong. While companies incorporated in China can issue three types of shares, they mostly issued A-shares, which account for about 65% of the domestic stock market. Since they were launched in 1993, H-shares now comprise half of the market capitalization of the Hong Kong stock market.

Note that Chinese A-shares have historically been inaccessible to foreign investors, but the Chinese government has opened its capital markets in recent years. It has launched a series of initiatives to allow institutional investors to purchase A-shares, such as the qualified foreign institutional investors (QFII) scheme. QFII was launched in 2002 with the purpose to attract foreign investments in A-shares. Recently, more and more A-shares have been added to the MSCI emerging markets.[6]

The B-shares were open to Chinese citizens on February 19, 2001. Due to the availability of H-shares and the accessibility of A-shares to foreign investors, the B-shares market has been diminishing in recent years.

3.2 Special Features of the Chinese Stock Market

Since its emergence in the early 1990s, the Chinese stock market has grown rapidly, in terms of both the number of listed companies and market capitalization. Here we present some important features of this stock market.

3.2.1 Rapid Growth

When the Chinese stock market started in 1990, it had only 8 public companies, with a market value of CNY1.234 billion. Thirty years later, it has grown into the second largest stock market in the world, with more than 3500 listed public companies and a total market capitalization of CNY 44 trillion (Fig. 3.3).

The rapid growth of the Chinese stock market is a clear reflection of the country's economic growth. The plots in Fig. 3.4 display the values of GDP (right axis) overlaid with market capitalization. We see that the two lines align with each other for the

[6]MSCI announced on February 28, 2019, that it will increase the weight of China A-shares in the MSCI Indexes using a three-step inclusion process beginning with the May 2019 Semi-Annual Index Review.

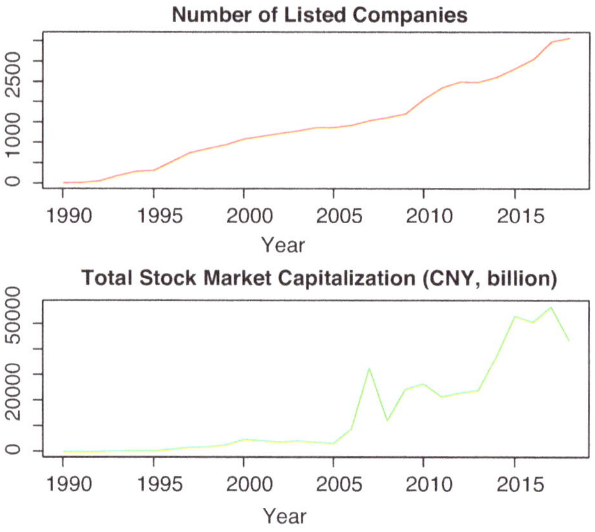

Fig. 3.3 The number of listed companies (top plot) and market capitalization (bottom plot) in the Chinese stock market from 1990 to 2018

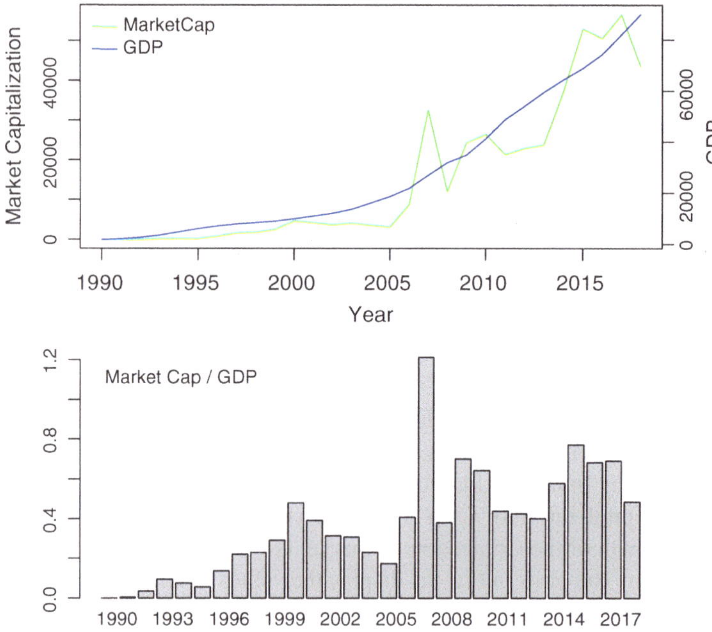

Fig. 3.4 The market capitalization of the Chinese stock market and GDP from 1990 to 2018, with the y-axis in billions of local currency. The bottom plot displays the ratio of market cap to GDP

entire period, with only a few years of outliers. The plot at the bottom shows the ratio of market cap to GDP at each year end. We see that the ratio stays at about 20–40% except for the early years, when the stock market had just started, and with a slight upward trend in later years. There are a few years with exceptional values, such as 2007, when the ratio jumped to 120% when the stock market was in a bubble and then collapsed in 2008.

3.2.2 *Market Participants: State-Owned Public Companies and Individual Investors*

Another important feature of the Chinese stock market is that more than half of China's A-share companies are state-owned enterprises (SOEs) whose executives are appointed by the government. In China, the initial incentive to have a stock market was to raise capital for SOEs. Note that an SOE still controls a tremendous number of shares even after it goes "public," resulting in only a small fraction of shares floating on the market, while the rest are held by the company as a legal representative of state capital. As such, public policy objectives may override the profitability of these companies. Research (e.g., Xie 2019) has found that SOEs are generally less efficient, less profitable, and more leveraged than their peers.

Regarding investors, there are four types of investment groups: legal investors, individual investors, institutional investors, and the central government. Legal investors are the legal holders or representatives of shares not floated on the market, such as state-owned public companies. Institutional investors are the big players in the stock market. They include pension funds, mutual funds, insurance companies, investment banks, and some private equity investors. In general, most institutional investors do not actually own the money they manage but rather invest for other people. Institutional investors are generally considered sophisticated and knowledgeable. On the other hand, retail or individual investors do not invest on someone else's behalf; rather, they manage their own or their family's money, usually driven by personal goals.

In the early years of the Chinese stock market, most investors were individuals. This was similar to the early years of the US stock market, where individual investors were the majority. The chart below (Fig. 3.5) presents the holdings of market share value across different types of investors in the Chinese stock market from 2006 to 2017. We see that individual investors constituted about 50% of the market in the early 1990s, then dropped gradually over time to about 40% in recent years. On the other hand, the percentage of institutional investors has been rising steadily during the past 20 years. It would not be surprising if eventually, perhaps not long from now, institutional investors will take the majority seat in the Chinese stock market.

In general, as an equity market evolves with more public companies and financial products, more institutional investors will emerge, such as mutual funds, pension funds, and government funds. These institutional investors are usually more

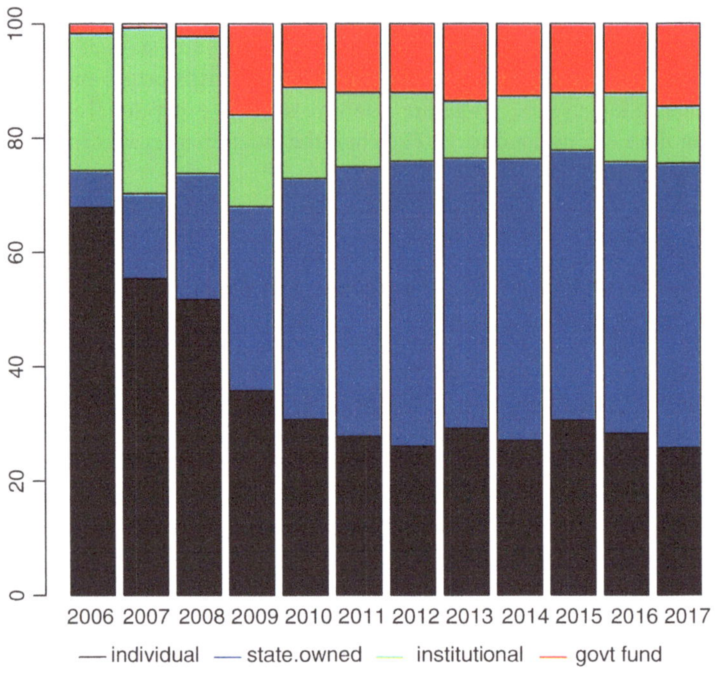

Fig. 3.5 The percentages of different investors in the Chinese stock market 2006–2017

sophisticated than individual investors due to their professional expertise, access to information, and ability to manage multiple strategies across different assets or markets with large amounts of money. This is a typical phenomenon across most stock markets in the DM. For example, as shown in Fig. 3.6, in the USA, the percentage of institutional investors was less than 20% in the 1940s. That percentage increased only after the pension system started to grow in the 1970s, when defined benefits and defined contribution plans became a large portion of assets. The percentage of institutional investors in the USA was about 70% in the 1990s and reached to about 80% in 2015.

If the Chinese stock market follows a similar trajectory, its institutional investors will continue to increase, albeit with some important differences due to its unique market structure. Institutional investors will take a larger share of the market if state-owned companies have more floating shares for the public. Currently, state-owned companies and government funds take up more than 60% of the market share. If this part transformed to market-based institutional investors, the trend in the percentage of institutional investors would accelerate in China. The professionalism of institutional investors will have positive impacts on the Chinese stock market in terms of, for example, efficiency and public information disclosure.

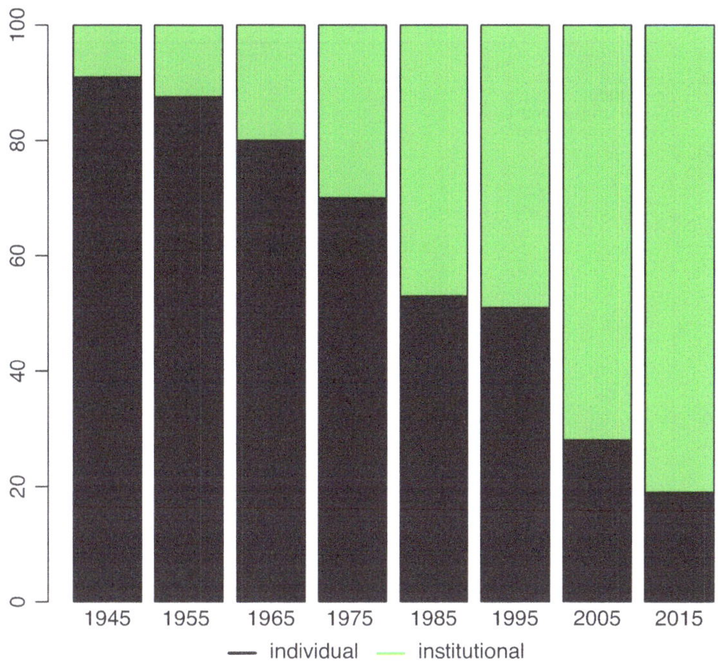

Fig. 3.6 The percentages of different investors in the US stock market, 1945–2015

3.2.3 Investment Behavior: Short-Term Speculation

In the Chinese stock market, individual investors focus on the short-term investment horizon, creating higher turnover than most stock markets in the world. The chart in Fig. 3.7 presents annual turnover of the Chinese stock market from 1992 to 2017, along with the USA, Indian, and South Korean markets. We see that compared with the USA, turnover in the Chinese market has been more than double. However, except for in recent years, turnover has been comparable with that in the South Korean stock market, perhaps due to cultural similarities. It is interesting that India is an EM, yet the turnover there is remarkably low.

Another notable feature of investment behavior in the Chinese stock market is that investors make "rational" decisions based on national policy and information from family and personal connections rather than professionals and public markets.

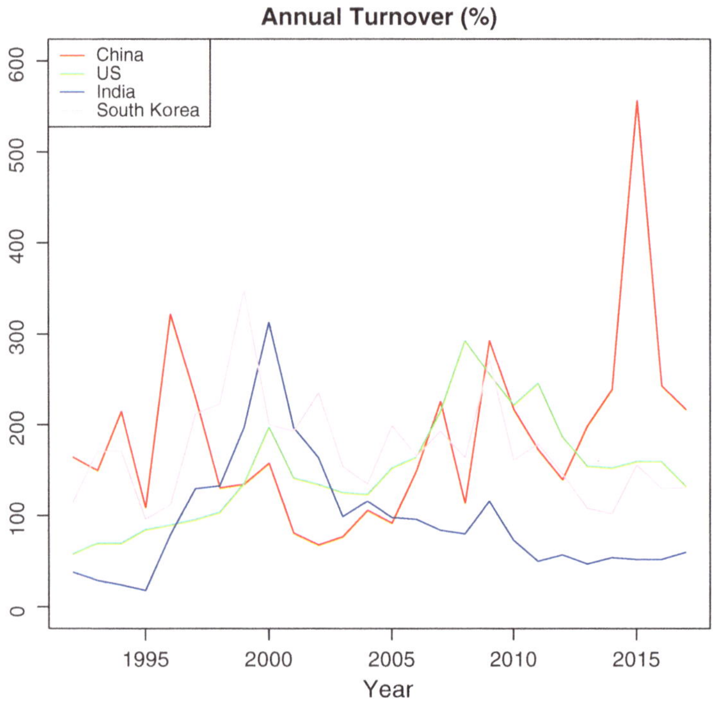

Fig. 3.7 The annual turnover (%) of the Chinese, the USA, Indian, and South Korean stock markets from 1992 to 2017

3.2.4 *Market Performance: Short-Term Bullish and Long-Term Bearish*

The Chinese stock market's performance has been very volatile for several reasons. First, China is an emerging economy, where high volatility is to be expected. Second, the central government's policy changes have caused large market movements. If we regard each policy change as a shock, each shock has tremendous impacts on the stock market over a short period due to its mandatory and efficient execution. Third, the Chinese stock market reacts strongly to the US stock market, which is exemplified both before and during the financial crisis in 2008.

The plot in Fig. 3.8 tracks daily price movements of the Shanghai Composite Index from 1991 to 2019. During this period, the SHCI had an annualized risk of 36.44% and an annualized return of 11.31%, resulting in a Sharpe ratio of 0.31. There are quite a few extremely volatile days, including 1 day when the index price doubled. Among 7136 trading days for this period, there are 235 days in which daily returns increased or decreased by more than 5% and 2870 days (40% of all trading days) in which returns increased or decreased by more than 1%.

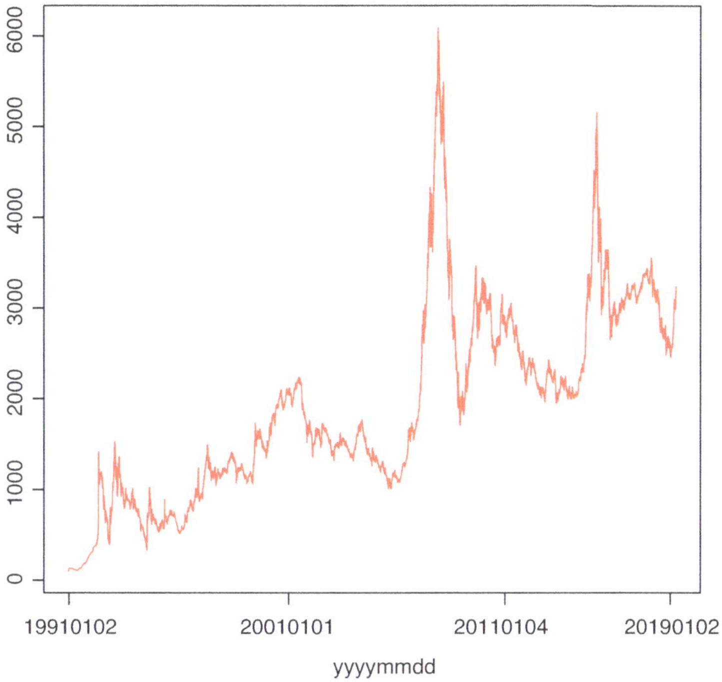

Fig. 3.8 The price movements of the Shanghai Composite Index from 1991 to 2019

Price movements in the Chinese stock market exhibit a short-term bullish and long-term bearish pattern. The market price climbed over a very short period and then dropped, followed by a long period of recovery. This distinctive pattern was largely driven by central government policy and dramatic changes in the US stock market, to which individual investors reacted. We mentioned above that in the Chinese stock market, individual investors have been very active. These individual investors—stay-at-home moms and retirees—prioritize the government's guidance (policy changes) over companies' performance. They know that even if some Chinese companies do not report any revenue, their stock prices may still soar as long as there is government policy support. Table 3.3 lists major events alongside policy changes and US stock market movements.

3.2.5 *Laws, Rules, and Regulation*

The supervisory and regulatory body of the Chinese financial market is the China Securities Regulatory Commission (CSRC). In 1998, China enacted its first comprehensive securities legislation, the Securities Law, which was effective on July 1, 1999. The law grants the CSRC authority to supervise the nationwide securities market. The CSRC has adopted many special rules and regulations for the Chinese stock market, such as stock halting, price change caps, and bans on day trading. All of these have had enormous impacts on stock prices.

3.2.5.1 Stock Halting

Stock halting refers to a situation in which a public company can have its stock trading stopped on an exchange. This can be due to an imminent earnings report, an ongoing business negotiation, or an investigation by the regulatory commission. Halting can last from days to 6 months.[7] Companies have often stopped trading of their shares to prevent their price from plummeting when the market sours towards the company. According to Yu and Shen (2018), in a report for Caixin Global, an average of 74 Shanghai-listed companies suspended trading of their shares each day in 2016, and 69 did so in 2017. Such suspensions accounted for about 5% in 2016 and 6% in 2017, of all listed company shares.

The original purpose of trade halting was to protect state-owned companies from value loss in the stock market. However, with more and more companies using the rule to suspend trading of their shares when they face the risk of a serious drop in their price, this introduces inefficiency and abnormal effects into markets.

In 2018, China announced policy changes aiming to cut down trading suspensions. The Shanghai Stock Exchange and the Shenzhen Stock Exchange strengthened the rules for share trading suspensions to curtail abuse of the practice. The suspension period for a major asset restructuring was shortened from 6 months to 10 days, and shares of companies whose major asset restructuring does not involve the issuance of new shares cannot be suspended. As a result, halted stocks dropped from 6% in 2017 to about 2% at the end of 2018 Yu and Shen (2018).

3.2.5.2 Caps on Price Changes

The practice of setting a limit on daily stock price changes has been widely adopted by many stock markets. Chinese stock markets also adopted the daily price limits. Chinese stocks are automatically halted for the rest of the trading day if there is a 10% price change. For stocks on the special treatment (ST) board, the cap is 5%. These rules apply to all stocks for all trading days. Thus, bear and bull markets are restrained to that 10% change per day. During extreme events, half of Chinese stocks may be halted!

The daily price limit was designed as a tool for market stabilization. This is particularly helpful in emerging markets where there is a large body of inexperienced investors and stock prices move wildly. There are many interesting studies about the pros and cons of daily price limits. For example, Kim and Rhee (1997) and Chan

---

[7]In the U.S. exchanges, halting is kept to less than an hour.

et al. (2005) find that price limit rules may induce imbalanced trading and barriers for market, while Cho et al. (2003) and Hsieh et al. (2009) showed a "magnet effect," asymmetric movements of stock prices towards thresholds. A recent study of Chen et al. (2017) confirmed that large investors Therefore, the price cap is a double-edged sword: on the one hand, it does reduce volatility, but the efficacy of this function depends on many other factors apparently.

3.2.5.3 Day Trading Ban, Maximum Order Size, and Short Selling

In addition to price caps, the CSRC imposes special regulations on trading. Day trading is not allowed for shares in the Chinese stock market. Therefore, shares of stocks bought on T-day can only be sold on and after $T + 1$ day. The maximum quantity for an order cannot exceed one million shares.

The CSRC also has policies on short selling. In March 2010, CSRC began to allow short selling, but only limited to fewer than 100 stocks. A few years later, the number of stocks allowed for short selling increased to about 700. In the middle of 2015, the Chinese stock market crashed and many Chinese trading companies stopped stock-shorting activities. Since the crash of 2015, while short selling remains legal in China, restrictions have been imposed. Currently, only shares of large companies are permitted to be lent out, though possibly at a high cost.[8] Chinese investors can hedge with domestic index futures.

There are other interesting features about the Chinese stock market. For example, few Chinese public companies have been delisted because local governments want to keep jobs and will not let companies go bankrupt. The colors associated with price movements are unique: green means down, while red, the favorite color of the Chinese culture, means up!

3.3 Performance of the Chinese Stock Market

To gauge the performance of the Chinese stock market, we use the Shanghai Stock Exchange (SSE) Composite Index. Although it is for the stocks listed in the SSE only, the SSE Composite Index reflects the overall Chinese stock market performance. In the following sections, we present an overview of the Chinese stock market's performance.

[8]The lending rate can be as high as 8.5%.

3.3.1 Overall Performance: Annual Return and Risk

To begin our analysis, we present the SSE Composite Index's performance from 1991 to 2019. Over the entire period, the annualized return of the SSE Composite Index was 12.68%, and the annualized risk was 38.45%. The plots in Fig. 3.9 display annual return and risk for each calendar year from 1991 to 2019.[9] We see there were dramatic ups and downs in returns, ranging from −65% to 167%, during this period. Even by emerging market standards, these returns are extreme. For example, among these nearly 30 years, there are 7 years with returns greater than 50% (Table 3.3). The tremendous price movements were usually associated with government policy changes. We briefly discuss these extreme periods below.

1991 and 1992 In the first 2 years of the stock market, the annual returns were 129% in 1991 and 167% in 1992. In 1991, the government encouraged investment in the stock market. In 1992, there were two big events that caused the Chinese stock market to skyrocket. On May 21, 1992, the SSE canceled the price cap of 1% and adopted the $T + 0$ rule (day trading). The closing price of the SSE rose from 667 to 1266, a 105% increase in a single day. The market dropped by 70% about 6

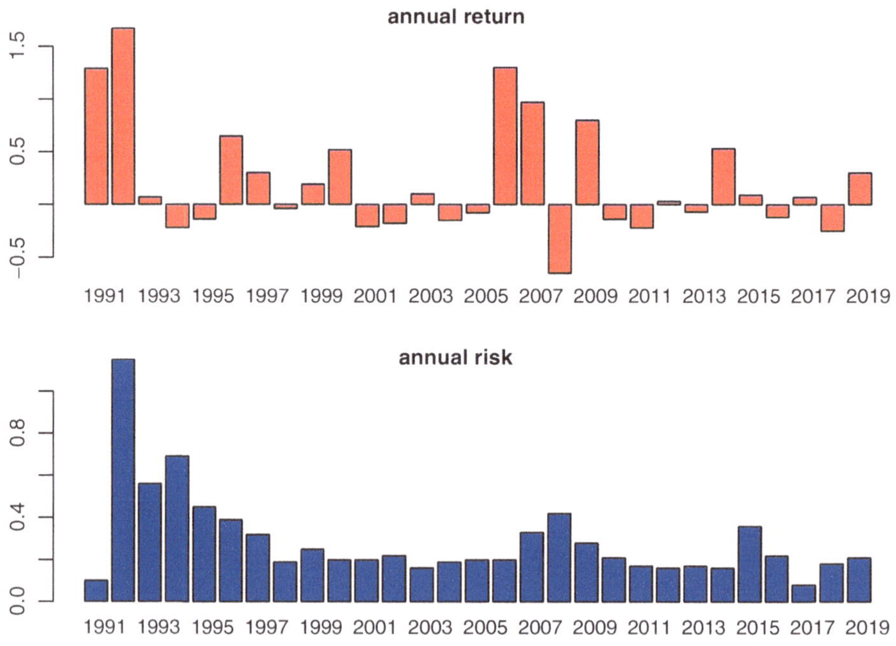

Fig. 3.9 Annual returns and risk of the Shanghai Composite Index, 1991–2019

[9]The annualized values for 2019 are based on data up to April 9.

Table 3.3 Annual performance of the SSE Composite Index from 1991 to 2019

yyyy	Return (%)	Risk (%)	Sharpe ratio	Note
1991	129	10	13.3	Stock mark first year
1992	167	115	1.45	Stock mark second year, Deng's talk
1993	7	56	0.12	
1994	−22	69	−0.32	
1995	−14	45	−0.32	
1996	65	39	1.67	Limited share supply
1997	30	32	0.96	
1998	−4	19	−0.21	
1999	19	25	0.75	
2000	52	20	2.64	
2001	−21	20	−1.04	
2002	−18	22	−0.8	
2003	10	16	0.63	
2004	−15	19	−0.82	
2005	−8	20	−0.42	
2006	130	20	6.51	Ownership reform
2007	97	33	2.94	Mutual funds expansion
2008	−65	42	−1.54	US Financial Crisis
2009	80	28	2.84	Money supply
2010	−14	21	−0.68	
2011	−22	17	−1.26	
2012	3	16	0.19	
2013	−7	17	−0.39	
2014	53	16	3.29	Economic reform theme
2015	9	36	0.26	
2016	−12	22	−0.57	
2017	7	8	0.81	
2018	−25	18	−1.34	Trade war
2019	30	21	1.43	

months later in November. Then, in late Nov 1992, the stock market was inspired by Deng Xiaoping's speech in the southern part of China, which encouraged openness and adopting new things in China from capitalism, such as the stock market. The SSE Composite Index increased from 386 to 1558 in about 3 months, an increase of 300%.

1996 The index had a return of 65% in 1996. New shares were not allowed to be issued after 1995. This control of supply and an increase in demand caused price increases. Investors started to focus on companies' business performance as a new concept called JiYouGu (referring to shares with good business performance) emerged.

2006 and 2007 The SSE Composite Index increased by 130% in 2006 and 97% in 2007. During these 2 years, there were three important policy changes that contributed to these extreme price movements: share ownership reform, appreciation of the Chinese currency RMB or CNY, and expansion of mutual funds.

2008 The US financial crisis had serious impacts on the Chinese stock market. The market was down by 65%, a greater decrease than in the US stock market. This was exacerbated by other events in China. For example, the central government called for a halt to mutual funds because their size doubled in a short period. The declining stock market was also partially due to the devastating earthquake in Sichuan province.

2014 The market was up dramatically after the summer, when the central government published many articles to promote the stock market with the theme that the stock market was undervalued and it should have reflected the success of China's economic development. People expected a strong stock market rally thereafter. Indeed, the market went up by about 50% in a single quarter. The rally continued into the early months of 2015.

2018 The Chinese stock market was down by more than 20% in 2018 due to trade wars with the USA. The new tariffs imposed on Chinese products and the sentiment of the trade war dragged Chinese economic growth down and had negative impacts on stock market performance.

While returns were very volatile, risk (measured by standard deviation) has been stable within the range of 20–40% each year after 1996 (Table 3.3 and Fig. 3.9). This is again due to price changes triggered by policy changes. The price would move in one direction for a while then reverse. Thus, except for the inflection points, the price movements have a momentum that causes the volatility to decrease between two consecutive events. Consequently, from the perspective of risk-adjusted return, the values of the Sharpe ratio were quite high for most years due to high returns and low risk within calendar years. For example, there are 9 years with Sharpe ratios greater than one and 4 years with Sharpe ratios less than negative one.

3.3.2 Performance by Sector

In China, stock market performance is very different across sectors. We present the growth rates of major sectors from May 2000 to April 2017 in Fig. 3.10. We see that Consumer Staples enjoyed a 550% growth, followed by Financials' 300%, then Consumer Discretionary's 140%. The companies in the Information Technology sector had a growth rate of 100%, Health had about 60%, and Industrials had the lowest growth rate of 40% during this period.

This pattern roughly reflects the growth of the Chinese economy across industries during the past 17 years. According to a research report by CaixaBank (2017), the service industry has increased its share by 11.8% per year over the past 16 years and accounted for 51.6% of GDP in 2016. The household consumption industry

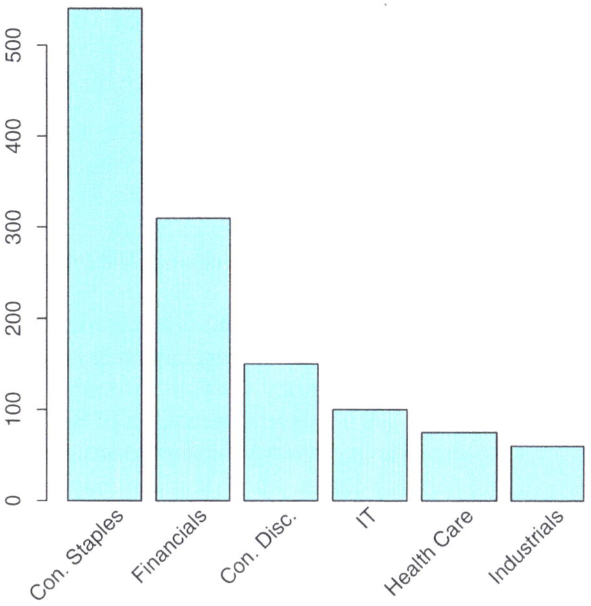

Fig. 3.10 The sector index growth of the Chinese stock market from May 2000 to April 2017. Data Source: Bloomberg

reached to around 37% of GDP in 2000s, with a slight upward trend thereafter. Given the huge size of the domestic market in China, e-commerce has developed rapidly during the last 10 years. Large e-commerce companies, such as Alibaba and Tencent, have made this industry bigger and stronger. As for the financials sector, national banks and insurance companies hold the monopoly power, they have grown faster than the economy, and their market caps have become much bigger than private financial firms. Meanwhile, there have been significant changes in the Chinese financial market. For instance, more financial companies have been launched and more financial products and services are now offered.

We see that in its early years, the Chinese stock market seemed "wild" due to both policy changes and irrational investment behaviors. The mixture of investment and speculation in the short term and a mentality to use the stock market as a wealth creation vehicle very much resemble the early years of the US stock market.

In the early years of the US stock market, before the Great Depression, the market was very volatile, and returns were very high. Securities trading in the latter nineteenth and early twentieth centuries was prone to panics and crashes. Investors used margin to "invest" in the stock market. This is the backdrop of the scientific research of Benjamin Graham, starting from the distinction between speculation and investment. In particular, he proposed a systematic method of value investing and how to manage risk through the margin of safety. We elaborated on Graham's contribution to investment in Chap. 2. Next, we present another very important figure who is closely related to Graham and his investment theories.

3.4 Investment Is a Science: Value Investing and the Buffett Factor

We introduced Benjamin Graham in Chap. 2. Investment is treated seriously as a science today in large part because of contributions from Graham and his student Warren Buffett at an early stage in the 1950s and 1960s (and continuing today). The former paved the theoretical foundation, while the latter practiced value investing in the real world. Buffett describes Graham's *The Intelligent Investor* as "the best book about investing ever written."

Buffett (Fig. 3.11) is one of the most successful practitioners of value investing.[10] His major contribution to quantitative investing is that he serves as a role model for scientific investing. Buffett has also published many articles, the most famous being his annual reports and letters to shareholders of Berkshire Hathaway. His investment success, public speaking prowess, and generous philanthropy have made Buffett an American business magnate whose impacts on investing are enduring.

> The basic ideas of investing are to look at stocks as business, use the market's fluctuations to your advantage, and seek a margin of safety. That's what Ben Graham taught us. A hundred years from now they will still be the cornerstones of investing. – Warren Buffett

3.4.1 Soundness of Stock Markets: The Buffett Factor

One example of value investing Buffett proposed at the macroeconomic level is how to judge the health of a stock market. In the long run, stock market performance generally reflects the economic health of a country. Moreover, the stock market represents the *expected* performance of a real economy. Intuitively, we can use the real economy and stock market value to gauge the equilibrium. There is a term for this measurement, called the Buffett factor (indicator), defined as follows:

Fig. 3.11 Warren Buffett, August 30, 1930–

[10]Photo credit: The India Today Group | Getty Images.

Fig. 3.12 The Buffett factor from December 1970 to December 2018. The top plot displays values of GDP and total stock market cap for the USA. The bottom plot displays the ratio. Source: www.gurufocus.com

$$\text{Buffett factor} = \frac{\text{real economic value}}{\text{stock market value}} = \frac{GNP}{\text{Stock Market Cap}},$$

where GNP is the gross national product (real not nominal), and the overall stock market value can be approximated by the Wilshire total market cap in the USA. In practice, due to data availability, people use GDP instead of GNP for the Buffett factor calculation. Buffett (2003) states that the percentage of total market cap (TMC) relative to the US GNP is "probably the best single measure of where valuations stand at any given moment."

In Fig. 3.12, we present the Buffett factor for the USA from Dec 1970 to Dec 2018. The top plot shows the values of GDP and TMC, while the bottom plot shows the ratio, the Buffett factor. We can see that during the past four decades, the Buffett factor had a wide range: it started from 75 in 1970, went down to 35% in 1982 (deep recession), then reached to 148% in 2000 (tech bubble). The ratio dropped during the 2008 financial crisis and went up thereafter. It is about 134% as of Feb 2, 2019 (the time of writing).

Compared with other stock markets, the US market is very efficient. The Buffett factor is expected to be around 100%, the equilibrium, with a range of 80–120%. Any major deviation from this range can be regarded as off the equilibrium, and the stock market will eventually adjust to reflect the real economy.

However, for other economies, such as emerging economies, the range of equilibrium may be between 60 and 80% as defined by the long-term ratio, based on specifics such as the economic and market structure of that particular country.

3.5 Bivariate Analysis: Correlation and Rank Correlation

In Chap. 2, we focused on analysis of a single variable. What if we have two variables and would like to know the relationship between them, such as the relationship between the US and Chinese stock markets? In quantitative investment, investors tend to explore the relationship between a variable and stock returns to analyze its forecasting power. In statistics, correlation is one way to define a quantitative relationship between two random variables. In this section, we introduce the concept of correlation and its variants.

3.5.1 Correlation

Recall that in Chap. 2, we defined variance: the variation in values of a random variable or how much they deviate from the average. Suppose the density function for a random variable X is $f(x)$, we express the variance σ^2 as follows:

$$F(X) = \int_{-\inf}^{x} f(t)dt$$

$$\mu = E(X) = \int xf(x)dx$$

$$\sigma^2 = Var(X) = E((X-\mu)(X-\mu)) - \mu^2 = \int x^2 f(x)dx - \mu^2.$$

Now, replacing the second term in the variance formula with another random variable Y, we get a new definition, $covariance$, which measures how closely two random variables follow each other.

Fig. 3.13 Karl Pearson (1857–1936). One of Pearson's major contributions to quantitative investing is correlation

$$\text{population: } Cov(X, X) = E[(X - \mu_X)(X - \mu_X)] = Var(X)$$
$$\text{sample: } Cov(X, Y) = \frac{1}{n}\sum(X - \bar{X})(Y - \bar{Y})$$

However, if, for example, we measure the movements of two elephants and two ants, both pairs move together, but the former would have a much larger value if we use the same unit. Of course, this does not make much sense. Karl Pearson (Fig. 3.13) proposed the *correlation* concept by scaling the covariance value by the standard deviations of both random variables, thus enabling apples-to-apples comparison (see Pearson 1895).[11]

$$\rho(X, Y) = \frac{Cov(X, Y)}{\sigma_X \sigma_Y}$$

Two special cases are as follows:

$$\rho(X, X) = \frac{Cov(X, X)}{\sigma_X \sigma_X} = 1, \quad \rho(X, -X) = \frac{Cov(X, -X)}{\sigma_X \sigma_X} = -1.$$

We see immediately that correlation exhibits symmetry, that is, the correlation between X and Y is the same as the correlation between Y and X. The values of a correlation range from -1 to 1. Another important property of the Pearson correlation coefficient is that it remains the same after location and scale shift. That is, $\rho(X, Y) = \rho(a + bX, c + dY)$, $b > 0$ and $d > 0$.

[11]This correlation value is also called the Pearson correlation coefficient.

However, it should be stressed here that, first, correlation measures a *linear* relationship between two variables; second, correlation does not present causal relationship; and third, zero correlation does not mean independence.

How do we make a judgment that two variables are highly correlated? For example, in quantitative investing, how do we tell if a factor has predictive power or not? This depends. When calibrating the relationship between a variable and forward returns, if the correlation is less than 1%, it is hard to say that the factor has predictive power, while if it is more than 20%, it is worth double-checking, as this usually arises from a contemporaneous relationship, the results of data mining, or simply outliers/errors in the data.

3.5.2 Variants: Rank Correlation and Moving Correlation

The Pearson correlation relies on the first two moments to gauge the linear relationship between two variables. However, we know from Chap. 2 that both the mean and standard deviation are very sensitive to outliers, so the Pearson correlation is not a robust measurement.

To obtain robustness, a simple way is to rank the variables first, then calculate Pearson correlation between the ranked values. This is called the Spearman rank correlation (Spearman 1904). Rank correlation has useful properties, such as robustness and monotonicity.

- Rank correlation is robust to outliers.
  For example, if X=1, 2, 3, 4, 5 changes to X=1, 2, 3, 4, 500, the rank correlation between variable Y with X stays the same.
- Rank correlation is invariant for monotonic data transformation, linear or nonlinear.
  Assuming both X and Y are positive, then the rank correlation between X and Y is the same as the rank correlation between $LOG(Y)$ and X^2.
- Rank correlation can be used for ordinal values.
  For example, if X = {"Very Unsatisfied," "Unsatisfied," "Satisfied," "Very Satisfied"} and Y= {"Low," "Medium," "High"}, then we can compute a rank correlation between X and Y.

In quantitative investing, one way to use rank correlation is to detect outliers: if the Pearson correlation is very different from the rank correlation, this usually indicates the existence of outliers.

In addition to rank correlation, another important variant used in quantitative investment analysis is a moving or rolling correlation. Rolling correlation can be used to examine how relationships between two assets change over time. In other words, while a once-for-all correlation measures the overall relationship between two assets, a rolling correlation reveals the dynamic features of such a relationship over time. In practice, the moving or rolling window could be a period of 20 days,

3 months, or 5 years, depending on the context of the study. One can apply both Pearson correlation and rank correlation to data within such windows. We will see applications in the next section when we explore the relationship between the US and Chinese stock markets.

3.6 Bivariate Analysis of the CSI 300 and S&P 500

In this section, we employ a major index, the CSI 300, to compare the performance of the Chinese stock market with the US stock market, represented by the S&P 500 index. The CSI 300 Index consists of the 300 largest and most liquid A-share stocks. It represents the overall performance of China's stock market. The index was launched on April 8, 2005, with the closing price of December 31, 2004, as the base of 1000. It is rebalanced semi-annually.

We compare the CSI 300 to the S&P 500 because the former comprises the largest and most liquid stocks, making it the most similar in the Chinese stock market to the S&P 500. Also, the index started in 2005, when the Chinese stock market became more developed and mature compared to 15 years ago, when there were only a few stocks.

We first compare the overall performance of the two indices, then run various correlations to analyze their quantitative relationship across different scenarios.

3.6.1 Stock Market Performance

We first present a scatter plot of returns for the CSI 300 and S&P 500 to see if we can visualize any relationship between the two stock markets (bottom plot in Fig. 3.14). The scatter plot shows that overall, it is difficult to discern any clear pattern. This is confirmed by the percentages for the count of direction of returns in each quadrant, where a positive return is counted as 1 and a negative return is counted as -1. The four quadrants represent four possibilities: both indices increase, the S&P 500 increases while the CSI 300 decreases, the S&P 500 decreases while the CSI 300 increases, or both indices decrease. We see that the percentages are around 22–25%, except for the first quadrant with 32% because of the upward trend of the two markets.

To see the relationship over time, we present the time series of price movements from 2005 to 2019 (top plot in Fig. 3.14). Overall, both indices had an upward trend, but the CSI 300 Index had much larger and more ups and downs than the S&P 500. Second, we see that there is indeed some relationship between the two indices' price movements: there is co-movement when both markets are down but not for other directions of price movements.

Table 3.4 displays annual performance for the two indices from 2005 to 2019. From the table, we see that the two stock markets had very different levels of

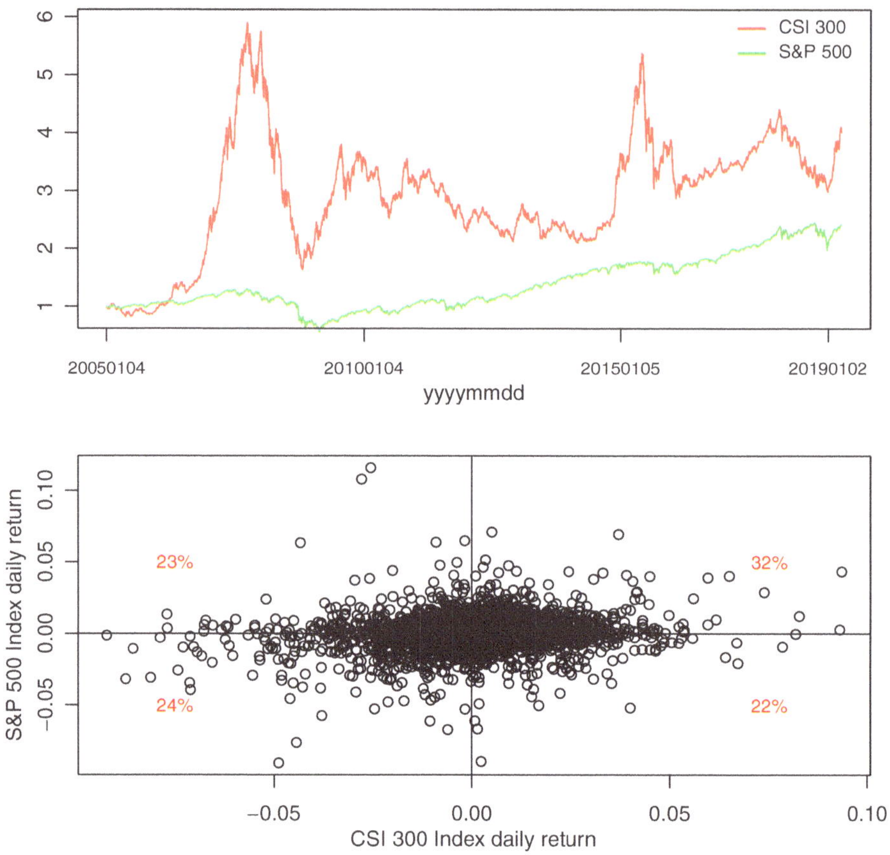

Fig. 3.14 The scatter (bottom) and time series (top) plots for S&P 500 and CSI 300 index prices from 2005 to 2019. In the bottom plot, the number for each quadrant is the percentage of the count of positive/negative returns for the two indices within that quadrant

performance each year. While in some years the two markets moved in the same direction—such as in 2008, when both markets declined dramatically during the financial crisis—in most years, the two markets moved in very different directions. For example, in 2007, while the US market was flat, the Chinese stock market went up by 152%, and in 2013 the reverse happened.

We present the distribution of daily returns for each index in Fig. 3.15. We can see that compared with the US stock market, the Chinese stock market had thick tails at both ends, especially the right one.

3.6.2 *Is There Any Relationship?*

We analyze the relationship between the Chinese and US stock markets using the correlation between the daily returns of the two major indices from 2005 to 2019.

Table 3.4 Annual performance of the S&P 500 and CSI 300 indices 2005–2019

yyyy	S&P 500 return	S&P 500 risk	S&P 500 Sharpe	CSI 300 return	CSI 300 risk	CSI 300 Sharpe
2005	0.05	0.1	0.5	−0.04	0.2	−0.22
2006	0.09	0.09	1.01	1.09	0.21	5.22
2007	0.01	0.15	0.05	1.52	0.34	4.45
2008	−0.31	0.38	−0.82	−0.62	0.45	−1.37
2009	0.17	0.25	0.65	0.87	0.3	2.85
2010	0.03	0.17	0.15	−0.08	0.24	−0.35
2011	−0.04	0.22	−0.18	−0.24	0.19	−1.24
2012	0.1	0.12	0.81	0.04	0.19	0.22
2013	0.31	0.1	2.99	−0.06	0.21	−0.29
2014	0.17	0.1	1.61	0.51	0.18	2.82
2015	−0.06	0.15	−0.41	−0.07	0.37	−0.2
2016	0.11	0.12	0.91	−0.13	0.19	−0.66
2017	0.19	0.06	2.97	0.2	0.1	2.07
2018	−0.04	0.16	−0.23	−0.24	0.2	−1.19
2019	0.15	0.12	1.18	0.36	0.23	1.61

The bold indicates that both markets have negative returns, which is addressed later in the information asymmetry section

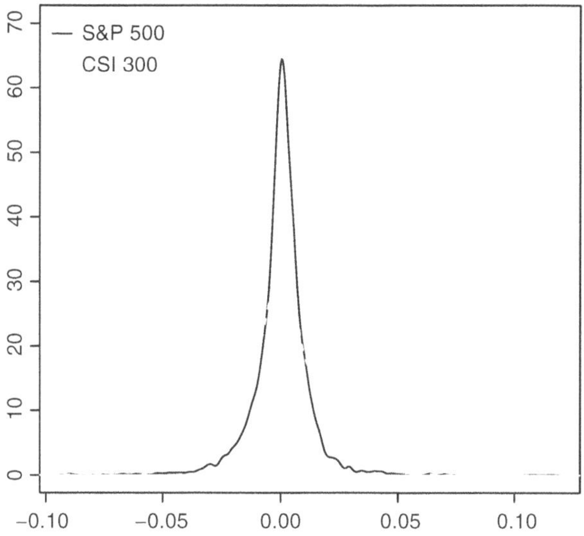

Fig. 3.15 Distribution of S&P 500 and CSI 300 daily returns from 2005 to 2019

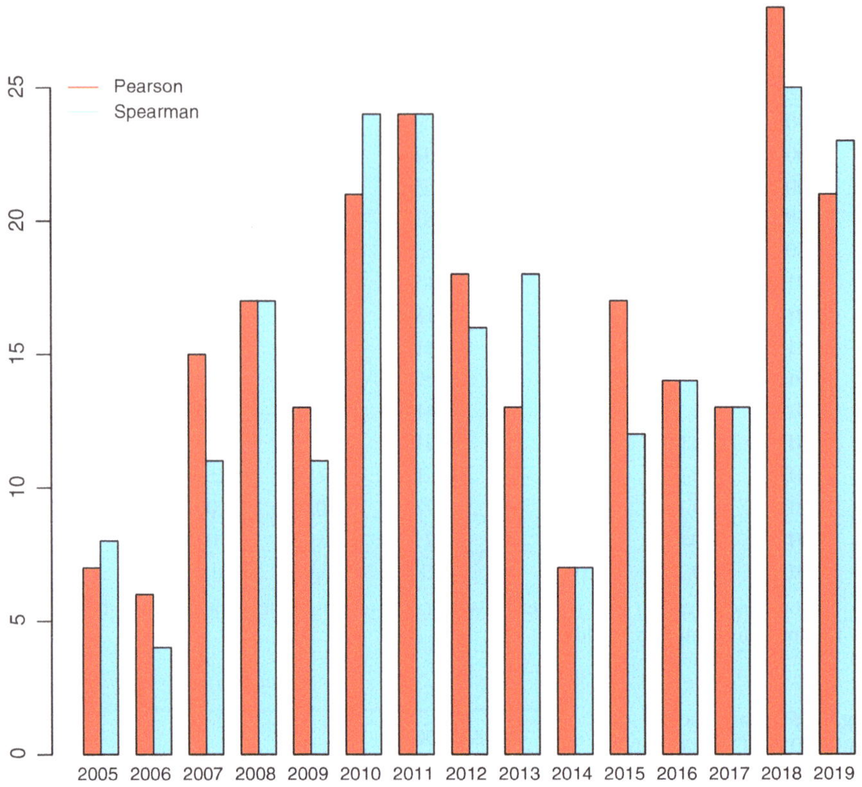

Fig. 3.16 Pearson and Spearman correlations between the returns of the S&P 500 and CSI 300 each year from 2005 to 2019

The regular (Pearson) and rank correlation values are 16% and 14%, respectively, their similarity indicating no extreme outliers.

Correlation between daily returns

```
> round(cor(sp5csi$return.sp5,sp5csi$return.csi),2)
[1] 0.16
> round(cor(sp5csi$return.sp5,sp5csi$return.csi,method="spearman"),2)
[1] 0.14
```

We then run Pearson and Spearman correlations between the daily returns of the S&P 500 and CSI 300 indices for each calendar year from 2005 to 2019 (Fig. 3.16).

To investigate a continuous dynamic relationship, we calculate rolling correlations between the daily returns of the two stock markets from 2005 to 2019 for 1-, 2-, and 3-year periods (Fig. 3.17).

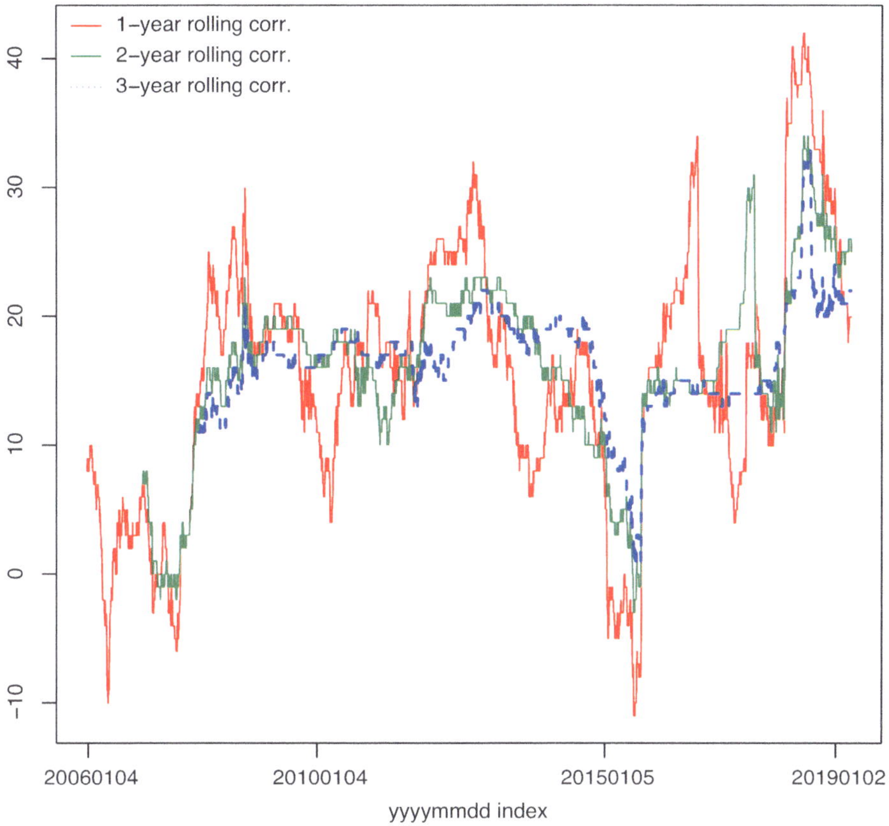

Fig. 3.17 Rolling correlations of 1-, 2-, and 3-year periods between returns of the S&P 500 and CSI 300 from 2005 to 2019

We also investigate whether either index follows a normal distribution. Instead of carrying out a rigorous test, we present the results of *QQplot* for the daily returns of each index in Fig. 3.18. First, we can see that neither index follows a normal distribution. This has very important implications: the first and second moments are not enough to characterize the returns of both stock markets. Second, the left sides of the distributions are more similar than the right sides.

3.7 Industry Insights: Information Decay and Asymmetry

The financial markets are intercorrelated. Given the size of the US economy and the dominant position of the US stock market among global financial markets, the impacts of the US stock market on the Chinese stock market are bigger than the inverse. In this section, we discuss two industry concepts—information decay and asymmetry—both of which are important for quantitative investment strategies.

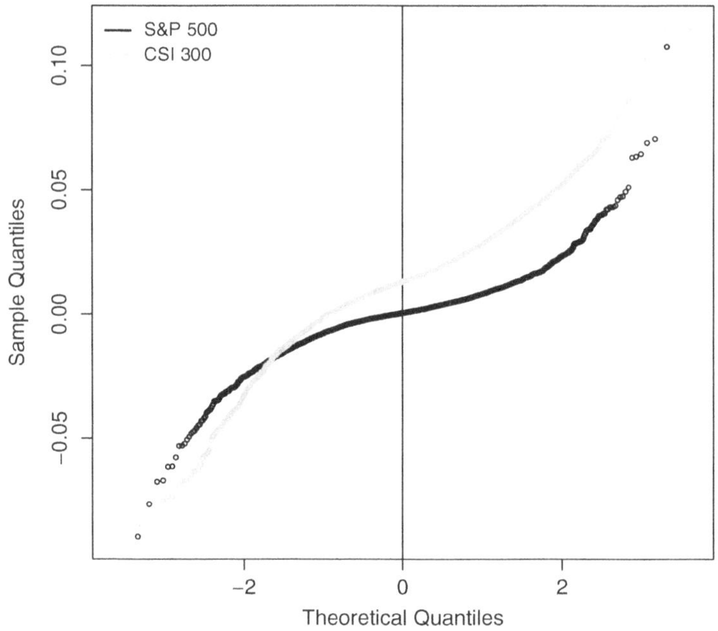

Fig. 3.18 The QQ plot of S&P 500 and CSI 300 daily returns from 2005 to 2019

3.7.1 Information Decay: Impacts of Events

Information decay, in an equity investment context, refers to changes in information over time, due to the effects of factors or events on stock returns. Information decay can be measured by the correlation between the event or causal factor and the values of impacted factors, using various lags and/or lead periods. For example, if the causal factor is A, and the response factor is B, we could use a series of correlations:

$$corr(A_t, B_t), \quad corr(A_t, B_{t+1}), \quad \dots \quad , corr(A_t, B_{t+10}),$$

with the expected decay showing in the series of values of correlations.

In quantitative investing, information decay is not just a concept; it has very important investment implications for factor efficacy, investment horizon, turnover, etc. The trend of information decay is faster nowadays due to factors such as technology enabling information to spread faster and money moving faster due to fewer barriers.

For illustration purposes, we carry out an information decay study between the US and Chinese stock markets. When the US market is up or down, what are the impacts on the Chinese market over time? We select a sequence of look-ahead periods of 1–10 business days, then 1, 3, and 6 months, so we can observe the decay of the impacts of the US stock market on the Chinese stock market. Information correlation

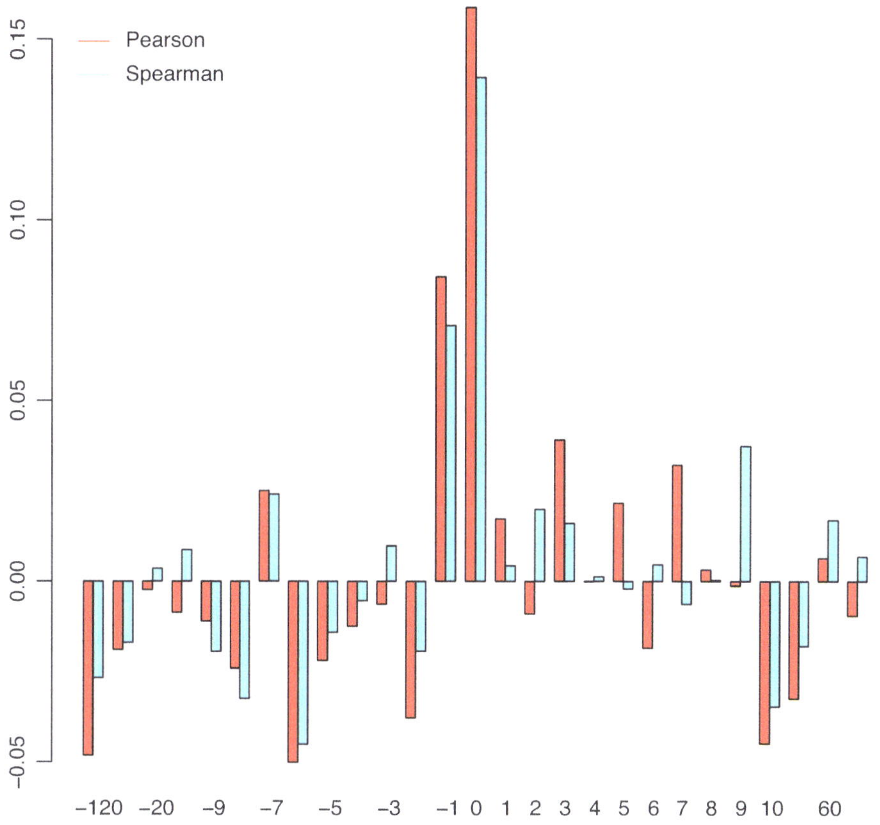

Fig. 3.19 An example of information decay: the price impacts of the S&P 500 on the CSI 300 using leads and lags of 1–10 days and 1, 3, and 6 months

is measured by both regular correlation (Pearson) and rank correlation (Spearman). For the purpose of comparison, we also provide the information for the leading periods.

The Shanghai Stock Exchange's hours are 9am–3pm Monday–Friday, and the NYSE's hours are 9:30am–3:30pm. Chinese time is ahead of US Eastern Time, with a 12-h difference during the summer and a 13-h difference at all other times. For example, when the NYSE closes at 3:30pm on July 16, 2019, it is 3:30am on July 17, 2019 in China, so a merging of data from the CSI 300 and the S&P 500 by date should use one trading date ahead for the CSI 300.

The plot in Fig. 3.19 shows that the correlation is about 15% on the same trading day (lag=0), then drops quickly to less than 5% for all other days. This implies that the impacts of the US stock markets on the Chinese stock market are rather short-lived, mostly lasting 1 day, after which there are barely any impacts on the Chinese stock market.

If we formulated a strategy based on this study, any time the S&P 500 dropped by more than 1%, we would short sell the CSI index the next day and then cover the position 1 day later. Would this be a profitable strategy? We will explore the answer in the next section after we have a better understanding of the relationship between the two stock markets.

3.7.2 Asymmetry: Bad News Travels Fast

Correlations for different parts of a return distribution can differ. For example, will the impacts of the US stock market on the Chinese stock market be the same when the US stock market is bullish and bearish? One simple way to measure asymmetric effects is to use correlation to measure different parts of a return distribution:

$$cor(A_t^+, B_t), \quad cor(A_t^-, B_t),$$

where A^+ is for positive values of A and A^- is for negative values of A. If we find that the correlation values are different for positive and negative parts of A, we would need more information to formulate a strategy. For example, we would need to carry out a test for the significance of the difference, and more importantly, to have a fundamental rationale for the difference. The former would be to ensure the difference is large enough, and the latter would be to ensure that the same difference may appear in the future.

We carry out an asymmetric impact study using the merged data of S&P 500 and CSI 300 returns by date. We select ten threshold points of daily returns, five positive and five negative. For the positive returns, we study the relationship when S&P 500 returns are larger than the threshold returns, For the negative returns, we study the relationship when S&P 500 returns are smaller than the threshold returns. Both Pearson and Spearman correlations are calculated. The number of observations is also listed in Table 3.5. We see that for the negative returns of the

Table 3.5 Asymmetric impacts of the US stock market on the Chinese stock market, 2005–2019

S&P 500 return	Num.obs	Pearson(%)	Spearman(%)
−0.05	12	46	46
−0.04	23	26	9
−0.03	51	−4	−12
−0.02	138	12	14
−0.01	407	13	10
0.01	416	1	2
0.02	105	−10	1
0.03	39	−31	−17
0.04	22	−46	−50
0.05	8	−40	−5

S&P 500, the correlation with the CSI 300 increases farther along the left tail: the Pearson correlation is about 10% when S&P 500 returns are less than (−1)–(−2)% and increases to 26% and 46% for extreme negative returns. The rank correlation (Spearman) values are of the same magnitude as Pearson correlations. However, for a bullish US stock market, the relationship is either very weak or the opposite: when S&P 500 returns are above 1%, the correlation with the CSI 300 index returns is 1%, but when S&P 500 returns are above 3%, the correlation with CSI 300 daily returns becomes negative, indicating the opposite direction of the Chinese stock market. Moreover, when the US stock market has positive daily returns, the Pearson correlation values and the rank correlation values are very different, indicating that the results are driven by a few outliers. For example, when the S&P 500 daily returns are above 3%, the Pearson correlation is −31%, while the Spearman correlation is −17%. When the S&P 500 returns are above 5%, the Pearson correlation is 40%, while the Spearman correlation is only −5%. It should be noted that in the latter case, there are only 8 trading days, so the results can be very random.

Asymmetric effects

```
> source("/.../springerLatex/bookQI/chapter2/chapter2.table.R")
> asymmetry.effects()
> sp5csi[1,]
  yyyymmdd return.sp5  return.csi
1 20050104   -0.01167     0.0099
> cc=which(sp5csi$return.sp5 < -0.05)
> length(cc)
[1] 12
```

Based on the results above, we can formulate an investment strategy. For example, we can short the CSI 300 index the following day if we know the S&P 500 was down the previous day. We leave details to readers for further exploration.

3.8 R Functions and Exporting Results of Tables and Plots

In this section, we show how to write a function in R and introduce methods for loops in R.

3.8.1 How to Write an R Function

One great strength of R is the user's ability to create functions. A function enables flexibility, reusability, and automaticity. We show here how readers can create a new function step by step.

1. Use the editor (right before the printer) in the bar line of R console.
2. The template for a function in R.

Template

```
functionName <- function () {

}
```

An R function has three mandatory parts: function name, inputs in ( ), and the contents in { }. We discuss each part below with examples. Note that the functionName is your own name for a function, although it can be anything just like your dog's name, but usually people give a name reflecting the contents or purpose of the function. For example:

meaningful function name

```
calculateRisk <- function ()
{
 x=1:10
 sigma=sd(x)
}
```

3. Add comments and use the R command *cat*.

A. To add comments in R, use "♯," R would ignore those comments in execution.

use ♯ for comments

```
calculateRisk <- function ()
{
  x=1:10       ### give values from 1 to 10 to x
  sigma=sd(x) ### calculate standard deviation of x
}
```

B. R command *cat* for comments in execution.

the command cat

```
calculateRisk <- function ()
{
        cat("This is an example to calculate risk \n")
        x=1:10       ### give values from 1 to 10 to x
        sigma=sd(x) ### calculate standard deviation of x
        cat("The risk is", sigma, "\n")
 }
```

Note that "\n" tells R to go to the next line for any new comments.

4. Add parameters in ( ). The major purpose of writing a function in R is to reuse it, this is achieved by using parameters in the ( ).

Add parameters

```
calculateRisk <- function (data.dir)
{
      ### import data
       my.data=read.csv(data.dir)
      ### calculate overall risk
      sigma=sd(my.data$return)*sqrt(220) # 220 trading days
      cat("The risk level is", sigma, "\n")
}
```

5. Create a matrix or table to store values. Note that objects in the function are local to the function.

Create matrix or table to store values

```
calculateSharpeRatio <- function (data.dir)
{
  ### import data
  my.data=read.csv(data.dir)

  ### calculate overall risk
  ann.risk=sd(my.data$return)*sqrt(220) # 220 trading days
  ### calculate return
  num.periods=dim(my.data)[1]
  ann.return=cumprod(1+my.data$return)^(220/num.periods)-1

  ###create a matrix
  sharpe.mat=matrix(NA, nrow=3,ncol=1)
  sharpe.mat[1,1]=ann.return
  sharpe.mat[2,1]=ann.risk
  sharpe.mat[3,1]=ann.return/ann.risk

  ###create a dataset or table
  sharpe.tab=data.frame(sharpe.mat)
  sharpe.tab$performance.name=c("ann.return","ann.risk","sharpe.ratio")
  names(sharpe.tab)[2]="performance.number"
  print(sharpe.tab)
}
```

Note: A matrix can contain either numbers or characters, but NOT both. A data set can have one column of numbers and another column of characters.

6. Export values and plots. A data frame can be exported by R command *write.table*, and a plot can be exported by R command *pdf*. See the R scripts below.

Create matrix or table to store values

```
calculateSharpeRatio<-function(data.dir,data.name, result.name, plot.name)
{
     ### same contents as the function  in item 5
     .....
     ### Now get the dataset (results out)
     data.out=paste(data.dir,result.name,sep="")
     write.table(sharpe.tab, data.out, sep=",")
```

```
    ### draw plot of return and risk in. a pdf file
    pdf(paste(data.dir,plot.name,".pdf", sep=""))
    barplot(sharpe.tab$performance.number[1:2])
    graphics.off()
}
```

A helpful practice when writing an R function is to add a "read.me" section at the top of the function to explain what the function does. An example is shown below.

Read me

```
###########################################
###
###         Purpose:        what does this function do
###         Input:          explain parameters, default values
###         Output:         specify what are produced and exported
###         Key note:       special attention or know how
###         Dependence:     pre requisite and context
###         Developer:      name, date
###
###########################################
```

3.8.2 How to Create Loops in R

Loops are very useful in R when the same method or procedure is used repeatedly. We first present basic R commands for loops (*for*, *while*, and *tapply*), then give an example of calculating annual returns with loops for each calendar year.

1. Loop by *for* , template: loop 10 times

Loop by for

```
for (i in 1:10)
{
     cat(" i is now  ", i, "\n")
}
```

2. Loop by *while*, template: loop 10 times

Loop by while

```
i=0
while (i < 10)
{
 cat(" the value of i is now: ", i, "\n")
 i = i+1
}
```

3. The *apply* family. In R, there are many useful built-in functions. Here we introduce one type of such functions, the *apply* type. The *apply* family includes *apply, tapply, sapply,* and *lapply*. We illustrate their use with *tapply*.

tapply function

```
> test.data=data.frame(name=c("a","b","c","a"), gpa=c(1,2,3,4))
> tapply (test.data$gpa,test.data$name,mean)
  a   b   c
2.5 2.0 3.0
```

4. An example: calculate annual returns for each calendar year.

Loop: calculate annual returns for each year

```
calculate.annualReturn <-function(data.dir,data.name,return.name)
{
 #### get the data
 data.in=paste(data.dir,data.name,sep="")
 my.data=read.csv(data.in)

 #### the data set has a column yyyymmdd
 ##make a new column with yyyy
 my.data$yyyy=as.numeric(substr(my.data,1,4))
 years=unique(my.data.yyyy)
 nyear=length(years)
```

```
### now set up a loop to calculate annual return for each year
for( i in 1:nyear)
{
  year.index=which(my.data$yyyy==years[i])
  my.data.year=my.data[year.index,]
  num.periods=dim(my.data.year)[1]
  annu.return=cumprod(1+my.data.year$return)[num.periods]-1
  cat("annual return for year ", years[i], " is ",annu.return, "\n" )
 }
 cat (" Dude, this is the end of the loop over years!")
}
```

Keywords, Problems, and Group Project

Part I: Keywords

Covariance, correlation, rank correlation, rolling correlation
CSI 300, information decay, Buffett factor, stock halting, price cap
asymmetric effects
R function, R loop

Part II: Problems

Problem 3.1 Compare the Chinese stock market (CSI 300) with the US stock market (S&P 500) with daily prices from 2005 to 2019.

(1) Calculate annualized return, risk, and Sharpe ratio for the CSI 300 and S&P 500 indices, for the entire period and each calendar year.
(2) Make scatter and time series plots for the price movement of the two indices.
(3) Calculate the four moments of the returns for both indices.
(4) For each index, plot the density of daily returns and see if it is Gaussian.
(5) For each index, identify extreme returns and explain what happened.
    (i) Extreme positive returns? When did they happen and why?
    (ii) Extreme negative returns? When did they happen and why?
(6) What are the major differences between the market performance of the two indices?

Problem 3.2 Run correlations between the two indices using the same daily price data as in Problem 3.1.

(1) Run Pearson correlation for daily returns between CSI 300 and S&P 500.
(2) Run rank correlation for daily returns between CSI 300 and S&P 500.
(3) Investigate differences between Pearson and rank correlations.

(4) Run rolling correlation (both Pearson and rank) for daily returns between CSI 300 and S&P 500 with windows of 1, 2, 3, and 5 years. Are these rolling correlations stable over time?

Problem 3.3 Investigate information asymmetry between the two indices using the same daily price data as in Problem 3.1.

(1) Run correlation (both Pearson and rank) for daily returns between CSI 300 and S&P 500, when S&P returns are positive and negative, respectively.
(2) Investigate differences when S&P 500 returns are positive and negative. Design an investment strategy.

Problem 3.4 Pick a stock in the CSI 300 index that is listed in the New York Stock Exchange and conduct fundamental analysis of the company.

(1) Company origination and development
(2) The IPO date, price, and market identifiers (ticker and cusip)
(3) Lines of business and core business
(4) Management team
(5) Peers or competitors in the industry
(6) Company performance during the last 3 years
   (i) business performance: total revenue, net income, total asset, cashflow
   (ii) EPS, DPS (if the company issues dividends)
(7) Stock market performance since public
   (i) market performance: return, volatility, and Sharpe ratio
   (ii) compare the performance with both CSI 300 and S&P 500 index and its peer companies in both China and the USA

Part III: Group Project

Problem 3.5 Collect data on macroeconomic indicators, GDP, total market cap, and population, for the USA, China, and another country for as long as possible.

(1) Calculate GDP growth rates and GDP per capita over time.
(2) Calculate the Buffett factor for each country and find the equilibrium value.
(3) Based on the equilibrium value, make a judgment of the soundness of the stock market in each country at the current stage.
(4) Economy decides the stock market performance. Stock market reflects investor's sentiment about the future economic situation. Provide examples for both arguments.

References

Buffett, W. 2003. "Preface to the Fourth Edition," in *The Intelligent Investor*, ed. Graham, B., 4th ed. New York: Harper & Row.

CaixaBank Research Report. 2017. *China's New Economy: Stock Market Transition and Sector Performance*, June 2017.

Chan, S.-H., K.A. Kim, and S.G. Rhee. 2005. "Price Limit Performance: Evidence from Transactions Data and the Limit Order Book." *Journal of Empirical Finance* 12: 269–290.

Chen, T., Z. Gao, J. He, W. Jiang, and W. Xiong. 2017. "Daily Price Limits and the Magnet Effect." Working Paper.

Cho, D.D., J. Russell, G.C. Tiao, and R. Tsay. 2003. "The Magnet Effect of Price Limits: Evidence from High-Frequency Data on Taiwan Stock Exchange." *Journal of Empirical Finance* 10: 133–168.

Hsieh, P.-H., Y.H. Kim, and J.J. Yang. 2009. "The Magnet Effect of Price Limits: A Logit Approach." *Journal of Empirical Finance* 16: 830–837.

Kim, K.A., and S.G. Rhee. 1997. "Price Limit Performance: Evidence from the Tokyo Stock Exchange." *Journal of Finance* 52: 885–901.

Pearson, K. 1895. "Notes on Regression and Inheritance in the Case of Two Parents." *Proceedings of the Royal Society of London* 58: 240–242.

Spearman, C. 1904. "The Proof and Measurement of Association Between Two Things." *American Journal of Psychology* 15(1): 72–101.

Xie, E. 2019. " Ten Things Investors Should Know About the China A Market." American Century Investments Research Report.

Yu, Z., and T. Shen. 2018. China stock markets tightens rules on when share trading can be suspended. *Caixin Global*. 28 Dec. 2018.

Chapter 4
How to Construct a Stock Selection Strategy: Multi-Factor Analysis

Abstract In this chapter, we introduce stock selection strategies and demonstrate how to employ a multi-factor model to build alphas for such a strategy. How can we forecast stock returns? To answer this critical question, we first discuss market inefficiency and identify sources of return anomalies. We then show how to transform these fundamental sources into a multi-factor alpha model. Regarding related finance theory, we introduce the capital asset pricing model (CAPM). On the quantitative side, we present the ordinary least squares (OLS) method. We explore estimation, inference, and properties and conditions of OLS estimates. Regarding industry insights, we show, using the Russell 1000 security level data, how to construct a multi-factor alpha model for a large-cap core stock selection portfolio. For R programming, we introduce commonly used utility functions in quantitative investing.

4.1 How to Forecast Stock Returns Using a Model

To build a quantitative investment strategy aimed at outperforming the market, we first need to understand the sources of stock returns. We can then model the source information by collecting data, constructing variables, and calculating the quantitative relationships between those variables and stock returns.

4.1.1 Stock Selection Strategy: Market Returns and Security Returns

In Chap. 3, we analyzed market returns for the US and Chinese stock markets using a major index in each market. The index return can be viewed as the overall market return for that stock market. It is helpful to know how the index return is calculated. For both the S&P 500 and CSI 300, the index return is the sum of weighted individual stock returns for all stocks within the index, where the weights derive from each stock's market capitalization.

L. Ma, *Quantitative Investing*, https://doi.org/10.1007/978-3-030-47202-3_4

$$\text{Index Return} = \sum_i^n w_i r_i, \quad w_i = \frac{mcap_i}{\sum_i^n mcap_i},$$

where *mcap* is market capitalization, which is simply the price times the number of shares.

If we allocate money just between the two indices, we are limited to only two choices. However, if we allocate money at the security level, we would have a much larger set of options. For example, if we use the S&P 500 as an investment universe, we have 500 stocks to select from the US market. Similarly, if we use the CSI 300, we have 300 stocks to select from the Chinese stock market. If we pick stocks with higher future returns than the market and put all these stocks in a portfolio, that portfolio will potentially deliver higher returns than the market. We have just formulated a stock selection strategy!

There are two requirements for a stock selection strategy to deliver higher than market returns: the ability to forecast stock returns and a process to select appropriate stocks into a portfolio. The former requires specialized skillsets and is often called *alpha* in the investment industry. The latter is a set of rules, approaches, and procedures designed to guide an investor's selections for an investment portfolio. We present below a brief structure of alpha and strategies used in the industry.

Definition 4.1 *Alpha*: refers to the excess market return of an investment strategy or skillsets needed to achieve superior performance; in quantitative investing, it also refers to the forecasted return (alpha model).

Theme: a field where sources of alpha can be identified and explored, such as profitability or price momentum. A theme can include many factors and is usually measured by a composite of similar factors. For example, a value theme can include factors such as P/E and P/B.

Signal: a variable that measures a subset of information in a theme; the term is interchangeable with factor, although the latter may include several signals.

Factor: a factor is considered a building block of a multi-factor model, it is usually an element of a theme and is sometimes called a signal, but can be a theme as well depending on the context.

A stock selection strategy, in simple terms, is an investment strategy for selecting desired stocks into a portfolio to achieve desired performance. Selection strategies for the equity asset class are usually classified into two types: passive and active.

- Passive strategies: follow the market or deviate little from the market, with most returns coming from exposure to the market. They are usually long-only strategies with long-term holding periods, low turnover, and low trading costs. These do not require as much in terms of talent and skillsets. Industry products of index, index plus, and structured portfolios all belong to this category.
- Active strategies: investors use specialized skillsets to pick stocks and deliver better than market performance. These can be divided into two different subtypes: total return and relative return. The former focuses on the total return regardless

of benchmark in terms of portfolio weights and stock selection, while the latter focuses on the excess return over a benchmark, with weight variation relative to a benchmark being very important.

- *Total-return strategy* A total-return strategy focuses its performance on total return without considering the benchmark in the portfolio. It is also called the absolute return since the strategy is independent of any standard. For example, a long-short market neutral stock selection strategy is a typical total return active strategy. A long-short strategy allows both long and short sales. For example, if an investor has USD10 million cash, such a strategy requires the investor to buy long stocks worth USD10 million and borrow USD10 million shares for short sale. This type of strategy aims to generate returns without directional exposure to the markets so that it is market neutral. Because its returns are not driven by broad market exposure, an absolute return strategy relies on the investor's ability to select stocks with positive returns in the long position and stocks with negative returns in the short position.
- *Active-return strategy* Different from a total-return strategy, an active-return strategy focuses on the performance of a portfolio relative to a benchmark (usually the investment universe), where value can be added by overweighting good securities and/or underweighting bad securities (relative to a benchmark). That is, both picking winners and avoiding losers within the target benchmark contribute to a portfolio outperforming its benchmark. Thus, a strategy need not actually short sell securities on an absolute basis to benefit from the ability to identify underperforming securities. Rather, such a strategy benefits (relative to its benchmark) by underweighting securities that underperform. A typical active-return strategy in the equity markets is a long-only strategy (buy and hold) with a 3–5% standard deviation of weights from the benchmark. A long-only stock selection strategy does not involve short selling (selling shares of a stock the investor does not own). All the portfolio weights are zero or positive, though active weights can be negative for underweighting stocks.
  Active weight: active weight for a security measures the difference between its portfolio weight and its benchmark weight. Positive means overweighted and negative means underweighted (relative to the benchmark).

In quantitative investing, alpha refers to the results of a return-forecasting model or an alpha model. The term "alpha" is derived from its counterpart "beta," which we discuss in detail in the following subsection.

4.1.2 William Sharpe: The CAPM and the Sharpe Ratio

In order to formulate a successful stock selection strategy, a natural question is, what kinds of stocks have the potential to outperform the market? Economist William Sharpe (Fig. 4.1) proposed the capital asset pricing model (CAPM, Sharpe (1964)) to explore the quantitative relationship between expected returns of individual securities

Fig. 4.1 William Sharpe (1934–). Two of Sharpe's major contributions to quantitative investing are the CAPM and Sharpe ratio

and the market return.[1] From the quantitative investing perspective, Sharpe's major contributions are the CAPM and Sharpe ratio. We introduce the CAPM first and then define the Sharpe ratio.

The CAPM explores the relationship between individual returns and market returns through risk. It states that the expected return of a security is the risk-free rate plus compensation for the over-market risk, where the over-market risk is measured by the term β. The rationale is that if we hold all the money in cash, we will get the risk-free return, such as an interest rate earned on a banking deposit. If we invest that money in stock markets, then we risk losing money. In general, a higher return is required for riskier investments. How much should the additional return be to compensate for the additional risk? This is defined by β. According to Sharpe, if a stock's price movement is more volatile than the overall market movement, then a higher return should be expected and vice versa. Using $E(x)$ as an expected value of x, we have

$$E(r_i) = r_f + \beta_i(R_m - r_f), \tag{4.1}$$

where r_f is the risk-free return (cash) and R_m is the market return. While it seems naive to assume a linear relationship between expected return and market return, the CAPM's fundamental contribution to quantitative investing is not how to derive the expected return but the introduction of a risk definition and a general relationship between risk and return.

$E(r_i) - r_f$: additional return for taking additional risk

$R_m - r_f$: market premium, i.e., stock market return over cash return.

According to (4.1), different stocks will have different values of β. High-beta stocks mean that the company is more sensitive to market changes; conversely, low beta

[1]Photo source: https://www.mediatheque.lindau-nobel.org/laureates/sharpe.

stocks mean that the company is less sensitive to market changes. Note that the market can go in both directions: a high-beta implies that a stock may have greater positive returns when the market is up and larger negative returns when the market goes down. This double-edged feature plays a big role in financial products such as low beta and "smart beta" since the financial crisis of 2008.

The introduction of the CAPM and its linking of risk with expected returns in a simple equation contributes significantly to the overall investment world, including both traditional fundamental and quantitative investment. The CAPM is generally considered a must-have part of any modern investment textbook. It should be noted here that in addition to Sharpe (1964), there are other studies that proposed the CAPM concept independently around the 1960s, they are Jack Treynor (1961, 1962), John Lintner (1965a,b), and Jan Mossin (1966).[2]

In addition to the CAPM, Sharpe introduced another very important concept: the Sharpe ratio. In practice, there is a need to compare the performance of similar portfolios and a portfolio's performance against its benchmark. A natural question is, how can we measure portfolio performance quantitatively? Sharpe proposed a widely accepted solution: risk-adjusted return,

$$\text{Sharpe ratio} = \frac{\text{portfolio return} - \text{cash return}}{\text{portfolio risk}},$$

where portfolio risk can be measured by the standard deviation of portfolio returns minus the cash returns. The Sharpe ratio measures the overall performance of a fund and is used widely in the investment industry, especially for hedge funds and total return funds. The normal range of the Sharpe ratio in the industry is about 0.7–1.2. A Sharpe ratio below 0.5 is usually regarded as poor, and a Sharpe ratio above 2 is usually not sustainable. One variation of the Sharpe ratio is information ratio (IR), which measures the relative performance of a fund over its benchmark, expressed as risk-adjusted active or excess return,

$$\text{Information ratio} = \frac{\text{portfolio return} - \text{benchmark return}}{\text{active portfolio risk}},$$

where active portfolio risk can be measured by the standard deviation of portfolio returns minus the benchmark returns.

[2]Regarding the history of CAPM, see Fama (1968), French (2003), Perold (2004), and Sullivan (2006). Modigliani and other scholars made indirect contributions by encouraging the CAPM research and introducing the authors to each other during that period.

4.2 Rationale for a Stock Selection Strategy: Sources of Return Anomalies

We show in Chap. 1 that even for a 5- to 10-year period, there are still about 10–25% of equity funds that outperform the market. This holds true across all regions, indicating that there must be systematic sources of abnormal returns in equity markets. Identifying those sources requires a deep understanding of the market as well as the proper methodology to capture that information. In the following subsections, we briefly review the market efficiency hypothesis and discuss its necessary conditions and assumptions. We then explore real-world financial markets and identify sources of inefficiency that provide fundamental support for stock selection strategies.

4.2.1 Market Efficiency: Long- Versus Short-Run

A financial market is efficient when it reacts to any new information, public or private, immediately and completely. The assumptions behind the efficient market hypothesis (EMH) are that (1) there is a competitive market, (2) information travels without cost, (3) information is symmetric, and (4) the behavior of market participants is rational. The definition of a weak version of market efficiency is that the market is a consistent estimate of the true value of an investment. This means that the market efficiently reflects new information in the long run as a self-correction mechanism. Regardless of weak or strong versions of market efficiency, the assumptions above are usually too strong to be true. This is the fundamental reason for investors' confidence in stock selection strategies.

A common understanding in the industry is that over the long run, the market may be efficient, but in the short run, the market price can be biased for several reasons:

- The market is not fully competitive
  - investment barriers, monopolies, and regulations
- Information might be very costly
  - market segmentation
- Information is asymmetric
  - agency problem, structural issues
- Investors may not be rational
  - behavioral biases

For equity markets in developed countries, people usually regard the market as less inefficient, meaning it has an inherent mechanism to adjust itself to be efficient, in a long run. In emerging economies, equity markets are less efficient, meaning there are many barriers preventing stock prices from reflecting information fully and in a timely manner. Recall that in Chap. 1, we showed that the percentage of equity funds outperforming their markets is about 10–40% in developed countries but as high as 40–60% in emerging economies.

Reflected in price movements, market inefficiency occurs when a security sells in the market for substantially more or less than its intrinsic value. We can take advantage of this inefficiency by buying the security (the former) and holding it until the market recovers; or by short selling (the latter) and buying it back later at a lower price. Due to market inefficiency, opportunities for profit exist and can be exploited. For example, when there is bad news for a public company, many investors tend to overreact, thus creating downward pressure on the stock price. A few days later, the stock price may recover. The challenge then is how to identify the overreaction price range and when to enter the market to take advantage of this non-rational behavior, or in other words, how to determine the intrinsic value. Conventionally, this hypothetical value is usually proxied by the average value (plus reasonable deviations) over a reasonable period for a certain financial product.

Clearly, it is important to identify the sources of inefficiency and explore them in a systematic way. To explore market inefficiency, it is worthwhile to describe a financial market in general. A financial market is a place where participants conduct transactions involving financial products and services. We discuss sources of market inefficiency from the perspective of market participants in the following subsection.

4.2.2 Sources of Inefficiency

We discussed in Chap. 3 that the global stock markets can be divided into three segments: DM, EM, and FM. In terms of market efficiency, the EM and FM are much less efficient than the DM, but even the DM can be inefficient for various reasons. Here, we present systematic sources of market inefficiency for all three segments. The sources are categorized according to market participants and their respective business focuses, intentions, and action consequences. We also present examples of how to develop variables or factors in each category to capture inefficiency.

Participant	Motivation	Effects
Company	*to increase share price and size*	*creative accounting/windy CEO*[3]

From the perspective of the issuing company, the higher the stock price, the better. More often than not, the compensation of the senior management team is tied to the stock price and company size. Therefore, the senior management team has all the motivation to boost stock prices, such as by pursuing aggressive accounting methods to make core business earnings look better or making a series of acquisitions to build the empire. The consequences of these actions may not be captured immediately by

[3]Being from Chicago, the *windy* city is termed due to braggings of politicians in Chicago.

the market but will eventually be reflected in the performance of the company and realized by the market.

To identify investment opportunities created by public company behaviors, we could construct an earnings quality factor as the difference between net income and cash flow and an asset growth factor defined by the asset growth rate over a reasonable period (say, 3 years). Companies with extremely high values relative to their peers in the same industry will eventually be penalized by the market.

Participant	Motivation	Effects
Market maker	*to provide liquidity*	*information asymmetry*

In the stock market, liquidity providers may have more information than other investors. Large institutional investors may have more information and better ways to process information than others. Information asymmetry can create many issues in the market.

One variable in this category is short interest or borrowing cost. Many hedge funds borrow stocks from large brokers for short sales. The borrowing cost and changes in the short positions will indicate the direction of the price change of a stock. Of course, information is not symmetric because it is not public. Large institutional investors or pension fund managers will have more leverage to access those data.

Participant	Motivation	Effects
Investors	*to profit from price changes*	*over/under reaction*

In theory, investors do not care about the direction of price changes. They only need price changes to create opportunities for profit. This is especially true for long/short portfolios. When something bad happens, investors tend to walk away, and when something *very* bad happens, investors get scared and run off. This herding effect drives the price even further down in the short run. When things calm down, the price will steadily rise again.

A potential variable in this category is price reversal. Investors tend to chase winning companies and dump losing companies in their portfolios. Because of investors' overreactions to news and the market, increased buying or selling pushes the price up or down even further. Studies (e.g., DeBondt and Thaler 1985, French and Roll 1986, and Kaul and Ninmalendran 1990) find that for securities that undergo extreme price changes, the price reverses into the opposite direction (reverts to the mean) after three to ten trading days.

Participant	Motivation	Effects
Professionals	*to earn fees*	*self serving*

To illustrate this, we use an example of a trading desk and sell-side research department, both belong to an investment bank. Usually, a trading desk works very closely with its associated sell-side research department. Sell-side research may seem to be free of charge to investment companies (who are usually clients of investment banks) on the surface, but in fact it is not free! The value of sell-side research is to generate trading volume for the trading desk, either directly or indirectly. Indirectly, when an analyst changes his/her forecast, related trading signals, such as earnings momentum and diffusion, will change in value, thus driving trades. Directly, research and strategy reports from the research department are the give and the trading desk is the take is a typical relationship between an investment bank and an asset management company. This relationship is more important for small investment firms that do not have a strong capability on their own to conduct research and industry analysis and therefore rely heavily on sell-side research.

One variable in this category is estimation diffusion. When professional research analysts do not agree with each other, this indicates great uncertainty, which usually creates downward pressure on the price. Analysts seldom revise their estimates down in order to maintain their relationships with public companies. When they do revise down, it is usually for a significant reason.

- Participant | Motivation | Effects
  Government | *to regulate markets* | *agency problem, policy lags*

For macro-level policy instruments, there are always lags between announcements and their actual effects. Moreover, the same policy usually has different impacts on different industries. Policy effects cluster more at the industry or country level than the company level. For example, interest rate sensitivity varies because different industries react differently to any changes in the central bank's monetary policy. Interest rate changes will have more impacts on banking than the technology industry. Changes in the price of oil will have more impacts on countries like Canada and Russia than Germany and Greece.

We have identified major sources of market inefficiencies. For a stock selection strategy, we can first collect data and then build variables to capture these inefficiencies in equity markets. We can also apply concepts from Chaps. 2 and 3 to study univariate effects of each variable on future returns of stocks and the bivariate relationships between these variables. However, given that each variable represents only one piece of information, is there a way to combine variables to formulate a model to forecast future returns? How can we estimate the impacts of each variable and all the variables jointly? To answer these questions, we introduce multi-factor models and estimation methods in the next section.

4.3 Introduction to Multi-Factor Modeling

We focused on analysis of a single variable in Chap. 2 and the quantitative relationship between two variables in Chap. 3. But what if there are many variables? Can we formulate a model to characterize the relationship between one variable of interest and several other variables? When we investigate something we are interested in, we must define that "something" as a variable in the quantitative sense in order to explore it. We ask:

- What does this variable look like? This is a *univariate* model.

- How is this variable related to another variable? This is a *bivariate* model.
- Are there any other factors related closely to the variable? Is there a causal relationship? This is a *multi-factor* model!

In the following subsection, we show how to build a multi-factor model using three examples, then we present a general form of a linear model.

4.3.1 How to Build a Multi-Factor Model with Fundamental Insights

To build a multi-factor model, fundamental intuition is the key. First, we need to identify the variable we are interested in and then other variables whose values might impact the values of the variable of interest. The deeper our understanding of the situation, the better the model we will build.

We illustrate a model building process through three examples: people's weights, stock returns, and the price of oil.

Example 1: What Impacts People's Weights?

Intuition Using common sense, we identify the following variables that contribute to a person's weight:

- Height: This represents how tall a person is. A taller person generally weighs more than a shorter person.
- Gender: This represents male or female. Generally, a male weighs more than a female, all else being equal.
- Hours of exercise: Generally, physical exercise consumes energy and leads to weight loss.
- Food quality: This reflects the quality of a person's diet. For example, a balanced diet with vegetables and fruits rather than junk foods generally results in lower weight.

How Can We Build a Weight Model? We are interested in what causes changes in weight. In particular, we want to stay fit! Therefore, we want to know what factors contribute to weight gain and by how much. Based on the intuition above, we build a weight model with identified factors as follows:

$$\begin{aligned}\text{Weight} = b_0 + b_1\text{Height} + b_2\text{Gender}\\ +b_3\text{GymHours} + b_4\text{FoodQuality} + noise,\end{aligned}$$

where "noise" is everything other than the four factors we specified, b_k is the impact of variable k on weight.

Example 2: What Affects S&P 500 Stock Prices?

Intuition Using common sense, we identify the following variables that may impact a stock's price:

- Profitability: People want to invest in "good" companies. One basic measurement of quality is profitability: if a company makes profits, then the profits will help increase the stock price.
- Value: Value defines what is considered a good bargain. The price may already reflect the profitability, but if the market over- or undervalues a company, there is potential for the price to catch up in an undervalued case and decline in an overvalued case.
- Momentum: This refers to following a trend. Some companies see their stock prices increase because of favorable industry policy or overall high market demand for products from that industry, and others perform well in the stock market for more valid reasons. We want to follow winners and avoid losers.

How Can We Build a Pricing Model? We want to know what factors impact stock returns and by how much. Using intuition, we identify the variables that may impact or signal price movement of stocks: profitability, value, and momentum.

$$\text{Return} = b_0 + b_1\text{Profitability} + b_2\text{Value} + b_3\text{Momentum} + noise.$$

Example 3: What Impacts the Price of Crude Oil?

Intuition We identify the following variables that we think may impact the price of oil:

- US dollar strength: First, all transactions involving crude oil must be done in USD. Second, the USD is a comprehensive signal reflecting the strength of the USA in global affairs, including economic, geopolitical, and military power. This can be measured by the USDX, a USD index, which is the weighted average of exchange rates of USD to currencies from other major economic entities.
- Oil supply change: This refers to any information on oil supply changes. More oil supplies will push the price of oil down.
- Oil demand change: This refers to any information on changes in oil demand. For example, strong economic growth implies an increase in demand for oil.
- War and social crisis: This can refer to any extreme events around the world, particularly in oil-producing regions. These events usually increase the price of oil.

How Can We Build a Model for the Price of Oil? The price of oil is the variable we are interested to understand and perhaps forecast. We identify the USD strength, changes in oil supply and demand, and extreme events as major factors contributing to changes in the price of oil, resulting in the following oil pricing model:

$$\text{oilPrice} = b_0 + b_1 * \text{USDX} + b_2\text{Supply} + b_3\text{Demand} + b_4\text{Crisis} + noise$$

We have shown through these three examples how to build a multi-factor model to capture the forces that impact the variable we are interested in. How can we express these examples as a general linear model? Using the first example, a linear weight model can be rewritten as

$$y = b_0 + b_1 * x_1 + b_2 * x_2 + b_3 * x_3 + b_4 * x_4 + e,$$

$$\begin{aligned} where\ \ & y : weight, \\ & x_1 : height \\ & x_2 : gender \\ & x_3 : gymHours \\ & x_4 : foodQuality \\ & e : noise \text{ or error term}, \end{aligned}$$

We can apply the same equation to the other two examples and any other linear model. Here, linearity means being linear in parameters, $b = (b_0, b_1, b_2, b_3, b_4)$. We describe the components of a general linear model in the next section.

4.3.2 The Four Parts: The Known and the Unknown

Suppose we have a general linear model with two factors,

$$y = b_0 + b_1 x_1 + b_2 x_2 + \epsilon. \tag{4.2}$$

We see that there are four parts in this linear model: y, $X = (1, x_1, x_2)$, $b = (b_0, b_1, b_2)$, and ϵ. Each of them has a specific name in econometrics:

$$\begin{aligned} y :\ & \text{dependent or response variable} \\ X :\ & (1, x_1, x_2,),\ \text{independent variables, or factors in finance} \\ b :\ & (b_0, b_1, b_2,),\ \text{parameters or coefficients or weights} \\ e :\ & \text{residual, noise, or error term} \end{aligned}$$

We want to know how each variable in X impacts y. The relationship is characterized by the vector b, where b_0 is the constant (intercept) and b_k is the sensitivity of y to x_k. In order to derive the quantitative relationship, we first need to get data for X and y. Returning to the example of people's weights, we could conduct a survey to collect data on people's weight, height, food quality, hours of exercise, and gender. Usually, we would say that for a linear model, the known parts are the dependent and the independent variables because we can get data for these variables, while the unknown parts are the coefficients and the error.

Once we have values for (y, X), we need to estimate $b = (b_0, b_1, b_2)$ and then get $e = y - b_0 - b_1x_1 - b_2x_2$. How can we estimate b? We discuss estimation methodology in the next section.

4.4 Multi-Factor Model: Estimation for the Unknown

Suppose we have a linear model with just one factor,

$$y = b_0 + b_1x_1 + \epsilon. \tag{4.3}$$

We have data for (x_1, y), but how can we estimate parameters (b_0, b_1)? If we solve for (b_0, b_1), we can then define the line. There are many ways to fit a model to the data. How about an algorithm—could we find a line such that all data points are close to it? The question then is how to measure "closeness." One way to measure closeness is by distance from the line of best fit.

$$Distance = e = y - b_0 - b_1x_1.$$

We can then minimize the distance. For example, employing $g(e_i)$ as a function of e_i,

$$\min \sum_{i=1}^{n} g(e_i).$$

The simplest form would be $g(e_i) = e_i$; however, positive and negative distances would cancel each other out in this functional form:

$$e_i > 0 : y_i > b_0 + b_1x_1; \qquad e_i < 0 : y_i < b_0 + b_1x_1.$$

Apparently, $g(e_i) = e_i$ does not work because there are multiple solutions.

What if we square the distances, as we do when calculating variance?

$$g(e_i) = e_i^2 = (y_i - b_0 - b_1x_i)^2, \tag{4.4}$$

which makes all distances zero or positive! Or, can we just use the absolute value of e_i?

$$g(e_i) = |e_i| = |y_i - a - b * x_i|, \tag{4.5}$$

which again makes all distances zero or positive! The squared version (4.4) is called the least squares regression, while the absolute version (4.5) is the least absolute deviation regression. We introduce the least squares method (OLS) in the following section and the median estimation method as a special case of quantile regression in Chap. 8.

4.4.1 *Estimating a Multi-Factor Model by OLS*

Suppose we have a single-factor model (4.3) and N pairs of observations of

$$(x_i, y_i),\ i = 1, 2, \ldots, N-1, N.$$

We want to know how x impacts y, that is, the values for parameter $b = (b_0, b_1)$. As mentioned in the previous section, one way to obtain values of b is to employ (4.4),

$$g(e_i) = e_i^2.$$

We can add squares of errors together for all observations. Now, the optimality becomes finding (b_0, b_1) such that it minimizes the sum of the square of errors,

$$\begin{aligned} &\min \sum_{i=1}^{N} e_i^2 \\ &= \min \sum_{i=1}^{N} (y_i - b_0 - b_1 x_i)^2. \end{aligned} \tag{4.6}$$

This is called the least squares regression, where "least squares" means that the solution minimizes (hence "least") the sum of the "squares" of the errors of a linear model.

Carl Friedrich Gauss (Fig. 4.2) is accredited as the first scholar who proposed the least squares method in 1795 (Stigler 1981). But it was Adrien-Marie Legendre who first published a clear exposition of the least squares method in 1805 (Legendre (1805) in French). Using the draft note as evidence and his friends as witnesses, Gauss claimed that he developed the contents in Legendre (1805) many years before. According to Gauss, he did not bother to publish his research on the least squares method because he thought it was rather trivial: he happened to propose it while working on a major astronomical project.[4] Gauss (1809) claimed to have invented the method of least squares in 1795. A few years later, Gauss's method was used to locate the newly discovered asteroid Ceres. According to the back-then Hungarian

[4]Stigler (1981) provides an extensive account of the origins of OLS.

Fig. 4.2 Carl Friedrich Gauss (1777–1855). One of Gauss' major contributions to quantitative investing is the OLS methodology for linear multi-factor models. The German Mark featuring Gauss (1993, discontinued)

astronomer, Franz Xaver von Zach, among many methods, the least squares analysis is the only one that successfully located Ceres. Later, Karl Pearson used the word *regression* for least squares, "describing the regress to the mean." Equation of (4.6) is called the ordinary least squares (OLS) method.

An Example to Illustrate the OLS Method The least squares method is a statistical procedure designed to find the line of best fit for a set of data points, by minimizing the sum of the squared values of distances of those points from the fitted curve. We illustrate the OLS methodology through the following example.

Suppose we have five stocks with annual data on returns and profitability. What is the relationship between profitability and stock returns? More specifically, when profit increases by 1%, how much will the stock price change? We first make a scatter plot (Fig. 4.3) where visualization shows a positive relationship between the two. However, if we want to know the exact quantitative relationship between the two variables, we will have to seek some methodology, such as OLS:

$$\min \sum_{i=1}^{5} (y_i - b_0 - b_1 x_i)^2.$$

The solution from OLS, $\hat{b} = (\hat{b}_0, \hat{b}_1)$, is a straight line, with an intercept of $b_0 = -2.50$ and slope of $b_1 = 0.22$,

$$\hat{y} = -2.50 + 0.22x.$$

Now, let us add the line from the OLS estimates to the scatter plot. Recall that the line is derived with the estimates of b minimizing the squares of the error for each stock. We plot the distance or error for each stock with dotted lines. Any other line would produce higher values of the sum of the squared distances. In other words, the OLS line is the one with the overall minimum "distance." The values of the distances are listed in Table 4.1.

Note that points above the line and below the line produce positive and negative distances, respectively. They enter the regression with the same weight, but large deviations receive greater weights due to squaring. In our example, the second

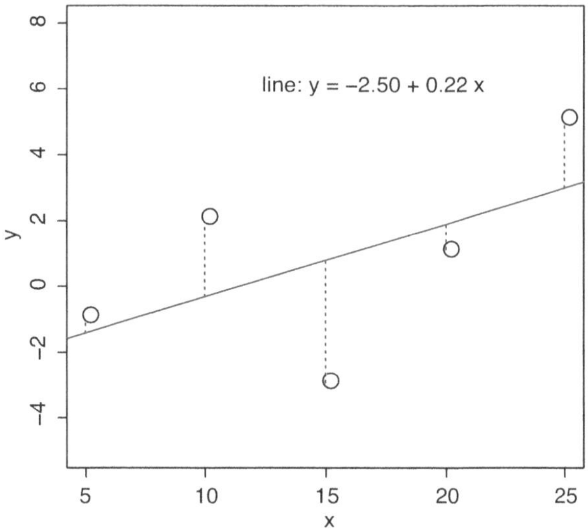

Fig. 4.3 A graphic illustration of the OLS methodology with five data points

Table 4.1 An OLS example with five data points

x	y	$\hat{y}$	Distance (e)	Distance squares (e^2)
10	2	−0.3	2.3	5.29
15	−3	0.8	−3.8	14.44
5	−1	−1.4	0.4	0.16
25	5	3	2	4
20	1	1.9	−0.9	0.81

We apply OLS to the model $y = b_0 + b_1 x + e$ and obtain $\hat{y} = -2.50 + 0.22x$, $\hat{e} = y - \hat{y}$

observation has an error of −3.8, yielding $e^2 = 14.44$, the largest value in the sum of squared errors (SSE). Based on the least squares definition,

$$SSE = 5.29 + 14.44 + 0.16 + 4 + 0.81 = 24.7$$

should be the smallest among all linear fitted lines.

Optimal Solution for OLS: First Order Condition The OLS optimality can be solved by the first order conditions. If the model has k factors, there should be k first order equations.

$$(\hat{b}_0, \hat{b}_1) = \text{argmin} \sum_{i=1}^{n} (y_i - b_0 - b_1 x_i)^2.$$

We define

$$G(b_0, b_1) = \sum_{i=1}^{n}(y_i - b_0 - b_1 x_i)^2 \tag{4.7}$$

and then solve for (b_0, b_1) using the first order conditions,

$$\frac{\partial G(b_0, b_1)}{\partial b_0} = -2\sum_{i=1}^{n}(y_i - b_0 - b_1 x_i) = 0$$

$$\frac{\partial G(b_0, b_1)}{\partial b_1} = -2\sum_{i=1}^{n} x_i(y_i - b_0 - b_1 x_i) = 0.$$

Let $\bar{y} = \frac{1}{n}\sum y_i$ and $\bar{x} = \frac{1}{n}\sum x_i$, the above equations can be simplified to

$$\bar{y} - b_0 - b_1\bar{x}_i = 0$$

$$\frac{1}{n}\sum_{i=1}^{n} x_i y_i - b_0\bar{x} - b_1\frac{1}{n}\sum_{i=1}^{n} x_i^2 = 0.$$

Solving for the two unknowns from the two equations yields immediately the OLS estimates:

$$\hat{b}_1 = \frac{\sum_{i=1}^{n}(x_i - \bar{x})(y_i - \bar{y})}{\sum_{i=1}^{n}(x_i - \bar{x})^2}$$

$$\hat{b}_0 = \bar{y} - \hat{b}_1\bar{x}.$$

Recall from Chap. 3 that $Cov(x, y) = \frac{\sum_{i=1}^{n}(x_i - \bar{x})(y_i - \bar{y})}{\sigma_x^2}$. We can rewrite b_1 as

$$\hat{b}_1 = \frac{Cov(x, y)}{Var(x)} = \rho_{xy}\frac{\sigma_y}{\sigma_x},$$

where ρ_{xy} is the correlation between x and y. Thus, we have the relationship between correlation and least squares regression estimates.

Formally, to verify whether $\hat{b} = (\hat{b}_0, \hat{b}_1)$ minimizes the sum of squared errors. We need to calculate the second derivatives with respect to b_0 and b_1, which constitute the Hessian matrix. The Hessian matrix needs to be positive definite to ensure that the OLS estimates the objective function globally.

We derive the Hessian matrix with values of the second-order derivatives based on (4.7),

$$H = \frac{\partial^2 G}{\partial^2 b} = \begin{bmatrix} 0 & \sum_{i=1}^{n} x_i \\ \sum_{i=1}^{n} & \sum_{i=1}^{n} x_i^2. \end{bmatrix}$$

The determinant value of the Hessian matrix is

$$|H| = \sum_{i=1}^{n} x_i^2 - (\sum_{i=1}^{n} x_i)^2 = \sum_{i=1}^{n} (x_i - \bar{x})^2 = n\, Var(x) > 0,$$

which completes the proof that, indeed, the OLS estimates minimize the sum of squared values!

Matrix Form: A "Real" Multi-Factor Linear Model So far, we have focused on the simplest linear model: a single-factor model. What if we have multiple factors, such as in Example 2 of Sect. 4.3.1, in which returns are impacted by profitability, value, and momentum? We can express Example 2 using a linear model in a mathematical format,

$$y = b_0 + b_1 x_1 + b_2 x_2 + b_3 x_3 + e,$$

where y is return, x_1 is profitability, x_2 is value, x_3 is momentum, and e is the error term. We can apply the same least squares method to derive OLS estimates.

To represent a general case, we now consider a linear model with K factors. We can formulate the model as

$$\begin{aligned} y &= b_0 + b_1 x_1 + b_2 x_2 + \ldots + b_{k-1} x_{k-1} + b_k x_k + e \\ &= (b_0, b_1, \ldots, b_{k-1}, b_k)(1_n, x_1, x_2, \ldots, x_{k-1}, x_k)^\top + e, \end{aligned}$$

where 1_n is a vector with the value of 1s and each x component contains a vector of values with observations. Note that before, we had only one factor, so we did not need to worry about relationships between factors. Now that we have multiple factors, what do we do if two of those factors have very similar values? We will revisit this issue when deriving estimates for b.

Defining $X = (1_n, x_1, x_2, \ldots, x_{k-1}, x_k)$ and $\beta = (b_0, b_1, \ldots, b_{k-1}, b_k)$, we can rewrite the general linear model above in a more compact matrix form:

$$y = X\beta + e,$$

where y is the response variable, X is the matrix of factors, and ϵ is the error term (noise). Regarding dimensions, there are n observations (rows) and k factors. Note that if we keep the intercept, we have the following dimensions: y is $n \times 1$, X is $n \times (k+1)$, β is $(k+1) \times 1$, and e is $n \times 1$. The full expression for y and X is as follows:

$$y = \begin{bmatrix} y_1 \\ y_2. \\ y_3 \\ \vdots \\ y_{n-1} \\ y_n \end{bmatrix} \qquad X = \begin{bmatrix} 1 & x_{11} & x_{21} & \dots & x_{k-1,1} & x_{k,1} \\ 1 & x_{12} & x_{22} & \dots & x_{k-1,2} & x_{k,2} \\ 1 & x_{13} & x_{23} & \dots & x_{k-1,3} & x_{k,3} \\ \vdots & \vdots & \vdots & \vdots & \vdots & \vdots \\ 1 & x_{1,n-1} & x_{2,n-1} & \dots & x_{k-1,n-1} & x_{k,n-1} \\ 1 & x_{1,n} & x_{2,n} & \dots & x_{k-1,n} & x_{k,n} \end{bmatrix}.$$

Now, let us consider a matrix form of Example 2. Putting all factors together in a matrix form, we have

$$\begin{aligned} y &= b_0 + b_1 x_1 + b_2 x_2 + b_3 x_3 + e \\ &= X\beta + e, \end{aligned}$$

where $X = (1_n, x_1, x_2.x_3)$, an $N \times 4$ matrix, and $\beta = (b_0, b_1, b_2, b_3)$. We can estimate parameters β using the OLS method. The least squares optimality equation is as follows:

$$\begin{aligned} \hat{\beta} &= \text{argmin}_\beta e^\top e \\ &= \text{argmin}_\beta (y - X\beta)^\top (y - X\beta). \end{aligned}$$

Using the first order condition, we get

$$\begin{aligned} 2X^\top (y - X\beta) &= 0 \\ X^\top y - X^\top X\beta &= 0 \\ X^\top y &= X^\top X\beta. \end{aligned}$$

Solving for β, we obtain the OLS estimate $\hat{\beta}$ immediately,

$$\begin{aligned} \hat{\beta} &= (X^\top X)^{-1} X^\top y \\ &= (X^\top X)^{-1} X^\top (X\beta + e) \\ &= \beta + (X^\top X)^{-1} X^\top e. \end{aligned} \tag{4.8}$$

The matrix solution (4.8) requires that $X^\top X$ be a full-rank matrix in order to have the reverse matrix. The full-rank requirement answers our earlier question about two factors having values close to each other: if this the case, it is better not to have both factors in a linear model. This is called the collinearity issue. Fundamentally speaking, adding an additional factor with values similar to one of the factors currently in the model does not contribute much value. Moreover, the collinearity issue creates potential problems for the solution. In quantitative investing, a linear model with factors representing diverse information is very important.

The last row in (4.8) also shows that the OLS estimates $\hat{\beta}$ are the population value β plus a factor-value weighted error term. This is a very useful equation for the analysis of OLS estimates. For example, we could derive the standard error for $\hat{\beta}$ as follows:

$$s.e.(\hat{\beta}) = \frac{\sigma_{\hat{\beta}}}{\sqrt{n}},$$

where $\sigma_{\hat{\beta}}$ can be obtained from

$$\begin{aligned} Var(\hat{\beta}) &= Var((X^\top X)^{-1} X^\top e) \\ &= (X^\top X)^{-1} Var(e) \\ &= (X^\top X)^{-1} \sigma_e^2. \end{aligned}$$

Now that we have established a general form of a linear model, it is convenient to discuss the interpretation of parameters,

$$\text{matrix form:} \quad y = X\beta + e.$$

Assuming that $E(e) = 0$, the expected value of y will be

$$E(y) = X\beta$$

and thus

$$\frac{\partial E(y)}{\partial X} = \beta.$$

This implies that β is the sensitivity of an expected value to the factors. In other words, β measures how changes in X impact the conditional mean of y. The "average" nature of the estimates contributes to the huge volume of applications of the OLS method, as many people seek average effects in the contexts of either estimates or forecasts. For quantitative investment, the implication of OLS is that the parameters β are the effects of factors on *expected* equity returns, or the conditional mean or *average* of stock returns. If the return distribution of a stock is not normal—for instance, having a fat tail or being highly skewed—this "mean" estimate may lose significance for stock selection.

Since its publication in 1805, the OLS method has become widely recognized and popular. In summary, it has two important features: (1) easy computation and (2) elegant properties. We have discussed the computation part: without large data sets and multiple factors, one could solve for OLS optimality with just a piece of paper and a pencil. With large data sets and/or multiple factors, OLS estimates can be derived quickly using a computer. This has important consequences. For example, the method can easily be applied to empirical studies, and interpretation is straightforward. However, ease of computation is only one reason. Another very important reason for OLS's popularity is that it has many elegant properties.

In the following section, we discuss properties of OLS estimates as well as the conditions required by those properties.

4.5 Multi-Factor Models: Properties of OLS

For a linear multi-factor model, OLS estimators have the following properties:

1. Linear: additive; OLS estimators are linear functions of the values of y.
2. Unbiased: the mean of the sample estimates is the same as the population mean.
3. Consistent: as the sample size increases, estimates will eventually converge to the population parameter.
4. Efficient (Best): the variance of the OLS estimator is the smallest among all unbiased linear estimators.

Because of the above properties, the OLS estimator is called BLUE: best linear unbiased estimator. Being BLUE is extremely important in econometrics, making OLS one of the strongest and most widely used estimators for unknown parameters. One should use the OLS method not only because it is unbiased but also because it has the lowest variance among the class of all linear unbiased estimators. However, there is no free lunch, as these properties hold only if the following conditions are met:

A1. β: The linear regression model is linear in parameters.
A2. X: There is no multicollinearity (or perfect collinearity).
A3. X and ϵ: The conditional mean should be zero.
A4. ϵ: Spherical errors: There is homoscedasticity and no autocorrelation.

How are these properties and conditions related to investment? In quantitative investing, OLS is employed widely in industry as well as academic research. This is partially because of the useful properties listed above and partially because of the complexity and uncertainty of asset price movements. A simple and reliable estimator helps investors understand quantitative relationships. Investors should avoid complicated methods because they require more assumptions about data, which finance data generally do not satisfy. As mentioned in the preface to this book, one of the four pillars of successful quantitative investment is mastery of quantitative methodologies: not only mastery of how to use them but also, and perhaps more importantly, understanding the properties and conditions of estimation methods and limitations of these methods if certain conditions are not met.

4.5.1 Properties of OLS: BLUE

We discuss each part of BLUE in detail in this section. Consider a multi-factor linear model,

$$y = X\beta + \epsilon.$$

Under assumptions A1– A4 above, we have the Gauss–Markov theorem, which states that the estimator $\hat{\beta}$ is the best linear unbiased estimator (BLUE) among all linear unbiased estimators. We present the properties of OLS estimates in mathematical notation below.

$$\begin{aligned} unbiased &: \ E(\hat{b}) = b \\ consistent &: \ n \rightarrow \infty, \hat{b} \Rightarrow b \\ efficient &: \ Var(\hat{b}) \leq Var(\tilde{b}). \end{aligned}$$

Next, we discuss each property of OLS estimates in detail and explore their relevance to quantitative investment. We start our discussion with linearity, the letter "L" in BLUE.

Linearity in y Linearity here does not mean linear in parameters, rather, it means that the OLS estimator $\hat{\beta}$ is a linear function of y_i with weights from regressors. This is shown clearly if we rewrite the OLS estimator as

$$\hat{\beta} = (X^{\top}X)^{-1}X^{\top}y = \left[(X^{\top}X)^{-1}X^{\top}\right]y,$$

the weighted $\left[(X^{\top}X)^{-1}X^{\top}\right]$ average of y. In a quantitative stock selection strategy, y is usually forward returns of stocks and X is a vector of factors. Thus, $\hat{\beta}$ from OLS is the factor-value weighted returns across stocks.

Unbiased Unbiased means that if we have M samples for a random variable X, then the average of the estimates from these M samples about X (suppose M is random and large enough) will be the same as the unknown population parameter. For example, if we have $M = 100$ samples of data from the S&P 500 and we would like to know how the value factor impacts stock returns. Applying OLS to each sample for the same model will produce an estimate for $\hat{\beta}_{value}$. Being unbiased means that the average of these 100 estimates can be regarded as the true value of value impacts on stock returns. In quantitative investment, no one knows the true or population value, but knowing that the estimate is not too far from the true value ensures that investors are estimating the "real" impacts, not something else.

We can demonstrate the property of unbiasedness with an example. Suppose we have a model with one regressor,

$$y = 10 + 5x + \epsilon, \tag{4.9}$$

with the true values for $b = (b_0, b_1) = (10, 5)$. Now, we generate data according to the model above with a sample size of 100. We then estimate b using the OLS method. We repeat the process 100 times, gathering 100 estimates of b, $\hat{b} = (\hat{b}_0, \hat{b}_1)$. Focusing on $\hat{b}_1$, we get the expected value $E(\hat{b}_1) = 4.995$, close enough as an unbiased estimator to the true value of $b_1 = 5$. The left plot in Fig. 4.4 shows the unbiasedness of the OLS estimate with the expected mean and density of $\hat{b}_1$. The density plot shows that values of $\hat{b}_1$ range from 4.8 to 5.2 with the average (the dashed line) very close to $b_1 = 5$. We present detailed computation and results via the R codes below.

OLS: R codes for the unbiased property

```
blue.unbiased <- function()
{
        ## model setup, sample size =100
        set.seed(99)
        x=rt(100,3)
        error=rnorm(100)
        b0=10
        b1=5
        y=b0+b1*x+erro
       ### biasedness, repeat estimate 100 times
       ntime=100
       bias.mat=rep(NA, ntime)
        for(i in 1:ntime)
        {
                bias.mat[i] = lm(y~x)$coef[2]
                x=rt(100,3)
                error=rnorm(100)
                y=b0+b1*x+error
        }
        return(bias.mat)
}
```

Consistency A consistent estimator is one that approaches the real value of the population parameter as the size of the sample, n, increases. This is very important for quantitative investing, where if we can get more data, such as enlarging a sample size from 500 stocks to 1000 stocks, the estimates will be closer to the population value. In the real world, it is impossible to have 100 samples for the same group of stocks at the same time. For example, there is only one set of observations of stock

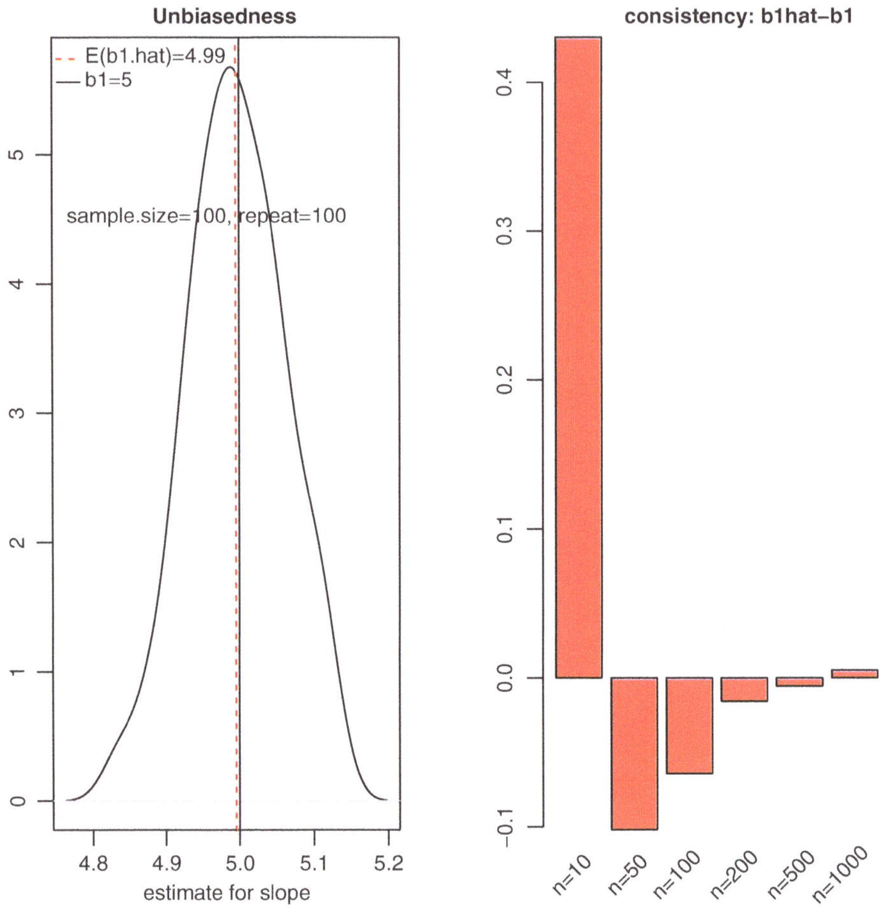

Fig. 4.4 The left plot demonstrates unbiasedness with the average of 100 sample estimates. The right plot demonstrates consistency as sample size increases. The simulation is based on model $y = 10 + 5 * x + \epsilon$ with the OLS estimator for $b_1 = 5$

closing prices for the S&P 500 from 1980 to 2010. However, we can easily increase the investment universe from 500 to 1000 stocks. For instance, this can be done by replacing the S&P 500 with the Russell 1000. However, we have to keep in mind that when we enlarge the investment universe, the newly added stocks may have different characteristics. For example, the additional 500 stocks from the Russell 1000 are smaller than those in the S&P 500. Nevertheless, we should still try to avoid a small sample size, which can yield estimates very far from the true value, because there is not enough data to be representative.

To illustrate this, we use the same model, (4.9), as we investigate the unbiased property,

Table 4.2 The OLS estimator of $b_1 = 5$ is consistent: as the sample size increases, $\hat{b}_1$ converges to b_1

	$n = 10$	$n = 50$	$n = 100$	$n = 200$	$n = 500$	$n = 1000$
$\hat{b}_1$	5.430	4.898	4.935	4.984	4.994	5.005
$\hat{b}_1 - b_1$	0.430	−0.102	−0.065	−0.016	−0.006	0.005

$$y = 10 + 5x + \epsilon.$$

Now we generate data sets with increasing sample sizes of 10, 50, 100, 200, 500, and 1000. We then apply OLS to derive the estimate for b_1 corresponding to each sample size. We present the consistency property of the OLS estimator in both Table 4.2 and the right plot in Fig. 4.4. It can be seen clearly that as the sample size increases, the OLS estimate, $\hat{b}_1$, converges towards the true value b_1. We present below the R codes with computation details and results demonstrating the consistency property of OLS estimates.

OLS: codes for the consistency property

```
ols.consistency <- function()
{
        ## model set up
        b0=10
        b1=5
        ## consistency, sample size=10, 50,100, 200, 500, 1000
        samples=c(10,50,100,200,500,1000)
        nsample=length(samples
        con.mat=rep(NA,nsample)
        for(j in 1:nsample)
        {
                set.seed(99)
                the.sample=samples[j]
                x=rt(the.sample,3)
                error=rnorm(the.sample)
                 y=b0+b1*x+error
                 con.mat[j]=lm(y~x)$coef[2]
         }
        return(con.mat)
}
```

Efficiency In the context of estimation, the term "best" means most efficient, and the term efficient in econometrics refers to a smaller error or, in ordinary language, greater accuracy. Efficiency complements unbiasedness. For example, $\hat{b}$ and $\tilde{b}$ can both be unbiased, but they may have very different errors:

$$E(\hat{b}) = E(\tilde{b}) = b, \quad Var(\hat{b}) = 10, Var(\tilde{b}) = 100.$$

Clearly, $\hat{b}$ is more accurate because it has a smaller error around the estimate. The efficiency property of OLS is usually a luxury because it needs to have the smallest variance among all unbiased linear estimates for a linear multi-factor model, which requires strict assumptions. We discuss this in the next section.

4.5.2 Conditions for Being BLUE

We have discussed the BLUE properties of OLS estimators, but it should be stressed that these properties require strict conditions:

A1. β: The linear regression model is linear in parameters.
A2. X: There is no multicollinearity.
A3. X and ϵ: The conditional mean of the error term is zero.
A4. ϵ: Spherical errors: There is homoscedasticity and no autocorrelation.

Before explaining in detail how each condition works, we summarize how they are related to the BLUE properties: A1 and A2 ensure the existence of an OLS solution; A3 ensures the properties of unbiasedness and consistency; and A4 ensures efficiency.

Now, recall the matrix format of the OLS estimator (4.8),

$$\hat{\beta} = \beta + (X^\top X)^{-1} X^\top e.$$

We now show in detail what conditions each property requires and discuss the implications in the context of quantitative investing.

Exogeneity ⇒ Unbiasedness and Consistency Regarding X and ϵ, the zero mean of ϵ conditioning on X can be interpreted as exogeneity between X and ϵ; that is, factors are exogenous to the error term. Exogeneity is the opposite of endogeneity.

$$\text{Zero mean error:} \quad E[\epsilon] = 0 \tag{4.10}$$

$$\text{Non-correlation between X and } \epsilon: \quad E[X^\top \epsilon] = 0 \tag{4.11}$$

$$\text{Exogeneity:} \quad E[\epsilon | X] = 0. \tag{4.12}$$

The above conditions are very restrictive for investment models, because intercorrelation between variables and events is very common, and identified factors are usually correlated with unidentified factors contained in the error term. For example, for quantitative stock selection strategies, forecasted returns (alpha) usually have a correlation of less than 15% with actual returns, and the value of R^2 for a multi-factor model is usually less than 5%, which implies that 80–90% of the information about returns is contained within the error term. Another reason is that many factors use stock price as a component (for example, value and momentum factors), so the price composition in a factor definition is naturally related to the response variable $return = \frac{P_t}{P_{t-T}}$, causing endogeneity mechanically.

Following the definition of unbiasedness, we have

$$E(\hat{\beta}) = \beta + E\left[(X^\top X)^{-1}X^\top \epsilon\right]. \tag{4.13}$$

If the second term is zero, then we have $E(\hat{\beta}) = \beta$. Condition A3 dictates that

$$E[\epsilon|X] = E[f(X)\epsilon] = 0,$$

where $f(X) = (X^\top X)^{-1}X^\top$ is a function of X. Hence, unbiasedness follows immediately from (4.13).

Now, let us consider the consistency of $\hat{\beta}$ to β,

$$\lim_{n\to\infty} \hat{\beta} = \beta + \lim_{n\to\infty} (X^\top X)^{-1}X^\top \epsilon.$$

The property of consistency requires that

$$\lim_{n\to\infty} (X^\top X)^{-1}X^\top \epsilon = 0.$$

Condition A2 states that there is no collinearity in X, which implies that $Q = n^{-1}(X^\top X)$ is a positive-definite matrix. With Condition A3, we obtain consistency as follows:

$$\begin{aligned}
&\lim_{n\to\infty} (X^\top X)^{-1}X^\top \epsilon \\
&= \left[n^{-1}(X^\top X)\right]^{-1} \lim_{n\to\infty} n^{-1}X^\top \epsilon \\
&= Q^{-1} \lim_{n\to\infty} E(X^\top \epsilon) \\
&= 0.
\end{aligned}$$

Spherical Errors ⇒ Efficiency For the error term, the homoscedasticity assumption means that all errors have the same variance, while no autocorrelation means the errors are not correlated with each other. This is a strict requirement of the error terms: they should be like identical twins who are strangers—the same but unknown to each other. A stronger version of condition A4 is that errors are independent and identically distributed (IID). In quantitative terms,

$$\text{Homoscedasticity:} \quad E[\epsilon_i^2|X] = \sigma^2 \tag{4.14}$$

$$\text{No autocorrelation:} \quad E[\epsilon_i \epsilon_j|X] = 0 \tag{4.15}$$

$$\text{Spherical:} \quad Var[\epsilon|X] = \sigma^2 I_n, \tag{4.16}$$

where I_n is the identity matrix with n dimensions. In the context of quantitative investing, condition A4 dictates that all other information, excluding factors, has the same impact for each stock and that information for each stock is not correlated. Obviously, this is not the case for asset returns in financial markets, where a shock, such as an interest rate increase, can impact all stocks at the same time (to different degrees). For example, a declaration of war on an important oil-producing country will impact oil production in that country and hence the world oil supply, which applies upward pressure to the price of oil that will affect many stocks, especially companies in the energy and airline industries.

If condition A4 is satisfied, then the OLS estimator will have the smallest variance. That is, if we have another linear unbiased estimator, $\tilde{\beta}$, from a different method, the following is always true:

$$\begin{aligned} Var(\hat{\beta}) &= Var(\beta + (X^\top X)^{-1} X^\top \epsilon) \\ &= (X^\top X)^{-1} Var(\epsilon) \\ &= (X^\top X)^{-1} \sigma_\epsilon^2 \\ &\leq Var(\tilde{\beta}). \end{aligned}$$

For a detailed proof, please see the Gauss–Markov theorem.

A linear multi-factor model describes the expected value of the dependent variable (outcome) depending on the known states of the independent variables (predictors). In regression, we always think about the "conditional" distribution of the dependent variable, i.e., the distribution of Y given $X = x$. Therefore, the conditional distribution of Y and the error term ϵ have the same distribution. If the error is normal, so is the conditional distribution $Y|X = x$. We should note here that for OLS estimates to be BLUE requires no assumptions about the distributions of independent or dependent variables. In theory, OLS does not require the distribution of error term to be normal. However, in practice, having the distribution being normal is important, such as to make the prediction intervals reliable.

In order for inference to work well (e.g., use of estimated coefficients and their standard errors), the estimates need to be normally distributed or very close to normally distributed (or a t-distribution with a large sample size). This can be met by either (1) normally distributed error terms (residuals) or (2) a large sample. If one has a large sample, the central limit theorem ensures normality of the estimates. The definition of "large" depends on the context. If there are many independent variables, the necessary sample size should be much larger.

4.6 Inference of OLS Estimates: Factor and Model Significance

In this section, we explore inference of OLS estimates, focusing in particular on significance testing of factors and models. We first apply hypothesis testing to show how to test the significance of a factor in a multi-factor linear model, then present a measure for model efficacy.

4.6.1 *Significance of a Factor*

How do we objectively evaluate whether a model is good or bad? An out-of-sample test should dictate the ultimate judgment. However, even before an out-of-sample test, is there a way to answer the following?

1. At the factor level: Is it significant? Do the data agree with the intuition?
2. At the model level: Can the model explain the data well overall?

We explore the answer to the first question in this section and address the model fit question in the next section. Recall that in Chap. 2, we used hypothesis testing to determine whether a statistic value from a sample is the same as the population value.

To investigate the significance of a factor, we use a one-factor model,

$$y = b_0 + b_1 x_1 + \epsilon. \tag{4.17}$$

In the worst-case scenario, x_1 has no relationship with y, or in other words, $b_1 = 0$. This can be tested by specifying the following hypotheses:

$$\text{Null hypothesis:} \quad H_0 : b_1 = 0$$
$$\text{Alternative hypothesis:} \quad H_a : b_1 \neq 0.$$

One way is to use a t-test, where the t-value is derived as

$$t\text{-}value = \frac{\hat{b}_1 - b_1}{s.e.(\hat{b}_1)} = \frac{\hat{b}_1}{s.e.(\hat{b}_1)} \sim \text{Student's t-distribution},$$

with $n - k$ degrees of freedom, where n is the number of observations and k is the number of factors. In (4.17), we have two unknowns ($k = 2$), b_0 and b_1. The denominator $s.e.(\hat{b}_1)$ is the standard error of $\hat{b}_1$. By building a confidence interval for a specified significance value of probability, we are able to test whether an OLS estimate is close to zero. Continuing the example of five observations, we get the OLS results from R as shown below.

Significance of a factor

```
> ols.example=lm(return~x,data=oo)
> summary(ols.example)

Call:
lm(formula = return ~ x, data = oo)

Residuals:
   1    2    3    4    5
 2.3 -3.8  0.4  2.0 -0.9

Coefficients:
            Estimate Std. Error t value Pr(>|t|)
(Intercept)  -2.5000     3.0094  -0.831    0.467
x             0.2200     0.1815   1.212    0.312

Residual standard error: 2.869 on 3 degrees of freedom
Multiple R-squared:  0.3288,Adjusted R-squared:  0.1051
F-statistic:  1.47 on 1 and 3 DF,  p-value: 0.3122
```

We see that the t-value for $\hat{b}_1$ is $0.22/0.1815 = 1.212$, with a corresponding p-value of 0.31. This is not significantly different from zero with a 10% acceptance rate. Of course, this is due to the small sample size. We will see real-world examples in the industry insights section.

We have just discussed how to make a quantitative inference about the significance of a single factor. In the next section, we explore the overall significance of a model.

4.6.2 Overall Fit of the Model to the Data

From the previous section, we see that there may be strong or weak factors in a model. A multi-factor linear model may have many weak factors, but these individual factors may work well jointly to explain values of the response variable. How can we know whether the model is good or bad overall? How can we make a quantitative judgment? In econometrics, this is termed goodness of fit and can be measured by the extent to which y can be explained by the fit of $\hat{y}$.

Equation (4.8) can be rewritten as

$$y = X\beta + e = X\hat{\beta} + \hat{e}.$$

Recall that OLS minimizes the squared values of the error term,

$$SSE = \sum_{i=1}^{n} e^2.$$

We now simply need to know how much of this value is in the percentage of total variance of y, that is,

$$SST = \sum_{i=1}^{n} (y - \bar{y})^2.$$

This can be reformulated and decomposed as follows:

$$\begin{aligned} SST = \sum_{i=1}^{n} (y - \bar{y})^2 &= \sum_{i=1}^{n} [(y - \hat{y}) + (\hat{y} - \bar{y})]^2 \\ &= \sum_{i=1}^{n} [\hat{e} + (\hat{y} - \bar{y})]^2 \\ &= \sum_{i=1}^{n} \hat{e}^2 + \sum_{i=1}^{n} (\hat{y} - \bar{y})]^2, \quad \sum_{i=1}^{n} \hat{e}\hat{y} = 0, \\ &= SSE + SSR, \end{aligned}$$

where $SSR = \sum_{i=1}^{n} (\hat{y} - \bar{y})]^2$. How much y can be explained by $X\beta$ can be expressed in mathematical notation as

$$R^2 = \frac{SSR}{SST} = 1 - \frac{SSE}{SST}. \tag{4.18}$$

Apparently, if the model explains everything in y, the goodness of fit for this perfect model would be $R^2 = 1$. In the other extreme case, where the model does not explain anything, y would be equal to the error term, so the goodness of fit would be $R^2 = 0$. Hence, the value of R^2 should be in the range of (0, 1).

Now, a natural question is, what value of R^2 implies a good or bad model? The answer is, "It depends." For example, in the weight model (Sect. 4.3, Example 1), the value of R^2 can be as high as 0.50 or 0.90. On the other hand, in the stock return model (Sect. 4.3, Example 2), the value of R^2 is usually less than 0.10. For stock return forecasting purposes, we should have the response variable as forward returns and the factors as the values known as of the most recent period. In other words, the relationship we seek to model is to predict future returns. In this context, the value of R^2 is usually less than 0.05. As a general rule of thumb, in the context of security return forecasting by a multi-factor model in developed and emerging markets, if

$R^2 > 0.05$, one should be very suspicious that something is wrong. In this case, we need to check the model, factors, data, and calculation carefully.

Another fact about R^2 is that its value will increase with the number of factors, assuming new factors have explanatory power. In quantitative investing, we usually do not rely on R^2 given its generally low values in that context. Therefore, the addition of a new factor or multiple factors should be based on fundamental judgment (for details, please refer to the new factor building section in Chap. 5).

4.7 Industry Insights: A Multi-Factor Alpha Model

The quantitative investment approach emerged with the development of econometrics, data, and computer technology. Some advantages that quantitative managers have relative to more traditional managers are a consistent and repeatable investment process, transparent risk control, and effective implementation.

In this section, we present a general process of alpha modeling in industry in the context of a stock selection strategy. We illustrate how to build a multi-factor model for security return forecasting. Recall from Chap. 1 that a typical quantitative investment process usually starts with an investment strategy within a particular investment universe, and an alpha model for return forecasting within that universe is a critical component.

Strategy with return/risk targets ⇒ Alpha model ⇒ Portfolio construction with constraints and risk control ⇒ Trading and rebalance ⇒ Performance attribution.

A common approach to the *Alpha model* part is to build a multi-factor model, with each factor capturing certain aspects of future returns of the assets in the selected investment universe. In the following subsection, we describe primary types of factors used by quantitative strategies in the industry today.

4.7.1 Where Does Alpha Come From

In this section, we show how to employ a multi-factor model to build alphas for a stock selection strategy. However, before any model building, we need to have a deep understanding of financial markets, especially the drivers and/or signals that impact price movements of stocks in an investment universe. Note that for different investment universes, such as country, industry, or market capitalization categories, the factors impacting stock returns can be very different. Here we discuss how to build a multi-factor alpha model in a general sense. For stock selection strategies with medium- or long-term investment horizons, buying a stock is like buying a piece of a company. Of course, we are interested in companies that have good business performance relative to their peers. The question is, how do we define, measure, and identify "good" business performance?

First, a company needs to make profits, which are important for the company's survival and growth. Furthermore, a company is more investable if it makes sustainable profits, not as just a one-time phenomenon. Second, we want to make sure that the company's profits are of good quality. For example, the profits should be generated from the company's core products, not because of creative accounting or aggressive sales through accepting too many payments on credit. Third, we need to understand the importance of the senior management team, who make strategic decisions that impact business performance in a profound way. For example, are the CEO and her team empire builders who make series of acquisitions because the size of the company is tied to their own compensation? Management quality is very important for a company and hence stock performance.

Now recall that equity markets, especially those in developed economic regions, can be efficient, meaning that information about profitability, earnings quality, and management quality may already be reflected in stock prices, as every investor values those characteristics. How do we know whether a company is overvalued or undervalued by the stock market? We can apply the principles of "value" investing. We learned from Chap. 2 that in 1934, Benjamin Graham proposed the concept of value investing based on the differential between a company's market and intrinsic values, an approach Warren Buffett went on to apply successfully. For example, for a long-only portfolio, we are interested in "undervalued" stocks with good business performance, increasing the likelihood that their prices will eventually increase.

In addition to profitability, earnings quality, and value, other factors are also very important. For example, price momentum, the "technical" aspect of price movements, is constructed based purely on pricing information about a stock. The investment logic is very simple: to follow winners and avoid losers. The fundamental intuition is that there must be reasons for a stock to be a winner or loser, and the situation will likely persist for a while. A "technical" approach is to identify winners and losers based on past price movements and then try to buy the winners and avoid (long only) or short (long/short) the losers.

Another set of factors is market sentiment, that is, how the market "perceives" a company. In the investment industry, there are sell-side research companies and consulting companies that have skilled teams evaluating companies on a continuous basis with a focus on earnings per share (EPS), recommendations to buy/hold/sell, etc. Note that these sell-side professionals are not investors themselves, but they are the ones who understand each company the best because they go over financial reports, visit companies, and talk with CEOs. They then publish their research results and forecasts. Investors can leverage the expertise of sell-side professionals. This set of factors is categorized as market sentiment in the industry.

Based on the fundamental insights above, we can categorize factors into several themes based on their fundamental definitions: profitability, earnings quality, management quality, value, momentum, and market sentiment. Suppose we have each variable for each theme, how can we then combine all the variables to forecast future stock returns? We discussed analysis of a single variable and two variables in Chaps. 2 and 3. Now we want to use multiple variables to forecast future stock returns.

The simplest way to start is to build a linear model, that is, to add all variables together in a linear form. For example, we can formulate a simple multi-factor alpha model as

$$\alpha = b_0 + b_1\text{PROF} + b_2\text{EQ} + b_3\text{MQ} + b_4\text{VALUE} + b_5\text{PM} + b_6\text{MS} + \epsilon, \qquad (4.19)$$

where α is future stock returns, PROF is profitability, EQ is earnings quality, MQ is management quality, VALUE is value, PM is price momentum, MS is market sentiment, b_k is the impact of each theme or factor on returns, and ϵ is everything else that impacts stock prices.

We have intuitively identified themes that impact stock returns. What are the signals in each theme and how can we construct them? The following section describes each theme with its fundamental intuitions and associated signals.

4.7.2 *Building Signals for Each Theme*

In this subsection, we show the building process for each theme in this section, including rationales, factor definitions, and formulas.

4.7.2.1 Profitability

How do we know a stock's return will outperform the market? At end of the day, a stock is simply a share in the ownership of a business. Stock prices can fluctuate due to a multitude of factors in the short term, but over the long term, the stock price and the value of a business tend to move in the same direction. Common sense suggests that the more profitable the business, the greater its ability to create value for investors. Unsurprisingly, statistical research has proven that companies with superior profitability levels tend to produce above-average returns for investors. If a company is consistently making above-average profits, this generally indicates that it has superior business qualities, such as a differentiated brand, better technology, and/or a more innovative management team.

Given the transparency and availability of public financial filings nowadays, it is fairly easy to identify profitable companies. For illustration purposes, we present three signals for the profitability theme.

Return on Equity This signal measures the value created for shareholders, where return is measured by net income from the income statement.

$$ROE = \frac{NI}{Equity} = \frac{\sum_{t=1}^{4} NI_t}{Equity_t}.$$

The numerator is called the trailing twelve-month (TTM) value or annualized value for net income.

Cash Flow to Assets This signal measures the capability of cash flow generation from business based on total assets, where cash flow from operation is from cash flow statements.

$$CFO2TA = \frac{CFO}{TA}.$$

Earnings per Share Growth This ratio measures the time series path of earnings growth over the most recent quarter and the same quarter in the previous year.

$$EPS_g = \frac{EPS_t}{EPS_{t-1}} + \frac{EPS_t}{EPS_{t-4}}.$$

4.7.2.2 Earnings Quality

All else being equal, it is desirable to invest in companies with higher profits. However, we need to answer several questions: Are those earnings "real"? Where do they come from? Are they sustainable? The quality of earnings is very important for stock prices. Simply put, earnings quality means that the earnings are generated from a company's sound and reoccurring business operations, especially its core business, rather than creative accounting or extraordinary items. The theme of earnings quality is often used to assess the accuracy and sustainability of historical earnings as well as the achievability of future projections. Evaluating earnings quality will help investors make judgments about the certainty of current income and the prospects for the future.

Earnings refers to sales minus costs. First, good quality earnings should be real (bona fide) earnings. Unfortunately, there are incentives to engage in creative accounting. Many public companies link earnings per share of the company to the compensation of the senior management team, either directly or indirectly. In quantitative investing, the following signals are used by professional investors to measure earnings quality.

Accruals This ratio measures the difference between cash flow and net income (Sloan 1996). Cash flow and net income should generally go in the same direction. For a public company, a gap with a high net income but negative cash flows from operations, deviating far from its peers or industry norm, may signal low earnings quality.

$$Accruals = \frac{NI - CFO}{TA}.$$

Sustainable Cash Flow This ratio measures the cash flow growth path over the last 3 years. Companies with stable and positive growth rates are preferred.

$$CFO_g = \frac{\frac{CFO_t}{CFO_{t-1}} + \frac{CFO_t}{CFO_{t-4}} + \frac{CFO_t}{CFO_{t-8}}}{sd(CFO_{t,t-1,\dots,t-12})}.$$

4.7.2.3 Value

Value signals measure the difference between market and accounting valuations of a public company. We can measure value signals from three different angles: net income, cash flow, and balance sheet. Note that the three angles will yield different information regarding how cheap the company is, although there is some overlapping: the denominator is the same, and there is a certain relationship between net income, cash flow, and book value.

Book Value to Market Cap This is the value signal from the balance sheet. It is measured by the accounting value of the company's common shares divided by the market value of those common shares (see Fama and French (1992)).

$$B/P = \frac{\text{Book Value}}{\text{Market Capitalization}} = \frac{\text{Bool value per share}}{price}.$$

Earnings to Price This is the value signal from the income statement. This reflects the earnings support per share for the stock price: the price comes from earnings and how the market evaluates per dollar of earnings.

$$E/P = \frac{NI}{\text{Market Capitalization}} = \frac{EPS}{Price}.$$

Cash Flow to Price This is the value signal from the cash flow statement. This reflects the cash flow support for the market price. Usually, cash flow is more difficult to manipulate and more transparent than earnings.

$$CFO/P = \frac{\text{CFO per share}}{Price}.$$

We see that the numerators of the above three value signals are derived from different financial statements: balance sheet, income statement, and cash flow statement. Since these three statements focus on different aspects of a public company, each value signal captures a different angle of valuation.

4.7.2.4 Management Capability

It is very challenging to evaluate companies' management teams. It is even more challenging to assign a rating score to the senior management team of a public company. Happy families are all the same: a great team is capable and honest and works to increase shareholders' value. However, in the real world, many management teams work for their own benefits, frankly speaking. How can we quantitatively evaluate public companies' senior management teams? While it is hard to tell who is really good, one way is to try to identify how far the team deviates from their peers. We can evaluate the CEO and her team from two perspectives: (1) How did they acquire money? and (2) How did they spend the money? Both aspects require strategic leadership and capability, which have significant impacts on the overall performance of the company.

The answer to the first question—where did the CEO get the money?—can be quantified using a ratio with external financing (externalFIN). External financing occurs when a public company issues either stocks or debts to finance their projects and spending. In general, organic internal financing is regarded as healthy, while too much external financing is regarded as unhealthy because it indicates the company does not have enough internal resources to leverage. Moreover, equity issuance will dilute per share metrics, and external debt issuance could have serious consequences when things take a turn for the worse (see, e.g., Loughran and Ritter (1995) and Richarson and Sloan (2002)).

externalFIN This can be measured by two signals. One is the current ratio, external financing scaled by total assets. The other is growth over time, the change in external financing.

$$exFIN = \frac{\text{external Financing}}{TotalAsset}$$

$$exFIN_g = \frac{exFin_t}{exFin_{t-1}}.$$

Another important decision the senior management team needs to make is how to invest money in projects. While investments are necessary for profitable projects, overspending is always a bad signal for the company. Regardless, many CEOs have a tendency to build a *bigger* company instead of a *stronger* company. Some CEOs go even further to build an empire in the industry or even go beyond their expertise to cross into other industries. The latter happens often when a company has been running well and becomes overconfident, undertaking large expansions into unfamiliar areas. While we cannot know each company's projects, we do know the monetary value of capital expenditure, which can be used to quantify overspending. If the capital expenditure ratio and growth are far greater than peer companies in the industry, it is usually a case of overspending and will eventually be penalized by the market due to low or even negative returns on investment – indicating that the senior management team has made bad decisions about projects. Classic academic studies on this topic include Jensen (1966), Titman et al. (2004), and Cooper et al. (2008).

Capital Expenditure This can be measured by the following ratios: total capital expenditure scaled by total assets and change in capital expenditure.

$$CAPX2TA = \frac{\text{Capital Expenditure}}{TotalAsset}$$

$$CAPX_g = \frac{\text{Capital Expenditure}_t}{\text{Capital Expenditure}_{t-1}}.$$

4.7.2.5 Momentum

Momentum signals are based on past price movements. In quantitative investing, for strategies with moderate or longer holding periods (ranging from months to years), the industry has employed momentum with different look-back periods, such as 6 or 12 months.

$$PM6m = \frac{P_t}{P_{t-T}}, \qquad \text{where T= 6 months.}$$

Why does the momentum strategy work? There are many academic explanations (e.g., Jegadeesh 1987, Jegadeesh and Titman 1993), such as behavioral overreaction/underreaction to news, the market payoff for taking greater risk, the extra push on price from herding effects of holding winners and avoiding losers, etc. Fundamentally speaking, if a firm's stock has been doing well for the past few months, it may indicate some business value, such as market share, new products, etc. These fundamental edges will last for a while, and learning and replication by peers in the same industry may take a while.

It is intuitive that momentum signals will work better for an up trending market. For example, the same momentum signal works effectively in the US stock market but not in the Japanese stock market because the former has been up and the latter has been flat. However, it should be noted that when the market turns around, the momentum strategy can cause huge losses. Given the frequent price changes in the market, momentum is a relatively short-term and high-turnover theme.

4.7.2.6 Market Sentiment

Market sentiment refers to the "mood" of the market. It can include (1) forecasts from massive professional analysts for public companies, such as estimates for EPS, and (2) investors' overall attitude towards a financial market. In the context of quantitative investing, for the former, the industry employs analyst estimates' consensus for momentum and diffusion; for the latter, one signal employed widely by the industry is shorting activities.

Sell-side research analysts estimate the performance of public companies. They are usually professionals who follow up with companies for years. They visit the companies, meet with senior management, and make informed evaluations. The key evaluation metrics are earnings per share (EPS), cash flow per share (CFPS), the net present value (NPV), and buy/hold/sell recommendations. There are numerous studies about analysts' estimates from both industry practitioners and academia. Some early studies include Zacks (1979), Brown et al. (1980), Abarbanell (1991) and Mikhail et al. (1999). Here we present two market sentiment signals: consensus and diffusion.

Three-Month Earnings Momentum This signal measures dynamic consensus of ups and downs of EPS revisions among analysts. The data is from IBES.

$$EM3m = EPS_t - \left[0.5\, EPS_{t-1} + 0.3\, EPS_{t-2} + 0.2\, EPS_{t-3}\right],$$

where EPS can be the average estimates for the upcoming fiscal years or quarters. Here, we treat all analysts the same and all estimates the same. Of course, there are more considerations one should make, such as lead analysts versus follower analysts, local analysts versus foreign analysts, etc. Another important aspect is accuracy versus timeliness.

Three-Month Earnings Diffusion This signal measures disagreement among analysts about the directional change of EPS for a public company.

$$EM3d = 0.7sd(EPS_t) + 0.2sd(EPS_{t-1}) + 0.1sd(EPS_{t-2}),$$

which is just a weighted average of disagreement about EPS over different periods. It has been found that more disagreement indicates more disappointing performance of the company's business, and hence downward pressure on its stock price. This is because analysts tend to have more agreement when the outlook is good, so disagreement usually indicates a bad situation.

Regarding the collective sentiment of investors towards a stock, we present short interest.

Short Interest Shares in short positions as the total number of shares floating. The data usually comes from a third-party vendor, brokerage, or custodian bank.

$$SHI = \frac{\text{short positions}}{\text{floating shares}}.$$

There are variations, such as dollar-value based or the change of SHI.

In addition to the signals mentioned above, there are other approaches measuring investor attention, such as survey-based sentiment indexes, textual sentiment from specialized online resources, internet search behavior, and non-economic factors.[5] Of course, the sentiment signal usually has high turnover as attitudes swing due to uncertainty. Sentiment signals are usually highly correlated with price movement

[5]For example, a consumer confidence index at the macro level.

signals because, for example, some analysts update their estimates based on price movements.

We have classified drivers and indicators for stock returns into themes and identified signals within each investment theme. We now need to combine those signals into a composite score as a factor or theme in a linear alpha model.

4.7.3 Signals, Themes, and Alpha: Data Treatment and OLS

In this subsection, we present an industry approach to return forecasting in the context of a stock selection strategy. Note that there may be large variations depending on investment strategy, universe, portfolio characteristics, etc.

We assume data availability and use constructed signals and themes as specified in the previous subsection. We present below a typical industry procedure for alpha building from a multi-factor analytical framework, starting from signals, then themes, and finally alpha.

1. Select factors for each theme.
2. Decide an investment horizon.
3. Use a proper industry classification.
4. Clean and treat raw signals.
    (a) missing data
    (b) errors
    (c) outliers
    (d) commensurability (compare signals apples-to-apples)
        - same range
        - demeaned by industry to remove industry bias
        - distribution
5. Before multi-factor analysis
    (a) univariate analysis: efficacy of each signal
    (b) bivariate analysis: correlation
6. OLS regression: multi-factor model

    Stage 1: form themes
    (a) coefficients, t-value, and R^2 value
    (b) weights for signals

    Stage 2: form alpha
    (a) match with features of investment strategy
    (b) critical for portfolio performance

We illustrate the above process using a stock selection strategy in the large cap segments of the US stock market. In the active institutional investment space, the large-cap part of the US stock market is usually represented by the Russell 1000 index, a conventional investment universe as well as a popular benchmark. There are several reasons that most institutional investors use the Russell 1000 rather than the S&P 500: First, the Russell 1000 has 500 more large companies and is therefore more representative of the large cap universe without sacrificing liquidity. This provides greater breadth for quantitative investing, which relies on the law of large numbers. Second, the Russell 1000's constituents are all US companies, whereas the S&P 500 includes foreign companies. Finally, the Russell 1000 has more transparent rules on index inclusion and rebalancing so that exiting and new entries are more predictable.

Universe and Signals We use Russell 1000 monthly data from January 31, 1995 to December 31, 2004. The data is available on the month-end trading date. Following the industry convention, we use *Cusip* as the id for the US companies. Any record will be identified by a cusip and date. We select the following factors for the six themes:[6]

VALUE:	S/P, B/P, E/P, CFO/EV
PM:	PM1m, PM6m, PM9m
PROF:	ROE, EBITDA/EV
EQ:	accrualsCF, CFOxInt/debt
MQ:	CAPXg, exFINg
MS:	EM9m, ED9m, EM12m, ED12m

Investment Horizon and Industry Classification Assuming the portfolio has a medium- to long-term investment horizon, we set the investment horizon to be nine-months. We use 9-month cumulative returns as the dependent variable for the multi-factor model. Regarding industry classification, we use GICS, which is widely used by institutional investors in the USA. It had ten sectors during the 10-year period from 1995 to 2004. Industry classification is very important because it defines industry characteristics, determines peer companies, and thus impacts stock selections within each industry. Accordingly, this should be reflected in signal treatment and alpha values. Industry classification is also important for portfolio construction because constraints are imposed at the industry level for risk management purposes.

For a large company in the Russell 1000, its industry exposure can be multiple and dynamic. First, a large company may have multiple lines of business across different industries. Second, when a company changes its major business, the industry classification will change. These should all be considered in investing.

[6]We use both names of S2P and S/P for the same ratio.

Raw Signals Treatment Following the procedure specified above, we clean the raw signals with error values, treat the raw signals with outliers, demean the signal values across industries, and standardize the final values to produce the same range.

- Cleaning raw signals
  - Missing values: if >30%, invalidate the signal; otherwise, fill in zeroes
  - Error values: remove and treat as missing
  - Outliers: truncation at 5 sigma and winsorization at 3 sigma
- Demean by industry
- Z-score with range −3 to 3
- Maintain the original distribution

We illustrate above using signals in the value theme. First, we show the raw signal values with the summary statistics generated by R scripts below. We see that the raw data are full of errors and outliers. If we use these raw values for investment the results will be totally meaningless. We need to process the raw signals based on the rules specified above. The values after the treatment are also displayed by R scripts below.[7]

Raw and treated signals

```
> # summary of raw value signals: S2P, B2P, E2P, CFO2EV
> dim(r1k.rawdata)
[1] 117777    280

> summary(r1k.rawdata$S2P)
    Min.   1st Qu.   Median     Mean  3rd Qu.      Max.  NA's
  0.0000    0.3202   0.6204   1.0439   1.1412  839.7784  1559

> summary(r1k.rawdata$B2P)
     Min.   1st Qu.   Median     Mean   3rd Qu.      Max.  NA's
-132.8115    0.2119   0.3638   0.4486    0.5617  113.2039  2628

> summary(r1k.rawdata$E2P)
     Min.   1st Qu.  Median   Mean   3rd Qu.      Max.   NA's
-50000.00      2.41    4.67  82.46      6.88  50000.00    751

> summary(r1k.rawdata$CFO2EV)
     Min.  1st Qu.   Median     Mean   3rd Qu.       Max.   NA's
-50000.00     4.43     7.50   230.92     11.21  50000.00  20448
```

[7]We discuss R treatment functions for signals and multi-factor model estimation in the next section.

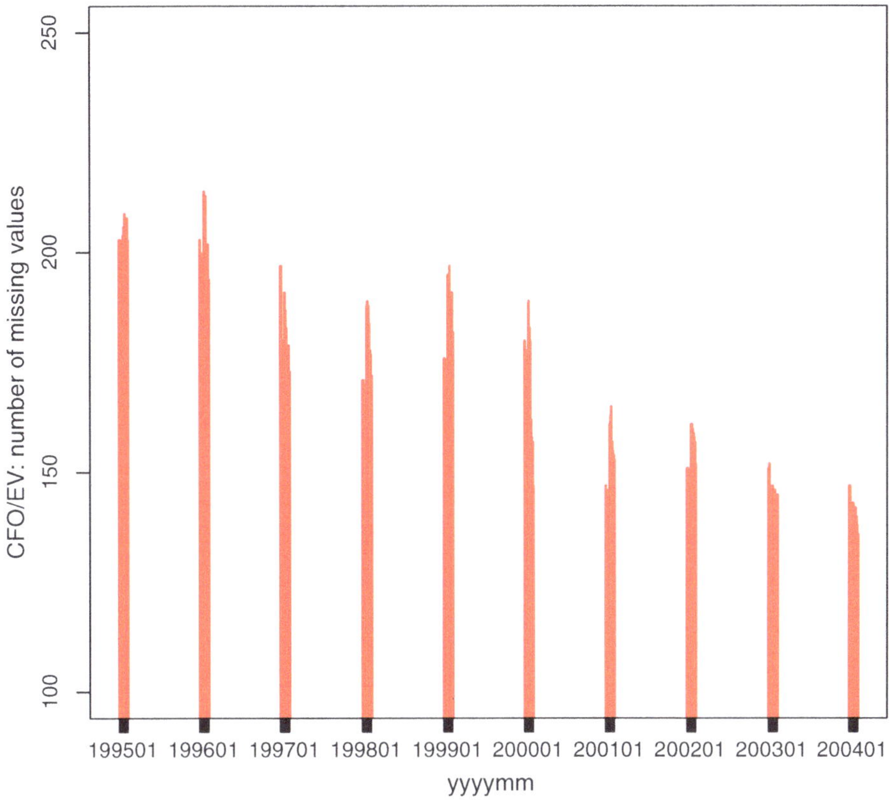

Fig. 4.5 The number of missing values for CFO/EV in the Russell 1000 universe decreased from around 200 on January 31, 1995 to around 150 on December 31, 2004

Note that pricing information is the most available and accurate, and there should be no missing values or errors, but there do exist outliers. However, accounting items are never short on such issues. Among the three financial statements, cash flow items generally have the most missing values because many items are not required to be reported. Using the value signals as an example, the CFO/EV signal has 17% missing values, while other value signals have only about 1–2% missing values. However, the good news is that the number of missing values for CFO/EV decreased dramatically from about 200 on January 31, 1995 to about 150 on December 31, 2004 (Fig. 4.5).

	S/P	B/P	E/P	CFO/EV
Missing Percentage:	*1%*	*2%*	*1%*	*17%.*

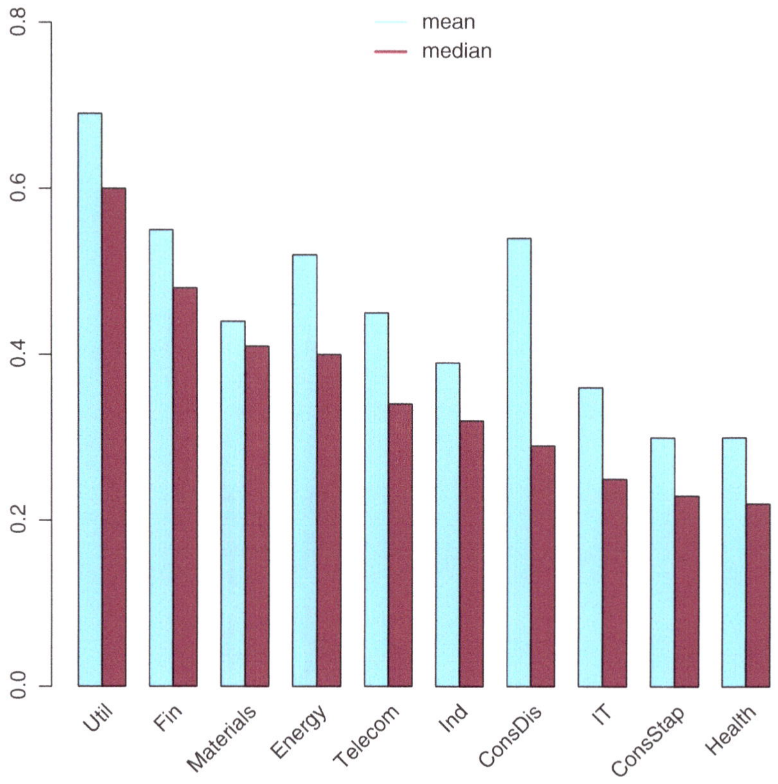

Fig. 4.6 The mean and median B/P values across GICS sectors (Russell 1000 universe) on December 31, 2004

Industries have special impacts on signal values. For example, cash flow items do not make much sense for the banking industry, and value factors are usually high for energy, banking, and utilities. This implies that without demeaning to remove industry effects, stock selection will be highly concentrated and skewed in some industries resulting in more of an industry selection than a stock selection. This is evidenced in Fig. 4.6, where plots display the mean and median B/P signal across ten GICS sectors during the period of 1995–2014. We see that, indeed, the values of B/P are very different across industries: the utilities, financials, and energy sectors have the highest median values of 0.40–0.60, while the IT, consumer staples, and health care sectors have the lowest median values of 0.20–0.25.

After dealing with the errors, missing values, and outliers in the data, and removing industry effects, we get the final treated values for each signal, with mean and median equal to zero, standard deviation equal to 1, and range equal to about −3 to 3. The box plots in Fig. 4.7 describe the distribution of final treated values for S/P, B/P, E/P, and CFO/EV on December 31, 2004 (right plot). For comparison purposes, we also have the box plots of the raw values for the four value signals (left

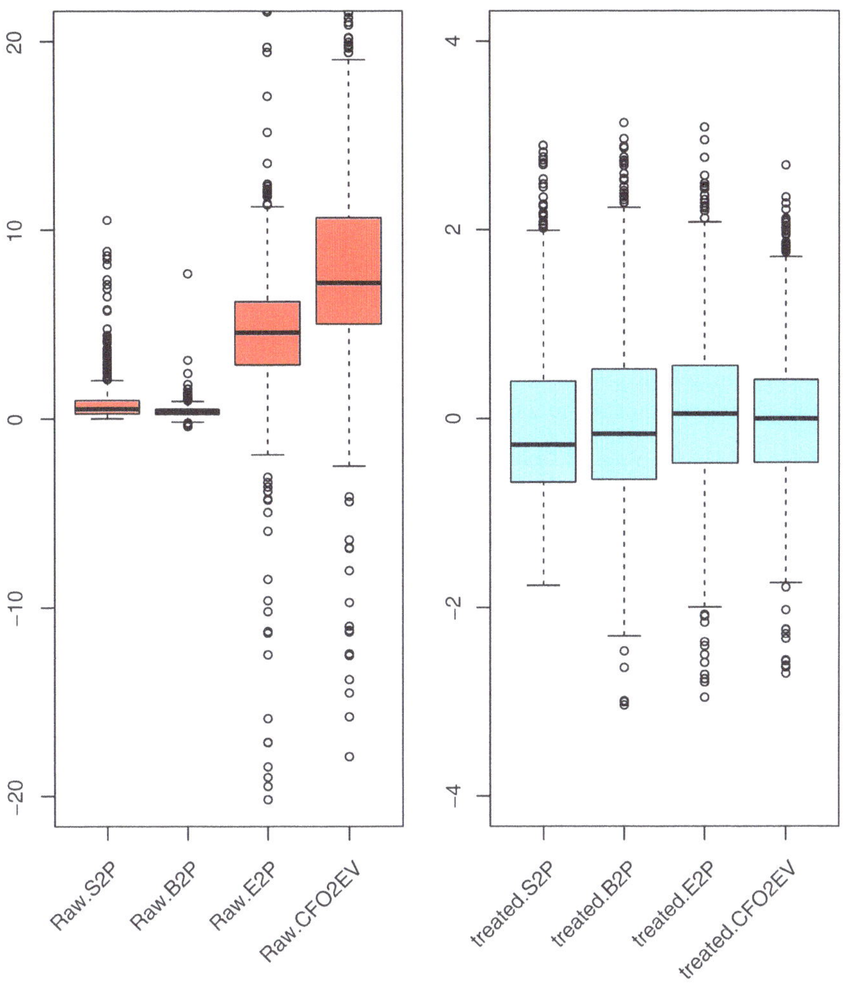

Fig. 4.7 Boxplots of value signals, raw values (left) and treated values (right), on December 31, 2004

plot). We see again that the raw scores are quite wild, the treated scores prepare us to compare signal values apples-to-apples for a stock selection strategy.

To see the factor values change after each step of the treatment, we present the correlations of values of B/P (B2P) in Fig. 4.8, where the lower triangle is for the rank correlation and the upper triangle is for the Pearson correlation. We see that the correlation is above 90%, indicating that, indeed, the signal treatment procedure removes or decreases some noise in the data while retaining the information contained in the factor.

	Raw.B2P	Truncate.B2P	Standardize.B2P	Windsorize.B2P	Neutralize.B2P	Standardize2.B2P	treated.B2P
Raw.B2P		1	1	0.98	0.93	0.93	0.93
Truncate.B2P	1		1	0.98	0.93	0.93	0.93
Standardize.B2P	1	1		0.98	0.93	0.93	0.93
Windsorize.B2P	1	1	1		0.94	0.94	0.94
Neutralize.B2P	0.92	0.92	0.92	0.92		1	1
Standardize2.B2P	0.92	0.92	0.92	0.92	1		1
treated.B2P	0.92	0.92	0.92	0.92	1	1	

Fig. 4.8 The Pearson and rank correlations between B/P values at each step of the treatment for stocks in the Russell 1000 universe on December 31, 2004

After building each signal with proper treatment of raw values, we now move on to the construction of a multi-factor model, which can be divided into two stages: stage 1—combine signals into a theme and stage 2—combine themes into an alpha.

Multi-Factor Model, Stage 1: Combine Signals into a Theme An immediate question is why we cannot put all signals in a multi-factor model rather than conduct a two-stage process. Well, the reason is simple. Because signals within a theme are highly correlated, combining signals into themes reduces the multicollinearity issue and produces more freedom in estimation. The conceptual meaning also makes sense: perhaps it is cleaner to have the factors at the theme level as the later reflect conceptual alpha sources.

Before running a multi-factor model, we investigate the efficacy of each signal with forward stock returns of 1, 3, 6, and 9 months. We continue using the value signals for illustration purposes, calculate correlations, and present them in Table 4.3. We see that overall, CFO/EV has the highest correlations, followed by E/P and S/P, while B/P has the lowest correlation. In terms of return forecasting power, in general, a correlation between signal values and forward returns around 0.01–0.05 is considered effective, 0.05–0.08 is considered very effective, and above 0.10 is suspicious and may be caused by errors, outliers, or a mistaken contemporaneous relationship. Based on the efficacy of CFO/EV, we see that cash flow is indeed the king in the US stock market, but this may not be true in other financial markets, such as in Japan, as we will see in Chap. 9. For value signals, one interesting feature is that they all have long-term effects, that is, the longer the investment horizon, the higher the forecasting power as measured by correlation. This implies that value signals are suitable for medium- to long-term investment strategies.

At this stage, we need to be extremely careful about the collinearity issue. If the correlation between signals is too high, we can just assign weights based on the results from univariate and bivariate analysis. Otherwise, we can apply OLS to the multi-factor model, use the t-value to make a judgment about the joint efficacy of factors, and then decide the weights based on univariate, bivariate, and multi-factor analysis. We give an example below for value factors. The correlations between value signals of S/P, B/P, E/P, and CFO/EV are listed in Table 4.4, where the upper triangle is the Pearson correlation and the lower triangle is the rank correlation. We see that for B/P and S/P, the Pearson and rank correlations are 0.46 and 0.50, respectively,

Table 4.3 Pearson and rank correlations of value signals with forward returns for stocks in the Russell 1000, based on monthly data from January 31, 1995 to December 31, 2004

	Retf1	Retf3	Retf6	Retf9		Retf1	Retf3	Retf6	Retf9
Pearson cor.					**Rank cor.**				
B/P	0.01	0.01	0.01	0.01	B/P	0.01	0.01	0.01	0.01
E/P	0.03	0.04	0.04	0.04	E/P	0.03	0.04	0.05	0.05
S/P	0.02	0.02	0.02	0.03	S/P	0.01	0.02	0.03	0.03
CFO/EV	0.03	0.04	0.05	0.05	CFO/EV	0.02	0.04	0.05	0.06

Retf1 is one-month forward stock returns

Table 4.4 Pearson and rank correlations between value signals for stocks in the Russell 1000, based on monthly data from January 31, 1995 to December 31, 2004

	B/P	E/P	S/P	CFO/EV
B/P	1	0.18	0.46	0.28
E/P	0.24	1	0.28	0.36
S/P	0.50	0.34	1	0.36
CFO/EV	0.34	0.38	0.42	1

Table 4.5 OLS coefficients (t-values) of value signals in the Russell 1000, based on monthly data from January 31, 1995 to December 31, 2004

	Retf9	y=Retf9	y=Retf9	y=Retf9	y=Retf9
B2P	0.0026 (2.10)				−0.0064 (−4.50)
E2P		0.0180 (14.32)			0.0122 (8.92)
S2P			0.0115 (9.19)		0.0054 (3.67)
CFO2EV				0.40219 (15.90)	0.0170 (10.88)

The first four columns are for each single variable and the last column corresponds to the multi-factor model (4.20)

which are high enough to cause collinearity in a multi-factor regression. Now, a natural question is, how high of a correlation is serious enough to cause collinearity? While the answer varies case-by-case, the rule of thumb is about 0.40 for OLS.

Now we run a multi-factor regression using different investment horizons.

$$R_F = b_0 + b_1 S2P + b_2\, B2P + b_3 E2P + b_4 CFO2EV + \epsilon, \qquad (4.20)$$

where R_F is forward returns at the stock level. Here we skip the time t as we apply OLS to the entire data set from January 31, 1995 to December 31, 2004. We present the estimates for coefficients in Table 4.5 with the t-values in parentheses. For comparison purposes, we have OLS estimates for each value signal in a single-factor model and then all value signals in a multi-factor model.

Connecting with theoretical explorations of OLS for multi-factor models in previous sections, we now have an industry model and real-world data to discuss in detail the OLS estimates for a multi-factor model for stocks in the Russell 1000 universe. We focus our discussion on coefficients, t-values, R^2, and BLUE conditions. The R scripts below produce the OLS regression results for the 4-factor model (4.20).

OLS, value signals

```
> #Summary of OLS of the 4 value signals:
>  summary(lm(formula = retf9 ~ S2P + B2P +E2P+CFO2EV,
                       data = r1k.signals))
Residuals:
    Min       1Q  Median      3Q     Max
-1.1468 -0.2085 -0.0114  0.1761 27.6524

Coefficients:
             Estimate Std. Error t value Pr(>|t|)
(Intercept)  0.097010   0.001242  78.135  < 2e-16 ***
S2P          0.005433   0.001482   3.666 0.000246 ***
B2P         -0.006449   0.001432  -4.503  6.7e-06 ***
E2P          0.012175   0.001364   8.923  < 2e-16 ***
CFO2EV       0.017002   0.001562  10.881  < 2e-16 ***
---
Residual standard error: 0.4094 on 108725 degrees of freedom
  (8162 observations deleted due to missingness)
Multiple R-squared:  0.003336,Adjusted R-squared: 0.003299
F-statistic: 90.97 on 4 and 108725 DF,  p-value: < 2.2e-16
```

Note that since we apply the same treatment process for all signals, the standard errors for coefficients are very similar. Because of collinearity issues, particularly between B/P and S/P, the coefficient for B/P is negative, so the effects are rendered by S/P because it has greater forecasting power as we observed in the correlation table. This also implies that univariate and bivariate analysis are very important before we run a multi-factor analysis. Note also that the coefficients are all significant, and the order of effects from the OLS results is the same as measured by correlation: CFO/EV has the highest coefficient, followed by E/P and S/P. The R-squared values are generally very low for a multi-factor model of a stock selection strategy, usually falling below 5% and often below 3%. Here we see that value signals together can only explain 0.3% of the nine-month forward returns, even though all signals are very significant based on t-values.

This also implies that over 99% of the information is contained in ϵ, the so-called noise in (4.20). Since the left-hand variable, the response variable–forward returns, is calculated based on prices, the error term must also relate to prices. The value signals all have denominators that are prices, so inevitably, the error term ϵ and signals x_k are correlated, therefore, the arise of the endogeneity issue. Based on what we learned from previous sections, in the presence of endogeneity, OLS estimates will no longer be unbiased or consistent.

Now, we discuss the homogeneity condition for efficiency. We obtain estimates for ϵ in (4.20). For the 10-year study period, there are 393 companies that appear in the Russell 1000 index each month. Of those 393 companies, we randomly select 100 companies and calculate standard deviation of residuals for each company and correlation between different companies. The results are presented in Fig. 4.9, where the top plot is for standard deviation and the bottom plot is for correlation. We see from the top plot that while many stocks have standard deviations of about 0.20, there do exist many stocks with standard deviations that differ significantly from 0.20, indicating the violation of error terms being identical. The violation of independence is more serious: the correlations in the bottom plot range from 10% to 40%, far from being zero! Clearly, the iid assumptions do not hold in this study. In fact, the iid condition barely holds for any financial market because securities are different and are related with each other. For example, Boeing and Bank of America are different companies, and their ϵ values will be very different, while American Airlines and United Airlines are peer companies belonging to the same industry, and their ϵ values will be highly correlated.

$$\begin{aligned} \text{violation of being identical: } & \sigma(\epsilon_i) \neq \sigma(\epsilon_j) \\ \text{violation of independence: } & cor(\epsilon_i, \epsilon_j) \neq 0. \end{aligned}$$

Apparently, the OLS estimator is far from being BLUE given the violations of exogeneity and errors being non-iid. However, we can still rely on OLS to estimate joint effects of signals and use this information with other analyses to make investment decisions.

We apply the same procedure for signals within each of the other themes, and summarize the results below. Note that this is only for illustration purposes. We derive all weights based on in-sample data and ignore many other aspects, such as turnover and risk characteristics.

$$\begin{aligned} VALUE &= 0.50CFO/EV + 0.30E2P + 0.15S2P + 0.05B2P \\ PM &= -0.20PM1m + 0.30PM6m + 0.50PM9m \\ PROF &= 0.20ROE + 0.80EBITDA/EV \\ EQ &= -0.55accrualsCF + 0.45CFOxInt/debt \\ MQ &= -0.20exFINg - 0.80CAPXg \\ MS. &= 0.20ER9m + 0.20ER12m + 0.30ED9m + 0.30ED12m. \end{aligned}$$

Multi-Factor Model, Stage 2: Combine Themes into an Alpha Now that we have themes, we focus on alpha construction. Before running the multi-factor model, we first conduct univariate and bivariate analysis.

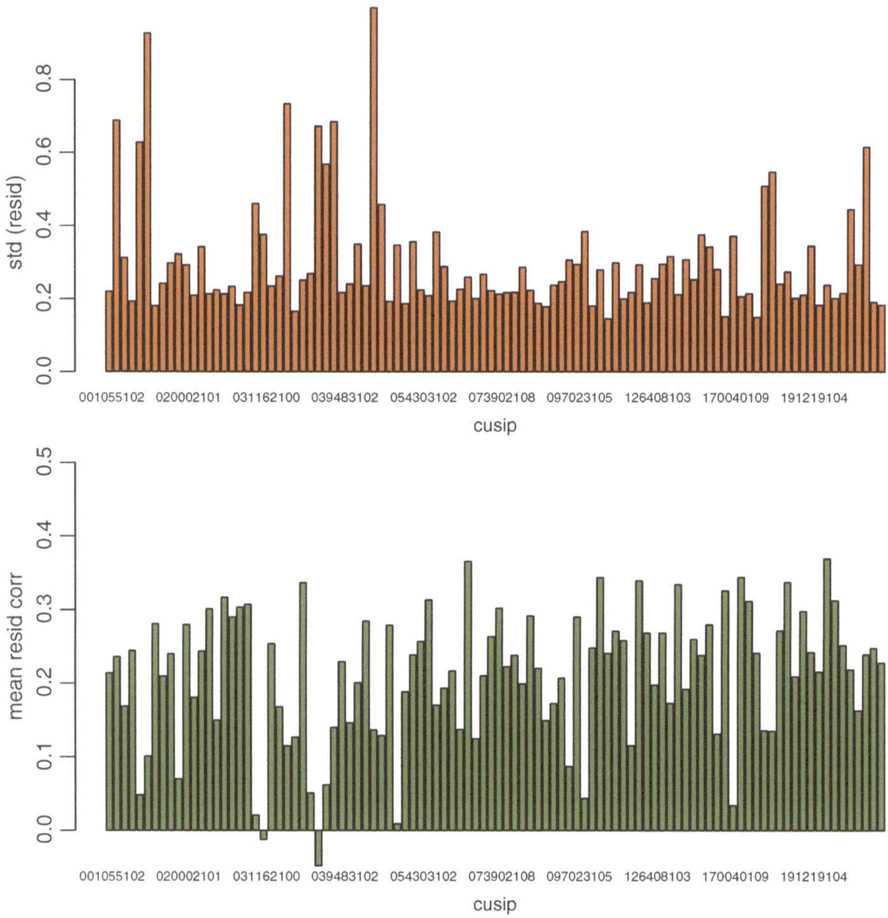

Fig. 4.9 Standard deviation of residuals (top plot) and correlation of residuals (bottom plot) of (4.20) for 100 randomly selected companies in the Russell 1000

The R scripts below yield summary statistics for each theme over the entire 10-year study period.

Univariate, summary statistics

```
##Univariate analysis, Summary of themes:
> summary(r1k.themes[,c("VALUE","PM","PROF","EQ","MQ,"MS")])

     VALUE                PM                 PROF                 EQ                   MQ
Min.   :-2.705059  Min.   :-3.577630  Min.   :-2.650443  Min.   :-3.1513605  Min.   :-2.447718
1st Qu.:-0.416829  1st Qu.:-0.450886  1st Qu.:-0.310891  1st Qu.:-0.3518615  1st Qu.:-0.330641
Median :-0.042301  Median :-0.016697  Median : 0.014486  Median :-0.0122225  Median : 0.060870
Mean   : 0.002291  Mean   : 0.002688  Mean   : 0.003099  Mean   :-0.0000649  Mean   : 0.001949
3rd Qu.: 0.351060  3rd Qu.: 0.439821  3rd Qu.: 0.363133  3rd Qu.: 0.3719101  3rd Qu.: 0.377287
```

```
Max.   : 3.043153  Max.   : 3.363608  Max.   : 2.738661  Max.   : 2.8282600  Max.   : 2.699267

       MS
Min.   :-3.466744
1st Qu.:-0.550293
Median : 0.000000
Mean   : 0.000887
3rd Qu.: 0.577463
Max.   : 3.505816
```

We apply necessary treatments to themes such as re-standardization, and then run correlations between each theme and a set of forward returns (Table 4.6) and between themes (Table 4.7).

As expected, the correlation scores with forward returns at the theme level are in general higher than at the signal level. Table 4.6 also shows that the rank correlations are of a similar magnitude to the Pearson correlations, indicating there are no effects from outliers. As measured by Pearson correlation, VALUE, PM, PROF, and MQ have correlation scores of 3–6%, while EQ and MS are a bit weaker, with correlation scores of 1–3%. MQ and EQ are more suited to long-term investment strategies, while MS and PM are more appropriate for short- to mid-term investment strategies. VALUE and PROF are in between.

We see in Table 4.7 that the Pearson and rank correlation scores are 74% and 68%, respectively, between VALUE and PROF; and 48% and 46%, respectively, between PM and MS. These are high enough to cause collinearity issues in a multi-factor model.

To overcome the multicollinearity issue, we employ a residual approach. We run OLS regressions of PROF on VALUE, then use residuals as a "cleaned" factor since they will still contain information of the response factor (PROF) but be clean of the independent factor (VALUE). We apply the same approach to MS and PM.

$$PROF = \gamma_0 + \gamma_1 VALUE + \mu$$
$$MS = \beta_0 + \beta_1 PM + \nu.$$

Table 4.6 Pearson and rank correlations of themes with forward returns for stocks in the Russell 1000, based on monthly data from January 31, 1995 to December 31, 2004

	Retf1	Retf3	Retf6	Retf9		Retf1	Retf3	Retf6	Retf9
Pearson cor.					**Rank cor.**				
VALUE	0.04	0.05	0.05	0.05	VALUE	0.03	0.05	0.06	0.07
PM	0.03	0.05	0.06	0.05	PM	0.03	0.04	0.05	0.04
PROF	0.04	0.05	0.06	0.06	PROF	0.04	0.06	0.07	0.08
EQ	0.01	0.02	0.03	0.03	EQ	0.01	0.02	0.03	0.03
MQ	0.03	0.05	0.06	0.06	MQ	0.02	0.04	0.06	0.06
MS	0.01	0.01	0.02	0.02	MS	0.01	0.02	0.02	0.03

Retf1 is one-month forward stock returns

Table 4.7 Pearson and rank correlations of themes for stocks in the Russell 1000, based on monthly data from January 31, 1995 to December 31, 2004

	VALUE	PM	PROF	EQ	MQ	MS
VALUE	1	−0.13	0.74	0.14	0.22	−0.12
PM	−0.14	1	0.01	0.07	0.04	0.48
PROF	0.68	0	1	0.27	0.3	0.04
EQ	0.11	0.07	0.25	1	0.2	0.11
MQ	0.22	0.04	0.31	0.2	1	−0.01
MS	−0.12	0.46	0.05	0.12	−0.01	1

The R scripts below show the relationship between the pairs and also a way to derive the residuals.

OLS, use residuals as a proxy

```
> summary(lm(PROF ~ VALUE, data = r1k.themes))
Coefficients:
            Estimate Std. Error t value Pr(>|t|)
(Intercept) 0.001506   0.001304   1.155    0.248
VALUE       0.695197   0.001830 379.979   <2e-16 ***
---
Residual standard error: 0.4459 on 116890 degrees of freedom
Multiple R-squared:  0.5526,Adjusted R-squared:  0.5526
F-statistic: 1.444e+05 on 1 and 116890 DF,  p-value: < 2.2e-16

> summary(lm(MS ~ PM, data = r1k.themes))
Coefficients:
              Estimate Std. Error t value Pr(>|t|)
(Intercept) -0.0006538  0.0023841  -0.274    0.784
PM           0.5732920  0.0030951 185.227   <2e-16 ***
---
Residual standard error: 0.8151 on 116890 degrees of freedom
Multiple R-squared:  0.2269,Adjusted R-squared:  0.2269
F-statistic: 3.431e+04 on 1 and 116890 DF,  p-value: < 2.2e-16
```

We now use residuals from the OLS regressions and apply OLS to a multi-factor model

$$Retf9 = b_0 + b_1 VALUE + b_2 PM + b_3 PROF.resid + b_4 EQ + b_5 MQ \\ + b_6 MS.resid + \epsilon.$$

OLS, multi-factor model with themes

```
> summary(lm(retf9 ~ VALUE + PM + PROF.resid + EQ
  + MQ +  MS.resid, data = r1k.themes))
Coefficients:
              Estimate Std. Error t value Pr(>|t|)
(Intercept)   9.705e-02  1.239e-03  78.357  < 2e-16 ***
VALUE         2.971e-02  1.820e-03  16.327  < 2e-16 ***
PM            2.642e-02  1.657e-03  15.947  < 2e-16 ***
PROF.resid.   1.313e-02  2.963e-03   4.431 9.38e-06 ***
EQ            3.733e-03  2.147e-03   1.739   0.0821 .
MQ            2.853e-02  2.188e-03  13.038  < 2e-16 ***
MS.resid.     5.728e-05  1.543e-03   0.037   0.9704
---
Residual standard error: 0.4084 on 108723 degrees of freedom
  (8162 observations deleted due to missingness)
Multiple R-squared:  0.008207,Adjusted R-squared: 0.008152
F-statistic: 149.9 on 6 and 108723 DF,  p-value: < 2.2e-16
```

Using the t-values of the multi-factor OLS results, we employ the following formula as a reference to calculate the weights for themes:

$$W_k = \frac{\text{T-VALUE}_k}{\sum_{k=1}^{6} \text{T-VALUE}_k}.$$

OLS, t-value based theme weights

```
Build Alpha now:
Theme weights based on OLS t-values:
VALUE   PM      PROF.resid   EQ     MQ    MS.resid
0.3169  0.3095    0.08601    0.0337 0.2530  0.0007
```

With consideration of univariate and bivariate results, we build alpha scores as follows:

$$ALPHA = 0.25\,\text{VALUE}+0.25\,\text{PM}+0.10\,\text{PROF}+0.10\,\text{EQ}+0.20\,\text{MQ}+0.10\,\text{MS}.$$

We then present simple summary statistics, correlations with forward returns, and an OLS regression model for ALPHA scores. The R scripts below show the results.

ALPHA

```
> summary(r1k.themes$ALPHA)
Statistics on the Efficacy of Alpha:
       V1
 Min.   :-2.227278
 1st Qu.:-0.221958
 Median : 0.006209
 Mean   : 0.002316
 3rd Qu.: 0.242810
 Max.   : 1.938843

> cor(r1k.themes$ALPHA, r1k.themes[, returns])
     retf1 retf3 retf6 retf9
[1,] 0.058  0.08 0.092  0.092

> cor(r1k.themes$ALPHA,r1k.themes[,returns],method="Spearman")
     retf1 retf3 retf6 retf9
[1,] 0.048 0.072 0.092 0.097

> summary(lm(retf9 ~ ALPHA, data = r1k.themes))
Residuals:
    Min      1Q  Median      3Q     Max
-1.1529 -0.2099 -0.0126  0.1751 27.6158

Coefficients:
            Estimate Std. Error t value Pr(>|t|)
(Intercept) 0.097047   0.001239   78.35   <2e-16 ***
ALPHA       0.090376   0.003042   29.71   <2e-16 ***
---
Residual standard error: 0.4084 on 108728 degrees of freedom
  (8162 observations deleted due to missingness)
Multiple R-squared:  0.008054,Adjusted R-squared: 0.008045
F-statistic: 882.8 on 1 and 108728 DF,  p-value: < 2.2e-16
```

As a brief summary, we present comments below about the results from R scripts for alpha diagnostics.

1. The summary statistics indicate that alpha values have mean/median zero and range from -2 to 2. Reflected in a stock selection portfolio, a stock with zero alpha indicates that the model does not have an opinion about its price movement (relative to the benchmark), and thus it should not be held in the portfolio from the pure alpha perspective. A stock with negative alpha should be avoided in a long-only portfolio or placed in short position for a long/short portfolio, Positive alphas should be associated with long-buy stocks.
2. The correlation values of ALPHA with forward returns are in the range of 5–10%, indicating high forecasting power of alpha values over the study period.
3. The OLS regression yields a t-value of 29.71, indicating a strong causal relationship between alpha values and nine-month forward returns. The R^2 is 0.8%, the highest we ever got comparing the regressions for signals and themes. The alpha can explain about 1% of all stock returns during this 10-year period.

So far, for a US large-cap-core stock selection strategy in Russell 1000, we have built signals, cleaned the data, applied proper treatments (such as removing industry biases), constructed themes, and built alpha values. We applied univariate, bivariate, and multi-factor OLS analysis during the process. We also made comments about the OLS properties (BLUE) and checked conditions for being BLUE using real-world data. However, we need to be cautious about several aspects of this alpha building process, including but not limited to:

- In-sample results. Note that all the analysis and results are based on an in-sample study. For a live portfolio, the future is unknown. An out-of-sample exploration is closer to a real-world investment.
- Industry-specific model. Each industry is different, but this does not necessarily mean we need to build a separate model for each industry. However, given the larger degree of difference between some industries, such as banking versus non-banking industries, an industry-specific banking model may be suitable to capture unique drivers and indicators for the stocks within the banking industry.
- Stationary versus dynamic. We apply the same procedure over the entire period and treat the signals, themes, and alpha all the same over time. However, we know that the relationships may change over time. Here we have ignored the dynamics.
- No risk involved in alpha construction. We either assume that risk information contained in signals is all the same or ignore the risk information during the alpha building process.
- Linear model. We adopt a simple linear approach for both themes and alpha construction. Of course, in the real world, the relationships between signals/themes/alpha are not linear. There are interactions among factors and nonlinear effects of factors on forward stock returns.

We will explore some of the issues above in detail in subsequent chapters with additional discussions, analysis, and possible solutions.

4.8 Commonly Used R Functions for Alpha Building

In this section, we introduce some simple R functions commonly used for alpha building in quantitative investing. We first introduce utility functions for data cleaning and signal treatment, then show multi-factor estimation for alpha construction.

4.8.1 R Functions: Data Cleaning and Signal Treatment

We discussed error and outliers in data in Chap. 2 and described the treatment procedure in the previous section. We now present simple functions for each step in the order of treatment.

$$Rawscores \Rightarrow Missing \Rightarrow Truncation \Rightarrow Standardization \Rightarrow Winsorization$$
$$\Rightarrow Industry\ demean \Rightarrow Re\text{-}standardization \Rightarrow Exclusion$$
$$\Rightarrow Missing\ values \Rightarrow Distribution \Rightarrow Treatedscores$$

The R functions for the above purposes are usually called utility functions as they can be used again and again in many cases of data cleaning and factor building. For a utility function, major inputs are typically data, ids and specific options for treatments. We will provide brief comments about them when appropriate.

Missing The presence of missing values can be serious if they exceed a certain percentage of the entire data. Unfortunately, this happens frequently. So, the first thing we need to do is to decide the maximum allowable percentage. For example, if the percentage of missing values is over 30%, we would simply drop the factor from the alpha building process.

Missing threshold

```
dataMiss <- function(data,varname,cut.point,theID="cusip")
{
   x=data
   cc=which(is.na(x[,varname]))
   if (sum(cc)>0)
   {
      ccc=length(cc)/dim(x)[1]
      x$missPercent=ccc
      if(ccc>cut.point)
      {
         x$missFlag=1  # date flag to avoid the factor
         x[,varname]=0  # assign zero for all names
```

```
      }
      else x$missFlag=0
      cat("Missing:", varname, ":", length(cc), "of",dim(x)[1],"\n")
      kkk=x[cc,c(theID,varname)]
      names(kkk)[2]="value"
      kkk$name=varname
      kkk$event="missing"
      print(kkk,row.names=F)
   }
   else
   {
     x$missPercent=0
     x$missFlag=0
     cat("Missing:", varname, ", no missing obs of ",dim(x)[1],"\n")
   }
   return(x)
}
```

Truncation, Standardization and Winsorization Once a signal passes the missing values threshold, we will clean up the data by removing errors and outliers and then standardize the scores and winsorize the outliers.

Regarding truncation, the criteria for removal can be based on either standard deviation or percentile. For example, we can set a rule that any stocks with values of more than 5 standard deviations (sigmas) will be removed or assigned NA (not available). Note that standard deviation itself is sensitive to outliers, we can overcome this issue by using a robust version of standard deviation calculation.

Truncation

```
dataTruncate <- function(data,varname,method="sigma",
                         LtruncPoint=5,RtruncPoint=5,theID="cusip")
{
  x=data

  ## deal with the left tail and right tail separately
  if(tolower(method)=="sigma")
  {
    tmp1<-mean(x[,varname],na.rm=T);
    tmp2<-sd(x[,varname],na.rm=T)
    minx=tmp1-LtruncPoint*tmp2
    maxx=tmp1+RtruncPoint*tmp2
    cc=which(x[,varname]>maxx | x[,varname]<minx)
    if (sum(cc)>0)
    {
      cat("Truncation: ",varname, "\n")
```

```
      kkk=x[cc,c(theID,varname)]
      names(kkk)[2]="value"
      kkk$name=varname
      kkk$event="truncation"
      print(kkk,row.names=F)
      x[cc,varname]<-NA
    }
    else cat("Truncation: ",varname," truncated ids = NA","\n")
  }
  else if(tolower(method)=="percentile")
  {
    minx<-as.vector(quantile(x[,varname],LtruncPoint/100,na.rm=T))
    maxx<-as.vector(quantile(x[,varname],(1-RtruncPoint/100),na.rm=T))
    cc=which(x[,varname]>maxx | x[,varname]<minx)
    if(sum(cc)>0)
    {
      cat("Truncation: ",varname, "\n")
      kkk=x[cc,c(theID,varname)]
      names(kkk)[2]="value"
      kkk$name=varname
      kkk$event="truncation"
      print(kkk,row.names=F)
      x[cc,varname]<-NA
    }
    else cat("Truncation: ",varname," truncated ids = NA","\n")
  }
  else  stop("Please select the correct truncation method! \n\n")
  return(x)
}
```

One option for standardization is to simply subtract the mean and divide by the standard deviation. This prepares for the winsorization step and eventually helps to compare factors apples-to-apples. Note that since standard deviation is sensitive to outliers, we add an option to use a robust version (R package *rrcov*) for the second moment calculation.

Standardization

```
dataStandardize <- function(data,varname,method="robust")
{
  # check the name in case of the repeated standar dization
  # robust version
  x = data
  ## If there are too few observations do not use robust
  if(method=="robust" & nrow(x) < 30) {
```

```
    method <- "simple"
  }

  if(method=="robust")
  {
   require(rrcov)
   #new.varname <- paste(varname,'.sdz',sep='');
   tmp <- CovSde(x[,varname]);
   x[,varname] <- x[,varname] - as.vector(tmp@center)
   x[,varname] <- x[,varname] / sqrt(as.vector(tmp@cov))
  }
  else
  {
     theMean=mean(x[,varname],na.rm=T)
     theStd=sd(x[,varname],na.rm=T)
     if(theStd != 0) x[,varname]=(x[,varname]-theMean)/theStd
  }
   return(x);
}
```

The winsorization is similar to truncation in the sense that both deal with outliers. The difference is that the former keeps outliers and shrinks them to a specified score, while the latter simply removes outliers.

Winsorization

```
dataWinsorize <- function(data,varname, Lwin=-3, Rwin=3,theID="cusip")
{
    x = data

    # deal with the left tail and right tail separately
    gt.index <- which(x[,varname] > Rwin);
    if(length(gt.index) > 0)
    {
       cat("Winsorization:",varname,"right tail:","\n")
       kkk=x[gt.index,c(theID,varname)]
       names(kkk)[2]="value"
       kkk$name=varname
       kkk$event="winsorization.right"
       print(kkk,row.names=F)
       x[gt.index,varname] <- Rwin
    }
    else cat("Winsorization: ",varname, "right tail = NA","\n")
```

```
    lt.index <- which(x[,varname] < Lwin);
    if(length(lt.index) > 0)
    {
       cat("Winsorization: ",varname,"left tail:","\n")
       kkk=x[lt.index,c(theID,varname)]
       names(kkk)[2]="value"
       kkk$name=varname
       kkk$event="winsorization.left"
       print(kkk,row.names=F)
       x[lt.index,varname] <- Lwin
    }
    else cat("Winsorization: ",varname, "left ids = NA","\n")

    return(x);
}
```

Industry Demean and Re-standardization This is a very important step. We explained the rationale in the previous section.

Industry demean

```
dataNeutralize <- function(data,varname, neutral.name)
{
   x=data
   # get the mean for each group
   cc=which(is.na(x[,neutral.name]))
   if(sum(cc)>0) x[cc,neutral.name] = "NA"
   aa=tapply(x[,varname],x[,neutral.name],mean,na.rm=T)
   theMean= data.frame(theMean = as.vector(aa), sss=names(aa))
   names(theMean)[2]=neutral.name
   x=merge(x,theMean,by=neutral.name,all.x=T)

   # get the demeaned score
   x[,varname]=x[,varname] - x$theMean
   mm=match("theMean",names(x))
   x=x[,-mm]
   return(x)
}
```

Re-standardization is employed to make sure factors have the same mean and standard deviation because industry demeaning may change the distribution of factor scores.

Exclusion Exclusion means to exclude a signal from a specified group. For example, cash flow factors do not make sense for banking, so we simply exclude cash-flow-based factors from themes and alpha building for banking companies.

Exclusion

```
dataExclude <-function(data, varname, col.exclude, exclude.name, theID="cusip")
{
  x=data
  # find the signal
  ss=match(varname,names(x))
  if(is.na(ss)) stop("varname does not exist in the data!\n")
  # exclude the group
  mm=match(col.exclude,names(x))
  if(is.na(mm)) stop ("col.exclude does not exist in the data!\n\n")
  nex=length(exclude.name)

  for(i in 1:nex)
  {
    cc=which(x[,mm]==exclude.name[i])
    if (sum(cc)>0)
    {
      x[cc,ss]=0
      cat("Exclusion: ",varname, "=", exclude.name[i], ": ",x[cc,theID],"\n")
    }
  }
  return(x)
 }
```

Missing Values and Distribution We deal with missing values again but now need to decide on how to treat the missing values: remove them or replace them with a preset score, such as zero, the mean, or the median? These preset values should be designed to associate stocks with missing values as neutral in a portfolio.

Missing replacement

```
dataReplaceMissing <- function(data,varname,theID="assetID",miss.fill="zero")
{
  x = data
  miss.index <- which(is.na(x[,varname]));

if(length(miss.index) > 0) {
 if(tolower(miss.fill)=="mean") x[miss.index,varname] <- mean(x[,varname],na.rm=T)
 if(tolower(miss.fill)=="median") x[miss.index,varname]<-median(x[,varname],na.rm=T)
 if(tolower(miss.fill)=="zero") x[miss.index,varname] <- 0
}
 return(x);
}
```

Distribution Matters This will depend on many considerations. For example, for an investment universe of small-cap companies or emerging markets, factor values are usually more wild than for large-cap companies in developed markets. The former may require ranking, while the latter can be normalized. We provide here a utility function with options for ranking, normalization, and maintaining the original distribution.

Distribution

```
dataDistribution <- function(data,varname, distrib)
{
  x=data
  if(distrib==1 |distrib==2)
  {
    #x$uniform=rank(x[,varname])/dim(x)[1]
 x$uniform=rank(x[,varname],na.last="keep",ties.method="average")/dim(x)[1]
    mm=match("uniform",names(x))
    names(x)[mm]=paste(varname,".unif",sep="")
  }
  if(distrib==2)
  {
    x$normal=pnorm(x[,mm],0,1)
    mm2=match("normal",names(x))
    names(x)[mm2]=paste(varname,".norm",sep="")
  }
  return(x)
}
```

Using the functions above, we clean the data and carry out treatment for the signal B/P for stocks in the Russell 1000 universe on December 31, 2004. We present the plot for the data at each stage in Fig. 4.10.

4.8.2 R Functions: Estimating a Multi-Factor Model with OLS

The previous section describes functions for data cleaning and factor treatment. In this section, we show how to estimate a multi-factor model with R scripts and functions. In particular, we show how to obtain OLS estimation results and output results using a function.

OLS and Its Attributes In R, the command to run OLS is *lm*. The protocol is

$$lm(y \sim x_1 + x_2, data = ABC),$$

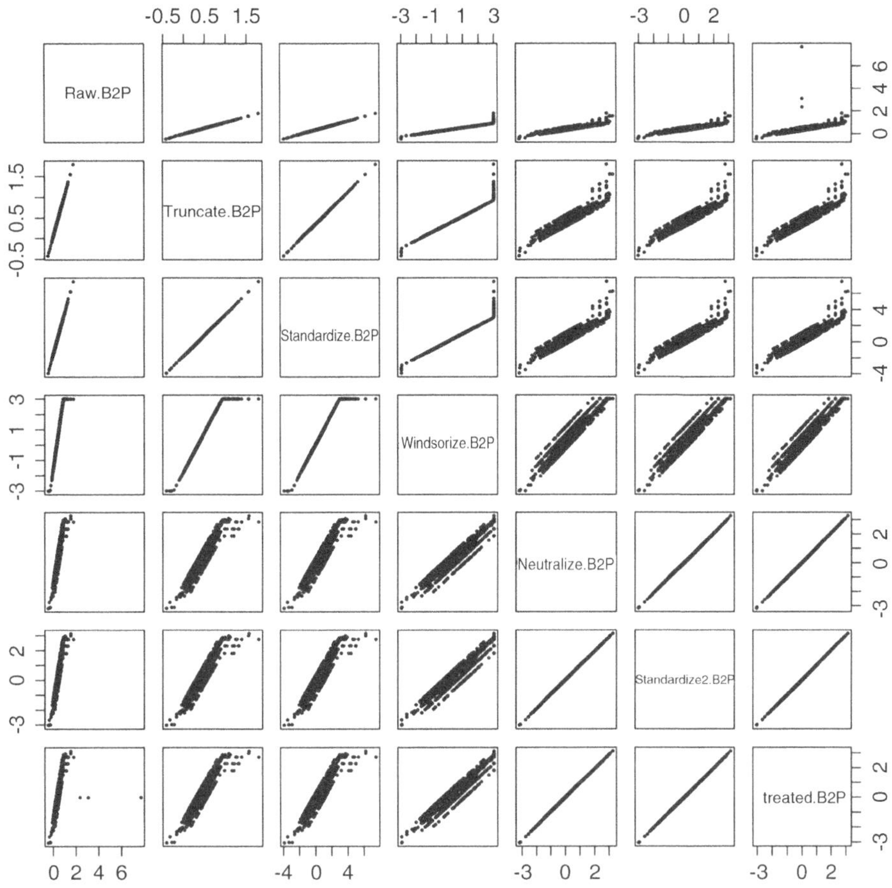

Fig. 4.10 The plot of B/P values at each step of the treatment for stocks in the Russell 1000 universe on December 31, 2004

where the data ABC contains columns of y, the response variable, and x_1 and x_2, the factors.

To get more information about the OLS results, we can use the R command *summary(ols.object)*. We use profitability signals as an example and show the R scripts below.

OLS example

```
> summary(lm(retf3 ~ FCF2EV + ROEFY0 + EBITDA2EV + roiCFO2CE,
          data = r1k.signals))

Coefficients:
             Estimate Std. Error t value Pr(>|t|)
(Intercept) 0.0327000  0.0006231  52.477  < 2e-16 ***
FCF2EV      0.0083439  0.0008187  10.192  < 2e-16 ***
ROEFY0      0.0036050  0.0007469   4.827 1.39e-06 ***
EBITDA2EV   0.0033314  0.0006872   4.848 1.25e-06 ***
roiCFO2CE   0.0022625  0.0008764   2.582  0.00984 **
---
Residual standard error: 0.2116 on 115355 degrees of freedom
  (1532 observations deleted due to missingness)
Multiple R-squared:  0.003079,Adjusted R-squared: 0.003045
F-statistic: 89.08 on 4 and 115355 DF,  p-value: < 2.2e-16
```

To obtain the components of the results above, we can use *coef* to get coefficients, *resid* to get residuals (the estimates for the errors), and *fitted* for fitted values $\hat{y}$. The R scripts below display the usage of these functions.

It is a good habit to start any R function with some explanatory comments. The "read.me" text for a function includes the purpose of the function and the prepared directories, folders and utility functions, etc.

Function, read.me part and preparation

```
##############################################################################
##
## Function: Quant Investmenting, Chapter 4, build alpha
##
##   Input:  data.dir -- directory of the data
##          data.name -- the name of the input data with theme values
##             themes -- the name list of themes
##            returns -- the name list of returns
##           industry -- industry name, in case it is needed
##
##  Output: sink file -- univariate, bivariate and multi-factor OLS results
##          csv files -- correlation of themes, themes with returns
##                    -- correlations of alpha with returns
##                    -- final data frame with alpha values added
##
```

```
##    Note: Alpha values are built with the t-values of OLS model
##          residual values are used for PROF~VALUE and MS~PM
##
##   Codes: Lingjie Ma, 20181128
##
#############################################################################

data.dir="/.../book/QuantInvesting/chapter3/factor.treatment/"
theme.list=c("VALUE","PM","PROF","EQ","MQ","MS")
return.list=c("retf1","retf3","retf6","retf9")

QI.chapter4.alpha <- function(data.dir,themes=them.list,
   returns=return.list, industry="sectorName")
{

}
```

We now present a sample function to illustrate the R commands for running OLS regression for a multi-factor model, obtaining components such as t-values and residuals, and calculating weights for factors based on OLS estimation results. Note that we also create a log file by command *sink* which is very helpful for debugging and recording purposes.

Sample R function: OLS for alpha model

```
QI.chapter4.alpha <- function(data.dir, data.name,
        themes=them.list, returns=return.list, industry="sectorName")
{
### get the data
r1k.themes=read.csv(paste(data.dir, data.name,sep=""), sep=",",header=T)

### start the log file
sink(paste(data.dir,"analysis/alpha.ols.txt",sep=""))

### Univariate results
cat("Univariate analysis, Summary of themes: \n")
print(summary(r1k.themes[,themes]))

### Bivariate, themes and with returns correlation
cat("Bivariate analysis, correlation of themes and with returns: \n")
## correlation with returns
 themes.corRetP=round(cor(r1k.themes[,themes],r1k.
              themes[,returns],use="complete.obs"),2)
 themes.corRetS=round(cor(r1k.themes[,themes],r1k.
                 themes[,returns],use="complete.obs",method="spearman"),2)

 write.table(themes.corRetP, paste(data.dir,"analysis/
              r1k.themes.corRetP.csv",sep=""),sep=",")
```

```
 write.table(themes.corRetS, paste(data.dir,"analysis/
              r1k.themes.corRetS.csv",sep=""),sep=",")

 ## correlation between signals
 themes.corP=round(cor(r1k.themes[,themes], use="complete.obs"),2)
 themes.corS=round(cor(r1k.themes[,themes],
      use="complete.obs",method="spearman"),2)

 write.table(themes.corP, paste(data.dir,"analysis/r1k.
                     themes.corP.csv",sep=""),sep=",")
 write.table(themes.corS, paste(data.dir,"analysis/r1k.
                     themes.corS.csv",sep=""),sep=",")

### Multi-factor model, OLS results: get resid
cat("Multi-factor, OLS of returns on themes: \n")
prof.ols=summary(lm(PROF~VALUE,data=r1k.themes))
print(prof.ols)
r1k.themes$PROFexVALUE.resid=prof.ols$resid

ms.ols=summary(lm(MS~PM,data=r1k.themes))
print(ms.ols)
r1k.themes$MSexPM.resid=ms.ols$resid

themes.ols=summary(lm(retf9~ VALUE+PM + PROF.resid
         + EQ+MQ+MS.resid,data=r1k.themes))
print(themes.ols)

### OLS, obtain t-values, calculate weight and build alpha scores
cat("Build Alpha now: \n")
theme.Tvalue=themes.ols$coef[-1,3]
theme.Tweight=theme.Tvalue/sum(abs(theme.Tvalue))
cat("Theme weights based on OLS t-values:\n")
print(theme.Tweight)

### Statistics on the Efficacy of Alpha
cat("Statistics on the Efficacy of Alpha: \n")
r1k.themes$ALPHA =  as.matrix(r1k.themes[,themes],
  ncol=6) %*% matrix(as.vector(theme.Tweight),ncol=1)
print(summary(r1k.themes$ALPHA))
print(round(cor(r1k.themes$ALPHA,r1k.themes[,returns],
                                  use="complete.obs"),3))
print(round(cor(r1k.themesALPHA,r1k.themes[,returns],
              use="complete.obs",method="spearman"),3))
print(summary(lm(retf9~ALPHA,data=r1k.themes)))

### Output the data with alpha values
write.table(r1k.themes, paste(data.dir,"treateddata/r1k.factors.
                alpha.csv",sep=""),sep=",",row.names=F)

sink()
}
```

Keywords, Problems, and Group Project

Part I. Keywords

Regression, Ordinary Least Squares (OLS), multi-factor model, t-value and R^2, collinearity, BLUE

CAPM, sources of market inefficiency, profitability, earnings quality, value, momentum, management quality, market sentiment, alpha model

CUSIP, GICS, investment strategies, market neutral, active weight

R treatment functions, run OLS in R and obtain results

Part II. Problems

Problem 4.1 Collect the data for the two companies you picked for the problems in Chaps. 2 and 3, and build signals.

(1) Collect data from quarterly financial statements filed with the SEC: balance sheet, income statement, and cash flow statement.
(2) Collect pricing data and calculate forward returns of 1, 3, 6, 9, 12 months.
(3) Construct quarterly ratios: price momentum, value, profitability, earnings quality, and management quality; merge with the return data by date and cusip.
(4) Apply proper treatments to raw signals.
(5) Conduct univariate analysis of the raw and treated values.
(6) Conduct bivariate correlation (both Pearson and rank) between signals (treated values), and between signals and forward returns.

 (i) Any correlations higher than 30 or 40%?
 (ii) Any return values outside the range of (−50%, +50&)?

(7) Conduct OLS regression using returns as the response variable and signals as factors for each theme.

 (i) Check coefficients, t-value, and R^2.
 (ii) Do the OLS regression results agree with bivariate results? Do the high-correlation signals cause collinearity issue?

Problem 4.2 Build themes with the treated signal data.

(1) Combine signals into themes with weights based on the univariate, bivariate and multi-factor analysis.
(2) Conduct univariate analysis for each theme, bivariate analysis between themes, and multi-factor analysis with all the themes.
(3) Use residual values in case of high correlation between themes.
(4) Change the weights by 5–10% and redo (2), investigate whether the results change significantly.

Problem 4.3 Build alpha values from themes using OLS.

(1) Run OLS for the multi-factor model.

$$R_F = b_0 + b_1 PROF + b_2 EQ + b_3 VALUE + b_4 PM + b_5 MQ + \epsilon$$

R_F is forward returns of T-month, $T = 1, 3, 9$.
(2) Build alpha with weights derived from t-values or coefficients.
(3) Conduct univariate analysis of alpha, correlation between alpha and forward returns, and OLS of returns on alpha. Evaluate the forecasting power of alpha scores.
(4) Plot distribution of return values and alpha scores. Do they follow a Gaussian (normal) distribution?

Problem 4.4 Calculate beta values for the two stocks (CAPM) in Problem 4.1.

(1) Write a function in R to use a loop to calculate beta.
(2) For the Chinese stock, calculate two beta values, one with the CSI 300 and the other with the S&P 500 as the market. Are the betas different?
(3) Calculate beta values when the market is positive and negative. Are the two sets of beta values the same?

Part III. Group Project
Problem 4.5 Quantitative investing is based on the law of large numbers, that is, the performance comes from breadth. Form a team with at least five people. Now with at least 10 stocks (the more the better), repeat the multi-factor analysis above to build the alpha for these stocks. Each team is expected to write a 5–10 page report and do a group presentation for 10–30 min.

(1) With more stocks, do you see the t-value and R^2 increase?
(2) Construct a long-only portfolio with stocks in the top 10% alphas for each quarter. Calculate portfolio performance over the entire in-sample period. Does your portfolio outperform the benchmark?
(3) Calculate beta for the portfolio. Are there any diversification benefits?

References

Abarbanell, J.S. 1991. "Do Analysts' Earnings Forecasts Incorporate Information in Prior Stock Prices Changes?" *Journal of Accounting and Economics* 14: 147–165.
Brown, L.D., and M.S. Rozel. 1980. "Analysts Can Forecast Accurately!" *Journal of Portfolio Management* 6: 31–34.
Cooper, M., H. Gulen, and M. Schill. 2008. "Asset Growth and the Cross-Section of Stock Returns." *Journal of Finance* 100(4): 1609–1650.
DeBondt, W., and R. Thaler. 1985. "Does the Stock Market Overreact?" *Journal of Finance* 40: 793–805.
Fama, E. 1968. "Risk, Return, and Equilibrium: Some Clarifying Comments." *Journal of Finance* 23: 29–40.

Fama, E., and K. French. 1992. "Cross-Section of Expected Stock Returns." *Journal of Finance* 47(2): 427–465.
French, C.W. 2003. "The Treynor Capital Asset Pricing Model." *Journal of Investment Management* 1(2): 60–72.
French, K., and R. Roll. 1986. "Stock Return Variances: The Arrival of Information and the Reaction of Traders." *Journal of Financial Economics* 17: 5–26.
Gauss, C.F. 1809. "Theoria Motus Corporum Coelestium in Sectionibus Conicis Solum Ambientium (Theory of the Motion of the Heavenly Bodies Moving About the Sun in Conic Sections)," the Original Work Published in 1809 and Original Translation Published, 1857, translated by Davis, C.H. (ed.) Mineola: Dover Publications, 2004.
Jegadeesh, N. 1987. "Predictable Behavior of Security Returns and Tests of Asset Pricing Models," Ph.D. dissertation, Columbia University.
Jegadeesh, N., and S. Titman. 1993. "Returns to Buying Winners and Selling Losers: Implications for Stock Market Efficiency." *The Journal of Finance* 48(1): 65–91.
Jensen, M. 1966. "Agency Costs of Free Cash Flow, Corporate Finance, and Takeover." *American Economic Review* 76: 323–329.
Kaul, G., and M. Ninmalendran. 1990. "Price Reversals." *Journal of Financial Economics* 28: 67–93.
Legendre, Adrien-Marie. 1805. Nouvelles méthodes pour la détermination des orbites des comètes (New Methods for the Determination of the Orbits of Comets) (in French). Paris: Courcier.
Lintner, J. 1965a. "The Valuation of Risk Assets and the Selection of Risky Investments in Stock Portfolios and Capital Budgets." *The Review of Economics and Statistics* 47: 13–37.
Lintner, J. 1965b. "Securities Prices, Risk, and Maximal Gains from Diversification." *Journal of Finance* 20(4): 587–615.
Loughran, T., and J. Ritter. 1995. "The New Issues Puzzle." *Journal of Finance* 50: 23–52.
Mikhail, M., B. R. Walther, and R. H. Willis. 1999. "Does Forecast Accuracy Matter to Security Analysts?" *The Accounting Review* 74: 185–200.
Perold, A.F. 2004. "The Capital Asset Pricing Model," *Journal of Economic Perspectives* 18(3): 3–24.
Richarson, S., and R. Sloan. 2003. " External Financing and Future Stock Returns." Working paper, The Wharton School.
Sharpe, W. 1964. "Capital Asset Prices: A Theory of Market Equilibrium." *The Journal of Finance* 19(3): 25–442.
Sloan, R. 1996. "Do Stock Prices Fully Reflect Information in Accruals and Cash Flows About Future Earnings?" *The Accounting Review* 71(3): 289–315.
Stigler, S.M. 1981. "Gauss and the Invention of Least Squares." *Annals of Statistics* 9(3): 465–474.
Sullivan, E.J. 2006. "A Brief History of the Capital Asset Pricing Model." APUBEF Proceedings.
Titman, S., K.C. Wei, and F. Xie. 2004. "Capital Investments and Stock Returns." *Journal of Financial and Quantitative Analysis* 39(4): 677–700.
Treynor, J.L. 1961. "Market Value, Time, and Risk." Unpublished manuscript, dated as Aug 8, 1961.
Treynor, J.L. 1962. "Toward a Theory of Market Value of Risky Assets." Unpublished manuscript, dated as fall 1962. A final version was published in *Asset Pricing and Portfolio Performance*, ed. Korajczyk, R.A., 15–22. London: Risk Books (1999).
Zacks, L. 1979. "EPS Forecasts? Accuracy Is Not Enough." *Financial Analysts Journal* 35(2): 53–55.

Chapter 5
More on Stock Selection Strategy: Alpha Hunting, Risk Adjustment, and Nonparametric Diagnostics

Abstract In the previous chapter, we introduced a general procedure and multi-factor framework for alpha construction of stock selection strategies. In this chapter, we continue to explore stock selection strategy with more advanced topics. In particular, we focus on alpha (new factor) hunting, risk adjustment, and nonparametric diagnostics. Regarding new alpha discovery, we present the guidance of IPARE. From a methodological perspective, we introduce the weighted least squares (WLS) method, which provides a tool to integrate risk into a multi-factor alpha model. We then introduce nonparametric approaches as a complement to parametric analysis. In the industry insights section, we provide a nonparametric diagnostics package used in the industry to investigate a new factor. The last section on R programming shows how to refine plots with parameters.

5.1 Alpha Hunting: IPRAE

In Chap. 4, we introduced general themes and signals for a stock selection strategy. Those factors have been used by industry professionals for many years in various ways. Over time and with technological advances, some "alpha" information contained in those factors has been explored away and gradually become "beta." In the world of quantitative investing, professionals are constantly searching for new information and building new factors in order to outperform markets and their peers. We propose here five criteria to guide new factor construction and analysis: intuitive, predictive, robust, additive, and executable.

1. Intuitive

   A factor should be economically intuitive, i.e., a factor should be proposed based on fundamental insights, such as how a firm makes profits and how those profits are transformed into market prices.

   As quantitative approaches have become more popular and big data have become more accessible, there is a great danger to make up a factor mechanically that shows some predictive power for future stock returns. However, if such a factor is not based on fundamental logic, the effect derived from the data might just be the result of data mining.

L. Ma, *Quantitative Investing*, https://doi.org/10.1007/978-3-030-47202-3_5

The intuition should be based on fundamental insights as well as in-depth understanding and analysis. For example, under what conditions does the factor make sense? Does the factor make sense in both recession and boom periods? Does the factor drive returns for all kinds of businesses?

2. Predictive
   If we think a factor is economically intuitive and tells a sound financial story, the next step is to ask, does the story hold up? Is the intuition credible or a misunderstanding? We need to check whether the factor indeed has predictive power for future stock returns. For example, how can we measure predictive power? Can we quantify relevant conditions to test them?
3. Robust
   There are numerous shocks in financial markets that come from different sources. A lot of factors are effective for predicting stocks' future returns but might not be robust. Robustness means that if there is some change in the definition of the factor or conditions of the firm or business, the factor is still predictive. Robustness is a very important "character" of a factor. How can we test for robustness in a quantitative way?
4. Additive
   This is especially important for a group of factors. A new factor may be intuitive, predictive, and robust, but does it contain any new information and thus add significant value to the investment? The contribution can be return enhancement, risk diversification, or cost reduction. How can we measure the value added or marginal contribution? One way is to employ a multi-factor model framework, with the new factor as the response variable and existing factors as independent variables, and see if there is any value in the residual.
5. Executable
   We need to consider implementation cost. A factor may look great on paper and satisfy all the rules above, but if it is hard to quantify or transaction costs are very high, then it is not executable.

In summary, when investigating a new factor, we can use these IPRAE criteria for guidance. The next section discusses how to realize these criteria with a quantitative approach, including risk adjustment and nonparametric diagnostics.

5.2 The Risk of Separating Risk from Alpha

In the previous chapters, we have discussed alpha and risk separately. However, both risk and alpha are based on price movements, so even by their definitions they are not isolated. One weakness of many quantitative investing strategies is that they separate risk from alpha. For example, a strategy may use an alpha model and a separate risk model, but the portfolio's performance is measured by the risk-adjusted return. Therefore, we cannot treat risk and return separately from the outset. One immediate consequence of ignoring risk when building alpha is the Grand Canyon in the portfolio between model alpha and portfolio weight.

5.2.1 Defining Risk as Volatility

In previous chapters (and most classical approaches), risk is defined as standard deviation, the volatility of price movements or the second moment as described in Chap. 1. Usually, "risk" is regarded as a bad thing. However, we should keep in mind that risk also means opportunity. If prices are constant, then the risk defined by the second moment is equal to zero, so there is no opportunity for investment. Thus, even by the definition of risk as standard deviation, alpha is an integral part.

However, we have to be cautious when linking risk to volatility. The two plots in Fig. 5.1 illustrate the drawbacks of using volatility as a risk measurement. In the left panel, securities A and B both start at 10 dollars per share and change at the same

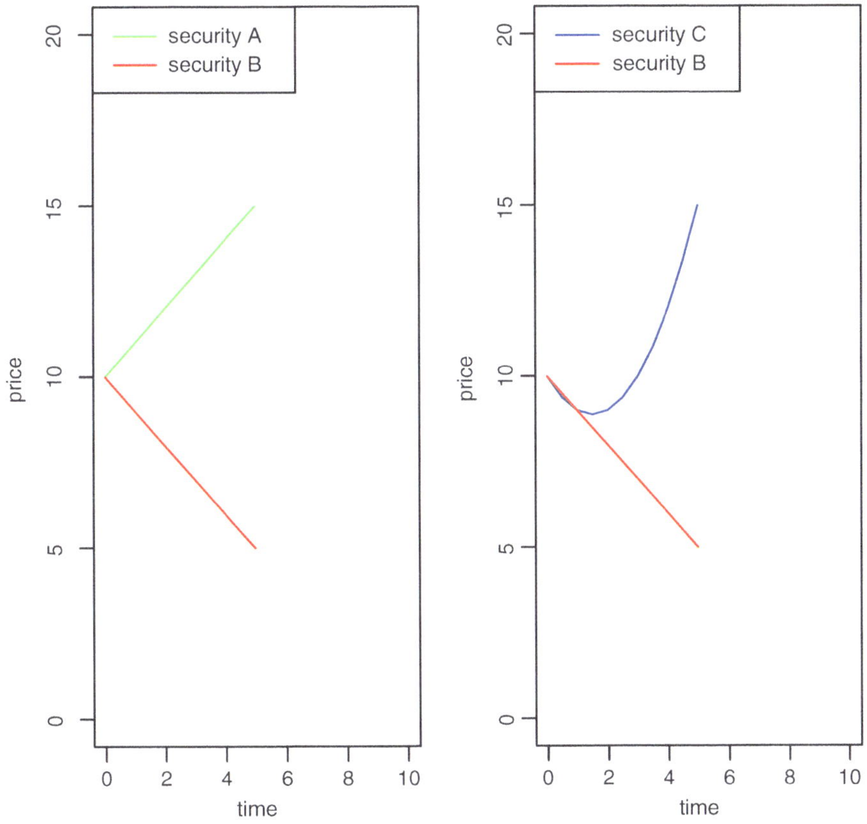

Fig. 5.1 In the left panel, security A and security B have the same values of standard deviation, while on the right panel security C has a higher standard deviation than security B

rate but in different directions: 5 business days later, security A ends up at 15 dollars, while security B ends up at 5 dollars. The standard deviations of price changes are the same for securities A and B, even though the prices went in completely different directions! If we were to use standard deviation to measure risk, securities A and B would have exactly the same risk level. However, we know that security A's price increased, and security B's price decreased. How absurd this is to define the two securities as having the same risk! Now, let us take a look at the right plot, which compares securities B and C. Security C's price starts at 10 and ends at 5, but with some up and downs. By just looking at the price paths, we would believe that security C would be "safer" than security B, because security C was up and security B was down during the same period. However, by the definition of risk as measured by volatility, security C would be regarded as the more risky asset than security B![1]

5.2.2 *Alpha Construction Without Risk Adjustment?*

So far we have separated risk and alpha: we have discussed alpha models, and we have discussed risk definitions. The alpha model is the core of quantitative investing, and risk is only a second-class citizen, the two meet with each other only at the portfolio construction stage, where alpha is maximized with some given level of risk or vice versa. However, separating alpha and risk in the very first place when building an alpha model is totally inappropriate.

In quantitative investing, alpha is an input in the mean-variance optimization for optimal portfolio weights. A simplified version of the mean-variance optimization is as follows:

$$\max W^{\top}\alpha - W^{\top}\Omega W, \tag{5.1}$$

where W is the optimal portfolio weights, Ω is the risk matrix that can be measured by variance of security returns. Solving the above mean-variance problem, we get

$$W = \Omega^{-1}\alpha,$$

so the optimal portfolio weights are derived as the risk-adjusted alpha. If we just use the variance part and ignore the covariance between security returns, we have

$$\Omega = \{\sigma_i^2\}.$$

[1]In addition to return volatility, there are other definitions of risk in investment, such as margin of safety and value at risk. We described the former in Chap. 2. Due to its subjectivity and ambiguity, the concept of the margin of safety is not often used in quantitative investing. We explore value at risk in Chap. 8.

Hence, we have

$$w_i = \frac{\alpha_i}{\sigma_i^2}$$

where σ_i^2 is used as the risk measurement for security i. Clearly, if α and σ are not aligned, the ratio can be very different from α in terms of both rank and magnitude. An extreme case is that we build a wonderful α that has high predictive power for forward returns, but once it is adjusted by risk, the forecasting power suddenly disappears! This is the *risk* of separating risk from alpha. Consider the following example:

$$Alpha = -3, -2, -1, 0, 1, 2, 3$$

$$Risk = 10, 5, 2, 1, 2, 5, 10$$

$$\frac{Alpha}{Risk} = -0.3, -0.4, -0.5, 0, 0.5, 0.4, 0.3$$

That is, after the risk adjustment, the stocks with alpha values of 1 and -1 become the long and short target positions instead of the original alpha values of 3 and -3.

One solution is to incorporate risk into the process of alpha construction. We discuss a methodology for this in Sect. 5.3 and applications in Sect. 5.4.

In early classical works, risk and return are integrated, such as the CAPM and Sharpe ratio from William Sharpe and the concept of the efficient frontier and mean-variance portfolio construction from Harry Markowitz. It is only since the heavy use of multi-factor models to forecast returns (or active return, aka alpha), where factors are identified and put in a (linear) model, that the alpha model seems to be the total focus and the risk model is forgotten. We should not blame academic research on multi-factor analysis for this problem. Rather, this is the result of the habits of many investment companies who had biased practices in the industry.

This is illustrated in a multi-factor alpha model we discussed in Chap. 4. Using an example of a stock selection strategy, a general alpha model is

$$R = b_0 + b_1 F_1 + b_2 F_2 + b_3 F_3 + \epsilon, \tag{5.2}$$

where R is security returns, F_j is the jth factor. For instance, F = {PROF, VALUE, EQ}. Is there a position for "risk" here? No, as we saw in Chap. 4. The caveat for this approach is that separating risk from alpha makes the alpha forecast weak. Recall that additional return should be compensation for taking additional risk. If we would like to add risk to the multi-factor model, how do we integrate them?

For a multi-factor framework, we could regard the factors that drive returns as potential sources of systematic risk, as they are for all securities; and the residual or error part as individual risk as the error is different for each security. In this sense, a multi-factor alpha model does have risk contents. Recall assumptions about the error term in order to satisfy the properties for the model estimates. For example, for

Fig. 5.2 Stephen Ross (1944–2017). Ross's major contributions to quantitative investing are arbitrage pricing model and binomial distribution for option pricing. Photo: Erica Ferrone for MIT Sloan School of Management

the OLS estimates for (5.2), the BLUE properties require exogeneity between F and ϵ and a spherical error term. These assumptions barely hold for real-world finance data because they derive from dynamic and extreme human activities rather than natural processes.

In the next section, we discuss each kind of violation for OLS estimates and then focus on the weighted least squares (WLS) method, which is very widely employed in quantitative investing. First, we introduce a multi-factor risk model that was a significant contribution to quantitative investing.

5.2.3 Arbitrage Pricing Theory (APT): A Multi-Factor Risk Model

The arbitrage pricing theory (APT) is about asset pricing that is based on a multi-factor framework with each factor serving as a proxy for a risk source. The coefficients for factors are the returns compensated for risks as represented by factor values. Thus, there is an equilibrium return level implied by the model. If returns for assets deviate from the equilibrium level, arbitrage will occur and bring them back. Stephen Ross (1976) proposed the APT in 1976. It was further developed by Roll and Ross (1980) and Chen et al. (1986) (Fig. 5.2).

By its original design, APT is used to identify and exploit mispriced assets by tracking a number of macroeconomic factors, which can be expressed as:

$$r = b_0 + b_1 F_1 + b_2 F_2 + \ldots + b_k F_k + \epsilon,$$

where F_k is a systematic factor, b_k is the sensitivity of r to F_k which is also called factor loading, and ϵ is the asset's individual random shock.[2] For example, many early studies about APT (e.g., Chen et al. 1986) focused on macro-level systematic factors comparable to the market premium of CAPM such as monetary policy indicators

[2]The error term, ϵ, is also called idiosyncratic risk in finance.

(interest rate), commodity price, and economic indicators (GDP). Unfortunately and expectedly, while these macro-level variables may help to explain structural changes in the economy or the overall status of a financial market, they often fail to explain asset returns at the individual level. This issue was solved by later research, such as the Fama–French three-factor model and other factor models, such as momentum and earnings quality, which seek to identify factors that can explain abnormal returns of individual stocks with statistical significance at the market level over time.

The contribution of APT to investment, especially quantitative investing, is that it proposes a general framework of a multi-factor linear model to analyze the relationship between expected stock returns and risk factors. Such a linear factor model structure is employed by many industry vendors as the basis for commercial risk products, for instance, the Barra Risk Model. These multi-factor risk products are used widely by asset managers in quantitative investing, especially at the portfolio construction stage.

Industry Risk Models During the period when Ross was publishing his research, the investment industry started to develop commercial risk models, designed to be used by practitioners for portfolio management. Barra was one of the first multi-factor based risk model vendors for the stock markets. It is still one of the largest in the world. Barra was founded by Barr Rosenberg.

> Since the factors can represent the components of return as seen by the financial analyst, the multiple-factor model is a natural representation of the real environment.
> –Barr Rosenberg (1974)

Currently, the major risk vendors in the US equity market are Barra, Northfield, and Axioma, in order of establishment. In Europe, one of the largest risk model vendors is APT, whose products of risk models are based on principle component analysis (PPA), a factor-fishing process.

Of course, risk and return are related, but not in a naive linear manner. In general, risk is compensated to the extent if the market is efficient; and risk does exist if market is not complete where there is mispricing. There is a wide range of returns for each risk level although the slope between return and risk is positive. We discuss a methodology for integrating risk and alpha in the next section.

5.3 What If OLS Conditions Are Violated?

We learned in Chap. 4 that for a linear multi-factor model, the parameters can be estimated by the OLS method, and under proper conditions, the OLS estimator is the best linear unbiased estimator (BLUE). However, those conditions, namely exogeneity and homoscedasticity, usually do not hold in the real world. What do we do then? Are there any treatments to rescue the BLUE from the gray world? In this section, we discuss refined methods that deal with cases where BLUE conditions are not satisfied. In particular, we discuss the consequences of each condition violation, the corresponding treatments, and the implications and applications for quantitative investing.

5.3.1 BLUE Is Great But in Reality It Can Be Gray

BLUE is the ideal status, but unfortunately the real world is usually gray.

The treatments for violations of BLUE conditions may enhance the properties of estimates such that they approach the BLUE. However, we have to keep in mind that the conditions, the data, and the model are all based on a theoretical framework. In the real world, we face many challenges in finance data,

1. data can be very dirty
   a. outliers
   b. incorrect data
   c. errors in measurement
   d. missing data
2. variables have different distributions
   a. can be very extreme
   b. correlations between variables can be very high
   c. omitted variables
3. linear model
   a. interactions among factors in part or overall
   b. relationships may be nonlinear
4. dynamic relationship
   a. spurious relationships
   b. contemporary and forecasting relationships are totally different

For all these practical issues, violations of exogeneity and homoscedasticity can be big and very serious. Even after treatment, we can never achieve the real BLUE, but the solutions we propose will make estimates less gray. This does not mean that the OLS method is not helpful. Rather, this deep exploration enables us to better understand the limits of modeling and be cautious in our interpretation and employment of statistical results in quantitative investing. From this perspective, we can say that a good quant is not the one who just knows how to build and estimate a model, but rather the one who understands deeply the weaknesses and limits of modeling.

In the next subsection, we discuss the consequences of such violations and proper treatments. We start with endogeneity and then move on to heteroscedasticity.

5.3.2 Unbiased: Endogeneity and 2SLS

We know from Chap. 4 that the unbiasedness of the OLS estimate depends on the assumption of exogeneity. Exogeneity, in simple terms, means that in a causal model,

$$y = X\beta + \epsilon,$$

the following hold:

1. X causes y, ϵ causes y
2. y does not cause X, ϵ does not cause X
3. Nothing that causes ϵ to also cause X

Exogeneity means that each X variable does not depend on the dependent variable y; rather, y depends on the $X's$ and on ϵ. Since y depends on ϵ, this means that the $X's$ are assumed to be independent of y and hence ϵ as well. These are standard assumptions of exogeneity we make in regression analysis. In mathematical terms, exogeneity can be expressed as $E(\epsilon|X) = 0$.

Any violation of at the above conditions would cause the endogeneity issue, the opposite of exogeneity. The endogeneity issue arises from multiple sources, such as simultaneity, correlation between explanatory variables, and measurement errors in factor values. There are serious issues of endogeneity in quantitative investment. Unfortunately, many sources of endogeneity in investment are not easy to deal with.

1. Simultaneity: Things happen at the same time and events are interrelated. This happens very often in financial markets. For example, changes in monetary policy impact security returns as well as companies' fundamental performance, such as profitability and earnings quality. This weakens the causal relationship from X to y.
2. Omitted variable: One or more important factors are omitted from the model. In financial markets, there are so many factors affecting the price of securities that there are always some factors that are not captured by a multi-factor model.[3]
3. Measurement error: The variable is measured with errors, and the measurement error may be correlated with the residual term of the model. In quantitative finance, many accounting-based items are estimates, such as net income or profit, which is derived from many components, assumptions, and estimate errors. This is unavoidable in finance.
4. Dynamic: autocorrelation of dependent variables. We will discuss this in detail in Chap. 6 on time series models.

No matter what causes endogeneity, the result is the same: x is not independent of ϵ. From Chap. 4, we know that if $E(\epsilon|X) \neq 0$, then neither unbiasedness nor consistency would hold.

Biasedness can be expressed as

$$E(\hat{b}) - b = E\left[(X^\top X)^{-1}X^\top \epsilon\right],$$

[3]In general, the R^2 for a multi-factor return forecast model is less than 10%. This does not mean that the other 90% is noise, but rather that much information is being ignored or not captured by the model. Examples include culture and the legal system at the macro-level and board member relationships at the micro governance level, which all impact public performance in the financial markets.

indicating that the higher the (absolute) correlation between x and ϵ, the greater the bias.

Consistency can be expressed as

$$\lim_{n\to\infty} \hat{b} - b = \lim_{n\to\infty} (X^\top X)^{-1} X^\top \epsilon,$$

implying less convergence for a higher relationship between x and ϵ.

The crucial issue here is that x is related to ϵ, which violates both the unbiasedness and consistency of the OLS estimator. One solution would be to find a variable that is fundamentally related to x but has no relationship with ϵ, such as a variable z. We can use z to approximate x, and since the function of z will not relate to the error term, the endogeneity issue is solved. In simple terms, say you (x) have a cousin (ϵ) from your father's side. To remove the relationship, we use your best friend and cousin from your mom's side (z) to replace you. Thus, z can replace x to some degree under the condition that z does not know ϵ! This is called the instrumental variable (IV) method, where z is the IV. The IV method was first proposed by Sewall Wright (1928) in the appendix of his father's book (Stock and Trebbis 2003).

In econometrics, we have the two-equation model, $E(\epsilon|x_1) \neq 0$,

$$y = b_0 + b_1 x_1 + b_2 x_2 + \epsilon \tag{5.3}$$

$$x_1 = \gamma_0 + \gamma_1 z + \nu, \tag{5.4}$$

where $E(\epsilon|Z) = 0$ and $E(\nu|Z) = 0$. We first estimate x_1 using z in (5.4) by OLS,

$$\hat{\gamma} = (Z^\top Z)^{-1} Z^\top x_1 = \gamma + (Z^\top Z)^{-1} Z^\top \nu,$$

where $Z = (1, z)$. Hence, the projected value for x_1 is

$$\hat{x}_1 = \hat{\gamma}_0 + \hat{\gamma}_1 z.$$

Replacing x_1 with $\hat{x}_1$ in (5.3), we have

$$\begin{aligned} y &= b_0 + b_1 \hat{x}_1 + b_2 x_2 + \left[b_1 (x_1 - \hat{x}_1) + \epsilon \right] \\ &= b_0 + b1 \hat{x}_1 + b_2 x_2 + \left[b_1 Z (\gamma - \hat{\gamma}) + b_1 \nu) + \epsilon \right] \\ &= b_0 + b1 \hat{x}_1 + b_2 x_2 + \left[b_1 (Z (Z^\top Z)^{-1} Z^\top + I) \nu + \epsilon \right]. \end{aligned}$$

Since $E(\epsilon) = 0$, $E(\epsilon|Z) = 0$, and $E(\nu|Z) = 0$, we can show that the OLS estimate for $b = (b_0, b_1, b_2)$ is unbiased and consistent. Since this involves two stages, we call the final estimates for b a two-stage least squares (2SLS) estimator. The term 2SLS is usually referred to as Heckman two-stage LS accrediting Heckman's contribution in this field (Heckman 1976, 1979). The statistical test for an IV was developed by Durbin, Wu, and Hauseman, called the Durbin–Wu–Hauseman specification test (Durbin 1954, Wu 1973, Hausman 1978).

While unbiasedness and consistency are very important properties, and the 2SLS (IV) method indeed overcomes the endogeneity issue as an econometric methodology, its applications in quantitative investing are very limited, perhaps for the following reasons:

- Endogeneity issue is too big an issue and very common in quantitative investing, so there is no way to solve it.
- There is no need to use 2SLS because the value added is too marginal.
- There are alternative solutions in quantitative investing. Sometimes, investors do use one factor to estimate another factor so as to forecast returns with less bias.

Why do not people employ the 2SLS or IV method in quantitative investing? There are several fundamental reasons. First, quant information that can be used to explain future returns is very limited. For example, quantitative models can explain only around 1–10 percent of price changes, implying that there are many factors or much information omitted. Given the dynamics of the market and uncertainty, contemporaneous relationships always exist and endogeneity is always an issue. On the other hand, more time and energy should be used to explore sources to identify fundamental factors, rather than exerting effort to address a big issue that yields little marginal value except in some special cases.

For example, profitability is important for future returns. Our focus should be on future profitability. If we could use historical profitability to forecast future profitability, then use the forecasted future profitability to forecast future returns, this makes much more sense in theory. However, in reality, future profitability is very hard to predict. Investors find it is more reliable to use historical profits to forecast returns. Instead of 2SLS, we can simply use z directly as an additional factor,

$$\begin{aligned} y &= b_0 + b_1(\gamma_0 + \gamma_1 z + \nu) + b_2 x_2 + \epsilon \\ &= b_0 + b_1\gamma_0 + b_1\gamma_1 z + b_2 x_2 + b_1 \nu + \epsilon \\ &= (b_0 + b_1\gamma_0) + b_1\gamma_1 z + b_2 x_2 + (b_1 \nu + \epsilon). \end{aligned}$$

In quantitative investing, investors really care about the forecasting results of y but not the value of b_1 or $b_1\gamma_1$. This is different from other fields, where the estimation of a factor x_1 is more important, such as when x_1 is a policy factor and y is the policy outcome.

However, we always need to keep in mind that in quantitative investing, for a linear model with OLS estimates, there are biases in the estimates! So we need to be very cautious in our interpretation and employment of estimates. In particular, we should be aware that a portfolio derived from a forecast of returns is based on biased estimates.

5.3.3 *Efficiency: Heteroscedasticity and WLS*

Every estimate deserves an error. Being efficient means that the standard error for the estimate is smaller, which requires the error term to be homogenous and not correlated with each other

$$Var(\epsilon|X) = \sigma^2 I,$$

where I is the identity matrix. If this is violated, then the OLS estimates will not be efficient any more. This is explained elegantly by Allison (1999): "The reason OLS is not optimal when heteroscedasticity is present is that it gives equal weight to all observations when, in fact, observations with larger disturbance variance contain less information than observations with smaller disturbance variance." Moreover, with the presence of heteroscedasticity, the estimated standard error will be biased which results in bias in related test statistics such as confidence intervals.

We discuss in detail about sources of heteroscedasticity in the context of finance data in the next section. Here, we show how to detect and treat heteroscedasticity from an econometric perspective.

Detecting Heteroscedasticity Heteroscedasticity can be detected by either plots or hypothesis tests. We show an example of residuals plotted against fitted values in Fig. 5.3. A rule of thumb is that if there is no clear pattern of distribution of the scatter plot, then it is in general safe to assume there is no serious issues of heteroscedasticity. However, if residual plot shows an uneven distribution across values of X, a formal test for heteroscedasticity should be considered.

There are many rigorous statistical tests for heteroscedasticity. Here we present one test used widely in economics and finance, the Breusch–Pagan test. Breusch and Pagan (1979) propose a statistical model to detect any linear form of heteroscedasticity. Suppose that we estimate a regression model

$$y = b_0 + b_1 x + \epsilon.$$

Step 1: Run OLS and obtain the values for $\hat{\epsilon}$, the residuals.
Step 2: Specify and estimate a derivative model:

$$\hat{\epsilon}^2 = \gamma_0 + \gamma_1 x + \nu.$$

If x helps to explain the variation in the least squares residual $\hat{\epsilon}$, then ϵ_i will not be the same across observations.[4] Now we need a test statistic to make a judgment. Since R^2 measures the proportion of variance in $\hat{\epsilon}^2$ explained by the x, it is a natural candidate for a test statistic. When $\gamma_1 = 0$ is true, the sample size N multiplied by R^2 has a χ^2 distribution with $k - 1$ degrees of freedom. That is,

[4]Otherwise, the model is not identifiable.

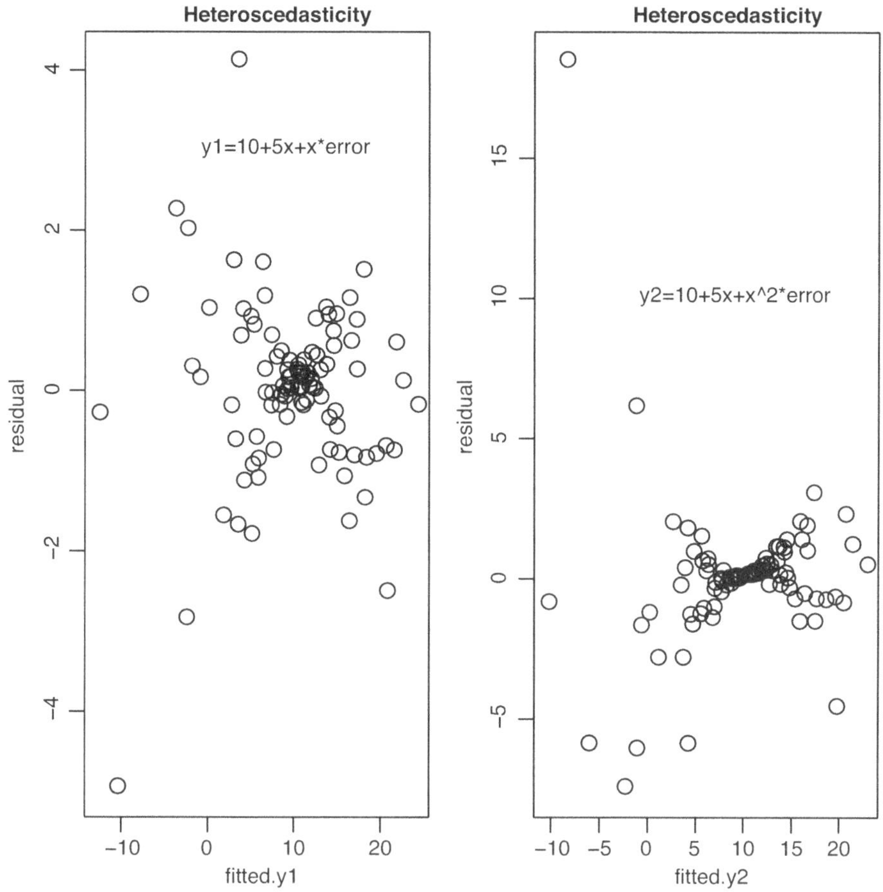

Fig. 5.3 In the left panel, the value of residuals decreases with x, and in the right plot, there is a bipolar relationship

$$\chi^2 = N \times R^2 \sim \chi^2_{k-1}.$$

Thus, we could specify a hypothesis test

$$H_0 : \gamma_1 = 0, \qquad H_a : \gamma_1 \neq 0.$$

Under H_0, the Breusch–Pagan test statistic follows a chi-squared distribution with k degrees of freedom, where k is the number of factors. We present an example below using a simulating model.

BP test for heteroscedasticity

```
## model setup
heter.test <- function()
{
     set.seed(99)
     x=rt(100,3)
     error=rnorm(100)
     b0=10
     b1=5
     y1=b0+b1*x+x*error
     y2=b0+b1*x+x^2*error

    fit1=lm(y1~x)
    fit2=lm(y2~x)
    #bptest(formula, varformula = NULL, studentize = TRUE, data = list())
    library(lmtest)
    print(bptest(fit1))
    print(bptest(fit2))
}

> source("/.../quantInvesting/springerLatex/chapter5/heter.test.R")
> heter.test()

studentized Breusch-Pagan test
data:  fit1
BP = 13.362, df = 1, p-value = 0.0002568

studentized Breusch-Pagan test
data:  fit2
BP = 15.28, df = 1, p-value = 9.267e-05
```

If the p-value is less than the level of significance (in this case, if the p-value is less than 0.01), we can reject the null hypothesis. In both cases, the BP tests reject homoscedasticity. In addition to the BP test, there are other tests, such as the White test (White (1980)), to detect heteroscedasticity.

Treatment of Heteroscedasticity The issue of heteroscedasticity can be treated in many ways. One way is to reformulate the model, for example, if there is an omitted factor that can be added back, or there are x^2 effects in the model, or the response variable should be logged or transformed, etc. Of course, this should be done in a fundamental way, not through data mining. Another treatment is to re-estimate the model by taking consideration of variation of errors with OLS. This is called weighted least squares (WLS) in a simple form and general least squares (GLS) in a general form: using residuals as weights to force the model to have uniform variance.

We now introduce the practical two step procedure of the WLS method for a multi-factor model. Consider a general multi-factor linear model with two factors,

$$y = b_0 + b_1 x_1 + b_2 x_2 + \epsilon.$$

For each observation i,

$$y_i = b_0 + b_1 x_{1,i} + b_2 x_{2,i} + \epsilon_i.$$

Since we know that ϵ_i is different from ϵ_j and $var(\epsilon_i) = \sigma_i^2$, we divide the above equation by σ_i

$$\frac{y_i}{\sigma_i} = b_0 \frac{1}{\sigma_i} + b_1 \frac{x_{1,i}}{\sigma_i} + b_2 \frac{x_{2,i}}{\sigma_i} + \frac{\epsilon_i}{\sigma_i}. \tag{5.5}$$

Renaming the model (5.5) with $\tilde{y} = \frac{y_i}{\sigma_i}$, we have

$$\tilde{y} = b_0 \tilde{x}_0 + b_1 \tilde{x}_1 + b_2 \tilde{x}_2 + \tilde{\epsilon} \tag{5.6}$$

and apply OLS to (5.6),

$$\min_b \sum_{i=1}^{n} (\tilde{y}_i - b_0 \tilde{x}_{0,i} + b_1 \tilde{x}_{1,i} + b_2 \tilde{x}_{2,i})^2,$$

where $b = (b_0, b_1, b_2)$, yielding the OLS estimate

$$\begin{aligned}
\hat{b} &= (\tilde{X}^\top \tilde{X})^{-1} \tilde{X}^\top \tilde{y} \\
&= (\tilde{X}^\top \tilde{X})^{-1} \tilde{X}^\top (\tilde{X} b + \tilde{\epsilon}) \\
&= b + (\tilde{X}^\top \tilde{X})^{-1} \tilde{X}^\top \tilde{\epsilon}.
\end{aligned}$$

The variance of $\hat{b}$ is

$$\begin{aligned}
Var(\hat{b}) &= Var\left[b + (\tilde{X}^\top \tilde{X})^{-1} \tilde{X}^\top \tilde{\epsilon}\right] \\
&= Var\left[(\tilde{X}^\top \tilde{X})^{-1} \tilde{X}^\top \tilde{\epsilon}\right] \\
&= (\tilde{X}^\top \tilde{X})^{-1} Var(\tilde{\epsilon}) \\
&= (\tilde{X}^\top \tilde{X})^{-1} Var(\frac{\epsilon}{\sigma_i}) \\
&= (\tilde{X}^\top \tilde{X})^{-1}.
\end{aligned}$$

Thus, we obtain efficiency again! This is because $\tilde{\epsilon}_i$ and $\tilde{\epsilon}_j$ have the same variance. However, we do not know the value of σ_i. Nevertheless, we could estimate σ_i. One way is to apply OLS to the original model and get the estimate, $\hat{\sigma}_i$, then weight the original model by $\frac{1}{\hat{\sigma}_i}$ and then apply OLS again.

Step 1: Run OLS and get $\hat{\epsilon}_i$.
Step 2: Estimate $\hat{\sigma}_i$ and run OLS again with weights $w_i = \frac{1}{\hat{\sigma}_i}$.

We illustrate the above procedure with $w_i = \frac{1}{|\hat{\epsilon}_i|}$ using the R scripts below.

Example of WLS

```
## continue the codes form the simulation model for the heteroscedasticity test
## run WLS
fit1.wls=lm(y1~x,weights=1/abs(fit1$resid))
fit2.wls=lm(y2~x,weights=1/abs(fit2$resid))

print(summary(fit1))
print(summary(fit1.wls))
Call:
lm(formula = y1 ~ x)
Residuals:
    Min      1Q  Median      3Q     Max
-4.9569 -0.3809  0.0505  0.4628  4.1150
Coefficients:
            Estimate Std. Error t value Pr(>|t|)
(Intercept)  9.88121    0.10817   91.35   <2e-16 ***
x            4.97695    0.08008   62.15   <2e-16 ***
---
Residual standard error: 1.081 on 98 degrees of freedom
Multiple R-squared:  0.9753,Adjusted R-squared:  0.975
F-statistic:  3863 on 1 and 98 DF,  p-value: < 2.2e-16

Call:
lm(formula = y1 ~ x, weights = 1/abs(y1resid))
Weighted Residuals:
    Min      1Q  Median      3Q     Max
-2.2417 -0.6276  0.1860  0.6684  2.0200
Coefficients:
            Estimate Std. Error t value Pr(>|t|)
(Intercept)  9.89061    0.02076   476.4   <2e-16 ***
x            4.97097    0.03955   125.7   <2e-16 ***
---
Residual standard error: 0.8477 on 98 degrees of freedom
Multiple R-squared:  0.9938,Adjusted R-squared:  0.9938
F-statistic: 1.579e+04 on 1 and 98 DF, p-value:< 2.2e-16
```

```
print(summary(fit2))
print(summary(fit2.wls))
Call:
lm(formula = y2 ~ x)
Residuals:
    Min      1Q  Median      3Q     Max
-7.4677 -0.4266  0.1014  0.4379 18.4612
Coefficients:
            Estimate Std. Error t value Pr(>|t|)
(Intercept)   9.9331     0.2575   38.58   <2e-16 ***
x             4.4862     0.1906   23.53   <2e-16 ***
---
Residual standard error: 2.572 on 98 degrees of freedom
Multiple R-squared:  0.8497,Adjusted R-squared:  0.8481
F-statistic: 553.8 on 1 and 98 DF,  p-value: < 2.2e-16

Call:
lm(formula = y2 ~ x, weights = 1/abs(y2resid))
Weighted Residuals:
    Min      1Q  Median      3Q     Max
-2.6305 -0.7465  0.1893  0.5907  4.3955
Coefficients:
            Estimate Std. Error t value Pr(>|t|)
(Intercept)  9.95506    0.02526  394.10   <2e-16 ***
x            4.59477    0.06583   69.79   <2e-16 ***
---
Residual standard error: 1.073 on 98 degrees of freedom
Multiple R-squared:  0.9803,Adjusted R-squared:  0.9801
F-statistic:  4871 on 1 and 98 DF,  p-value: < 2.2e-16
```

The OLS and WLS results for the simulating model are summarized below:

$$\text{Model 1:} \quad y1 = 10 + 5x + x\epsilon$$
$$OLS: \hat{b}_1 = 4.9769, \quad s.e.(\hat{b}_1) = 0.0800$$
$$WLS: \hat{b}_1 = 4.9709, \quad s.e.(\hat{b}_1) = 0.03955;$$

$$\text{Model 2:} \quad y2 = 10 + 5x + x^2\epsilon$$
$$OLS: \hat{b}_1 = 4.4862, \quad s.e.(\hat{b}_1) = 0.1906$$
$$WLS: \hat{b}_1 = 4.5947, \quad s.e.(\hat{b}_1) = 0.06583.$$

We see that after the application of WLS, the estimate for x stays almost the same, while the standard errors for x decreased dramatically in both model 1 (from 0.08 to 0.04) and model 2 (from 0.19 to 0.07). Also, R^2 increased for WLS significantly. The WLS methodology does increase the efficiency of OLS estimates with the presence of heteroscedasticity.

5.4 Applications of WLS in Finance: Risk-Adjusted Alpha

In classical finance, risk is measured by volatility expressed by standard deviation. In the WLS method, the weight is the standard deviation, the risk. If we have a multi-linear alpha model, then the sigma-weighted least squares method is just the risk-adjusted alpha model, with the sigma being the individual risk for each stock in a stock selection strategy. For this particular reason, WLS has been employed often in quantitative investing.[5]

For OLS estimators to be efficient requires the model error term to be identically and independently distributed. However, as we pointed out earlier, these assumptions or conditions are rather restrictive, and too restrictive to be true in finance. For example, for a stock selection strategy in the Russell 1000 investment universe.

1. Each company is different. The error term ϵ may have different distributions across different stocks, that is, ϵ_i and ϵ_j are different. For example, for Bank of America and Amazon, the residuals in a linear model will have very different characteristics for the two companies.
2. For a stock, price movements are continuous and related. That is, there may be a time path meaning $\epsilon_{i,t}$ and $\epsilon_{i,t-1}$ are correlated.
3. Stocks are related, especially those in the same industry. For example, the stock prices of United Airlines and American Airlines are highly correlated. This means that $\epsilon_{i,t}$ and $\epsilon_{j,t}$ are correlated.

Therefore, instead of $var(\epsilon_i) = \sigma_i^2$, in reality we have

1. Heterogeneity: $\sigma_i \neq \sigma_j$
2. Cross-section correlation: $\sigma_{i,j} \neq 0$
3. Time series correlation: $\sigma_{i,t} \neq \sigma_{i,t-1}$

Consequently, these violations will cause the OLS estimate to be inefficient, the letter "B" in BLUE is not valid anymore. In econometrics, are there ways to deal with such cases to make the OLS estimate efficient? Yes, there are. For cases 1 and 2, which are cross-sectional, the solutions are WLS or GLS. For case 3, which is a time series, we discuss a solution in Chap. 6. We now show how WLS can be employed to incorporate risk into alpha construction.

Recall that in quantitative investing, mean-variance optimization (5.1) yields optimal portfolio weights

$$W = I(\frac{1}{\sigma_i^2})\alpha,$$

or $w_i = \frac{\alpha_i}{\sigma_i^2}$ for each stock i. The optimal portfolio return is calculated as

[5]While 2SLS is not applied widely in quantitative finance.

$$R^P = W^\top R = \sum_{i=1}^{n} \frac{\alpha_i}{\sigma_i^2} R_i = \sum_{i=1}^{n} \frac{\alpha_i}{\sigma_i} \times \frac{R_i}{\sigma_i}, \tag{5.7}$$

where R is the vector of security returns. The maximization of portfolio return, R^P, can be obtained by the WLS method,

$$R = F\gamma + e, \quad w_i = \frac{1}{\sigma_i}, \tag{5.8}$$

where F is a vector of factors. We present detailed algorithms below which link multi-factor regression with portfolio weights in the context of contrasting WLS with OLS.

$$\begin{aligned} OLS\colon\ & R = F\beta + \epsilon \\ & \alpha_{ols} = F\hat{\beta}_{ols} \\ WLS\colon\ & R = F\gamma + \epsilon, \quad w_i = \frac{1}{\sigma_i} \\ & \frac{R}{\sigma_i} = \frac{F}{\sigma_i}\gamma + \frac{\epsilon}{\sigma_i} \\ & \alpha_{wls} = F\hat{\gamma}_{wls}. \end{aligned}$$

With the values of alpha in hand, we can get the optimal portfolio weights by following (5.1) corresponding to OLS and WLS estimates,

$$\begin{aligned} OLS\colon\ & w_{ols,i} = \frac{\alpha_{ols,i}}{\sigma_i^2} \\ & R_{ols}^P = \sum w_{ols,i} R_i = \sum \frac{F\hat{\beta}_{ols}}{\sigma_i^2} \times R = \sum \frac{F\hat{\beta}_{ols} \times R}{\sigma_i^2} \\ WLS\colon\ & w_{wls,i} = \frac{\alpha_{wls,i}}{\sigma_i^2} \\ & R_{wls}^P = \sum w_{wls,i} R_i = \sum \frac{F\hat{\gamma}_{wls}}{\sigma_i^2} \times R = \sum \frac{F}{\sigma_i} \times \frac{R}{\sigma_i} \times \hat{\gamma}_{wls}. \end{aligned}$$

Clearly, the WLS method provides a tool for risk-adjusted alpha. The "magic" touch here is simply because classical quantitative investing happens to use σ (of the error term) as the risk measurement.

Table 5.1 The correlation of OLS alpha with returns drops after risk adjustment

	Retf1	Retf3	Retf6	Retf9
Pearson cor.				
α_{ols}	0.052	0.079	0.107	0.122
$\alpha_{ols,i}/\sigma_i^2$	0.030	0.044	0.064	0.079
Rank cor.				
α_{ols}	0.043	0.066	0.089	0.098
$\alpha_{ols,i}/\sigma_i^2$	0.036	0.052	0.072	0.082

Now we carry out an empirical study applying WLS to the alpha model with the same data employed in Chap. 4. We have

$$OLS: \; R = VALUE + PM + PROF.resid + EQ + MQ + MS.resid$$

$$WLS: \; R = VALUE + PM + PROF.resid + EQ + MQ + MS.resid \quad w_i = \frac{1}{\hat{\sigma}_i}, \tag{5.9}$$

where the risk metric $\hat{\sigma}_i$ is estimated from the OLS regression.

Before showing the improvements brought by WLS, we first show the decreasing forecasting power of OLS alpha when applying risk adjustment (Table 5.1). We see that from alpha to risk-adjusted alpha, correlation scores (both Pearson and rank versions) with forward returns drop by one-third. Indeed, there is a risk in separating risk from alpha!

The R scripts below show the WLS model and results. For comparison purposes, we also have the OLS results.[6] After applying the WLS method, the value of R^2 increases dramatically from 0.02 for OLS to 0.12 for WLS. Note that we apply the robust version of residuals, which makes the standard errors of WLS estimates similar to those from OLS. Without the robust treatment, the R^2 is over 90%, and standard errors of WLS estimates are significantly lower than those from OLS. In sum, compared with OLS, WLS improves model performance with the data.

WLS: risk adjusted alpha

```
> ## WLS:  the weight is resid.sd measured by sd(residuals), estimated for each stock
> fit1=lm(retf9~    VALUE+PM+PROF.resid+EQ+MQ+MS.resid,data=r1k.themes)
> fit2=lm(retf9~    -1 + resid.sd + VALUE+PM+PROF.resid+EQ+MQ
              +MS.resid,data=r1k.themes, weights=1/r1k.themes$resid.sd^2)

> summary(fit1)
Residuals:
```

[6]Note that in the WLS model, we remove the intercept and add the inverse of weights, which plays the role of the intercept after the weights are applied.

```
     Min        1Q    Median        3Q       Max
-0.95460 -0.17704  0.01274  0.18679  0.85625

Coefficients:
               Estimate Std. Error t value Pr(>|t|)
(Intercept)   0.0633759  0.0009074  69.844   <2e-16 ***
VALUE         0.0326819  0.0013528  24.159   <2e-16 ***
PM            0.0194549  0.0012339  15.767   <2e-16 ***
PROF.resid    0.0280734  0.0022064  12.724   <2e-16 ***
EQ           -0.0025077  0.0015011  -1.671   0.0948 .
MQ            0.0258251  0.0016262  15.880   <2e-16 ***
MS.resid      0.0007949  0.0011345   0.701   0.4836
---
Residual standard error: 0.291 on 102920 degrees of freedom
Multiple R-squared: 0.01521, Adjusted R-squared: 0.01515
F-statistic: 264.9 on 6 and 102920 DF, p-value: < 2.2e-16

> summary(fit2)
Weighted Residuals:
      Min         1Q     Median         3Q        Max
-0.049767 -0.007470 -0.000603  0.006676  0.055194

Coefficients:
              Estimate Std. Error t value Pr(>|t|)
resid.sd     3.570e-03  3.381e-05 105.601  < 2e-16 ***
VALUE        2.240e-02  1.379e-03  16.238  < 2e-16 ***
PM           2.479e-02  1.322e-03  18.755  < 2e-16 ***
PROF.resid   3.893e-02  2.167e-03  17.967  < 2e-16 ***
EQ          -6.515e-03  1.534e-03  -4.246 2.17e-05 ***
MQ           2.242e-02  1.628e-03  13.771  < 2e-16 ***
MS.resid     8.109e-03  1.057e-03   7.670 1.73e-14 ***
---
Residual standard error: 0.01079 on 102920 degrees of freedom
Multiple R-squared:  0.1156,Adjusted R-squared:  0.1155
F-statistic: 1922 on 7 and 102920 DF, p-value: < 2.2e-16
```

Based on the two sets of coefficients of the OLS and WLS estimates, we build two sets of alphas, the weights for the six themes are specified in Table 5.2. The correlation between the two sets of alphas is about 95%.

We calculate raw portfolio returns using the formula specified earlier ($R^P = \sum \frac{\alpha_i}{\sigma_i^2} \times R_i$) and obtain

Portfolio return based on OLS weights: 1.75,
Portfolio return based on OLS weights: 1.92.

Table 5.2 The correlation of alpha with returns decreases after risk adjustment

	VALUE	PM	PROFresid	EQ	MQ	MSresid
OLS	0.35	0.23	0.19	-0.02	0.23	0.02
WLS.	0.23	0.27	0.25	-0.06	0.20	0.11

We see that there is a non-trivial improvement, about 10%, of portfolio return from WLS. Without the robust treatment, the raw return improves more than 50%. This shows that incorporating risk into the alpha construction does add value.

5.5 Nonparametric Analysis

We have so far explored parametric approaches to cases with one variable, two variables, and multiple factors. In this section, we introduce nonparametric approaches to quantitative investing. We explain reasons and benefits by employing a nonparametric approach and describe various nonparametric methods in the context of a stock selection strategy.

5.5.1 Why do We Need a Nonparametric Approach?

In previous chapters, we discussed data analysis with one, two, and multiple variables to explore the US stock market, the relationship between the US and Chinese markets, and the factors driving returns of individual stocks. In these kinds of analyses, we specify a model with parameters, such as the four moments of the US stock return distribution, the correlation between the S&P 500 and CSI 300, or the OLS estimation of a linear multi-factor model. The common feature of the above analyses is that we specify explicit parameters and assumptions about the data and model in an analytical framework. This type of analysis is called a parametric approach.

However, in the real world, there are often many cases where parameters are hard to specify and assumptions do not hold. For example, we learned in previous chapters that in finance, asset returns seldom follow a normal distribution; rather, they are usually skewed with fat tails.[7] That is, we cannot adequately describe a return distribution using just a few parameters. Instead, this type of data can be analyzed by a nonparametric approach for special parts of the return distribution, such as tail behavior. This is one of the main reasons that nonparametric approaches are applied widely in quantitative investing. In sum, methods that do not require us to make distributional assumptions about the data are called nonparametric methods.

We present the density plots of daily returns for two major indices in Fig. 5.4: the S&P 500 (top) and the CSI 300 (bottom). In the left panel, we also add the theoretical distribution using the same mean and standard deviation as the actual returns for each index. The right panel is the qqnorm plot, where a close-to-normal empirical distribution should produce a nearly straight line against a theoretical normal distribution. All plots indicate that the empirical distributions of actual returns for both markets are far from normal. In fact, it is difficult to characterize

[7] Although we may transform the data to make it suitable for parametric analyses.

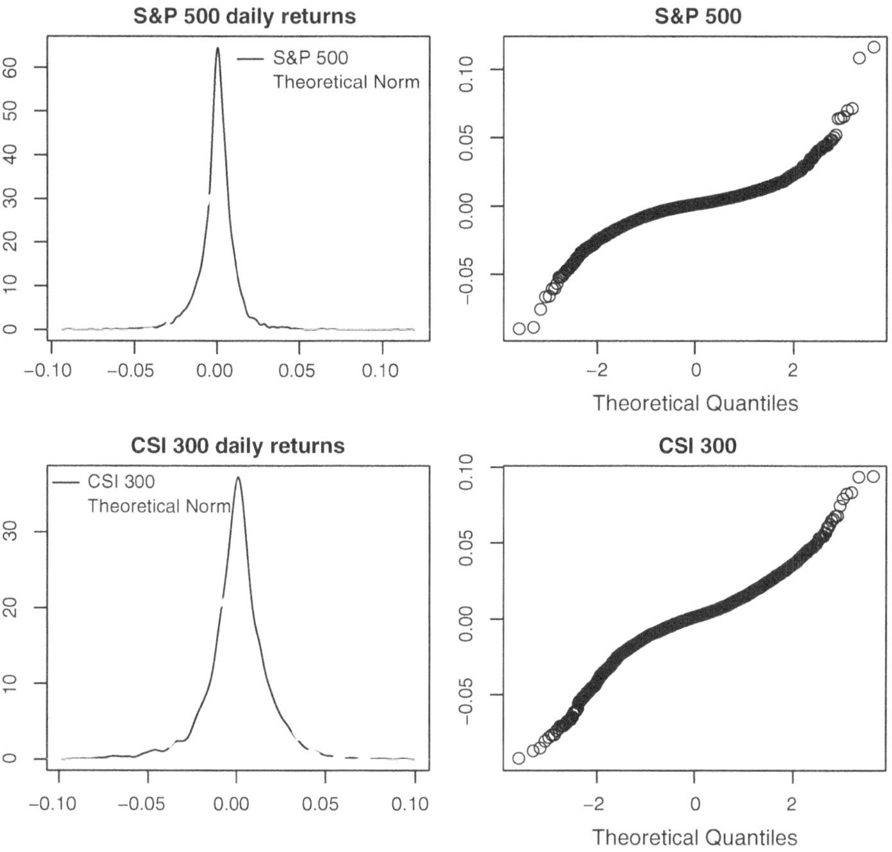

Fig. 5.4 The density and qqnorm plots for daily returns of the S&P 500 (top) and CSI 300 (bottom) 2005–2019

the distributions using just a few moments. The R codes below calculate summary statistics, moments, and the Jarque–Bera normality test for both indices. We see that the daily returns of both indices are skewed to the left with fat tails, and the test shows that both are thousands of miles from a Gaussian distribution! Thus, a nonparametric approach is necessary when we conduct quant analysis in terms of return forecasting power.

Test for Normality of Daily Returns Distribution

```
> library(moments)
> summary(sp5csi$return.sp5)
      Min.    1st Qu.     Median       Mean    3rd Qu.       Max.
-0.0903500 -0.0039000  0.0006900  0.0002901  0.0052650  0.1158000
```

```
> summary(sp5csi$return.csi)
      Min.    1st Qu.     Median       Mean    3rd Qu.       Max.
-0.0924000 -0.0072000  0.0009000  0.0005418  0.0088500  0.0934000

> skewness(sp5csi$return.sp5)
[1] -0.00557849
> skewness(sp5csi$return.csi)
[1] -0.3424697
> kurtosis(sp5csi$return.sp5)
[1] 14.79902
> kurtosis(sp5csi$return.csi)
[1] 6.643062
> jarque.test(sp5csi$return.sp5)

Jarque-Bera Normality Test

data:  sp5csi$return.sp5
JB = 19461, p-value < 2.2e-16
alternative hypothesis: greater

> jarque.test(sp5csi$return.csi)

Jarque-Bera Normality Test

data:  sp5csi$return.csi
JB = 1920.9, p-value < 2.2e-16
alternative hypothesis: greater
```

One major advantage of nonparametric approaches is that there are much less assumptions about the population than a parametric method. Many nonparametric methods are easy to understand and apply, such as rank correlation.

5.5.2 Nonparametric Methods

For almost any parametric approach, there is a corresponding nonparametric approach. The list below describes some commonly used nonparametric methods, many of which are used in quantitative investing, such as sign, count, rank, group, and contour.

1. Univariate estimation and inference
   a. Univariate kernel density estimation
   b. Goodness-of-Fit Tests: the Kolmogorov–Smirnov test
   c. Resampling and simulations for inference

   i. Bootstrap sampling and estimation
   ii. Monte Carlo simulations
   iii. Sign rank test

2. Bivariate relationship

   a. Spearman rank order
   b. Chi-squared tests for association for contingency table

3. Multi-factor model: quantile regression
4. Nonlinear models

   a. Cluster
   b. Neural network
   c. Local fitting

5. Others, for example

   a. Survival analysis: Kaplan–Meier curves
   b. Treatment effects: Nearest-neighbor matching

In quantitative investing, nonparametric approaches have been applied to for all financial markets such as asset pricing in the equity and bond markets, density estimation for the derivatives market (Fan 2005). Also very noticeable is the recent trend of artificial intelligence or machine learning. These tools depend heavily on nonparametric methods, because they do not presume conditions for finance data, which are chaotic and far from normal.

We apply some of these methods in the following sections when analyzing factor efficacy. For illustration purposes, we show two examples using simple but powerful nonparametric analysis in the context of factor effects on forward returns. We use the same data in Chap. 4, the monthly data from January 1995 to December 2004 for stocks in the Russell 1000 index. We focus on the B/P factor.

Example 1: Group Analysis Group analysis can be implemented with different number of groups depending on the number of stocks in a universe and how granular the effects we would like to see. For a stock selection strategy, group analysis is based on the rank of a factor's values (cross-sectionally) at a specified time. They are divided into a number of groups and then statistics are applied to returns (or any variable of interest) for stocks in each group. Of course, to ensure robustness, outliers and erroneous data are removed from variables.

We present a 5-group (quintile) analysis of B/P effects on one-month forward returns. For each quintile, we calculate average returns and then subtract the total average return for all the stocks in the investment universe for that period. Thus, we obtain the active average returns. The average returns are then annualized. The plots in Fig. 5.5 show annual active returns (left plot) and IRs (right plot). We see from the left plot that B/P has monotonic effects on one-month forward returns during the study period. If we buy the 200 stocks with the highest B/P scores each month, the active return is about 2% on an annual basis. We would obtain a similar return by shorting the 200 stocks with the lowest B/P scores each month. However, when

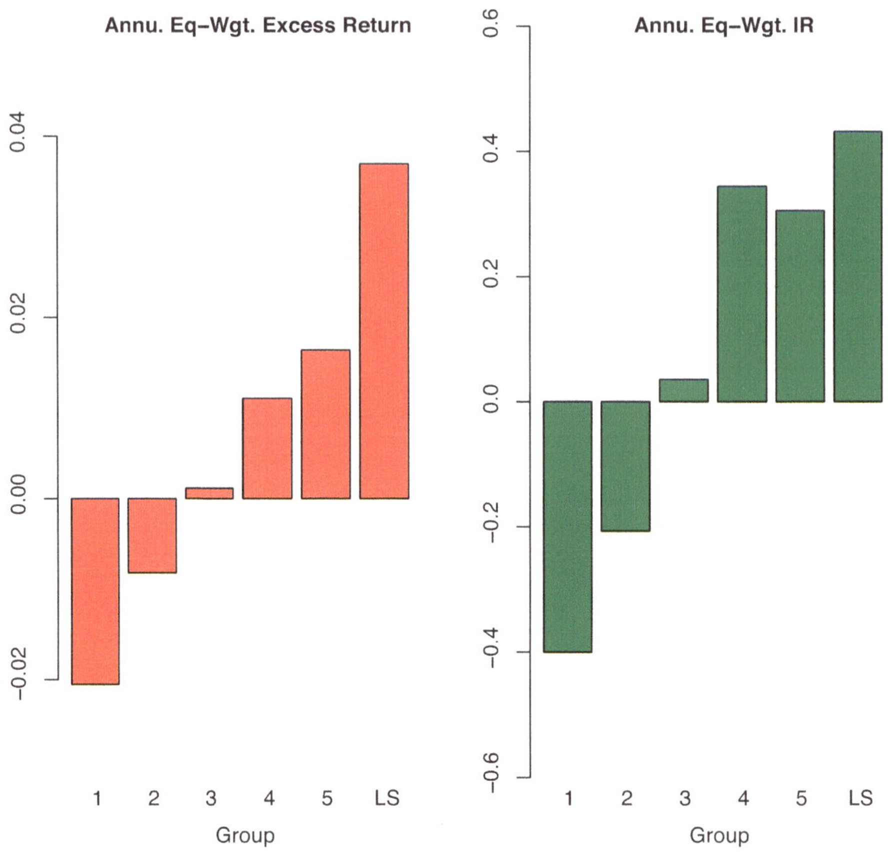

Fig. 5.5 The annual active returns (left plot) and IRs (right plot) by quintile of B/P for the data from January 1995 to December 2004

we add risk levels, the monotonicity changes: the 4th and 5th groups have about the same IRs, while the 1st group has a much higher IR (absolute value) than the 2nd group, indicating that the cheapest stocks have higher than average return volatilities.

In addition to the overall effects of B/P across all stocks, we investigate the effects of B/P on one-month forward returns across different sectors (GICS). The plot in Fig. 5.6 shows the annualized returns of a long-short (Q5–Q1) strategy for each industry. We see that B/P has very different effects across sectors, being most effective in telecommunication services and materials sectors, and most negative in utilities and IT sectors. This implies that we should perhaps treat B/P differently for each sector rather than using a one-size-fits-all approach.

Example 2: Nonlinear Effects Continuing with the same factor B/P, we now expand the group from quintile (5) to ventile (20). We are interested to see the monotonicity, in particular the tail behaviors—how deep value (cheapest) stocks perform. Each of

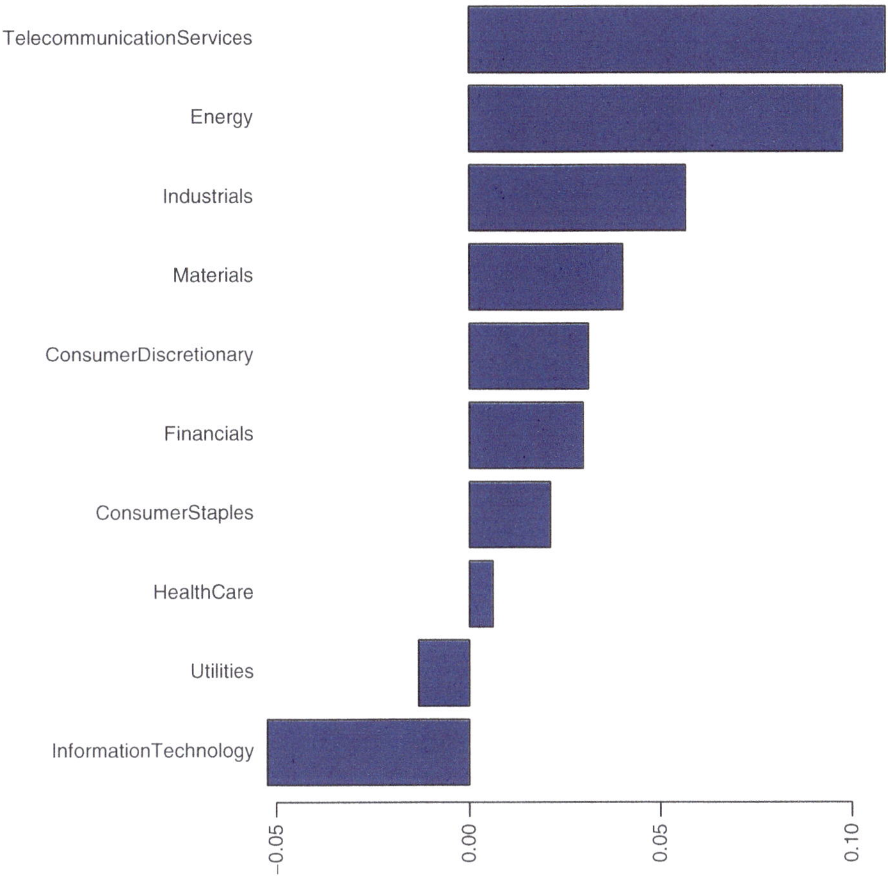

Fig. 5.6 The annual active returns of B/P by industry, January 1995–December 2004

the 20 groups has about 50 stocks so that the summary statistics are based on decent amount of data.[8] We find (Fig. 5.7) that while in general there is a roughly monotonic relationship between the 2nd and 18th groups of B/P values and stock returns during the study period, such a relationship does not hold at the tails, especially at the right tail. The deep value stocks are cheap—the last two groups have the lowest returns among all 20 groups. There is dramatic nonlinearity in the effects of B/P on forward returns.

Nonlinearity can cause serious issues for investments. For example, for a long-only portfolio, if we pick stocks in the 20th group, the portfolio will underperform;

[8]We also performed a decile analysis to check robustness and found similar results as for the 20 groups.

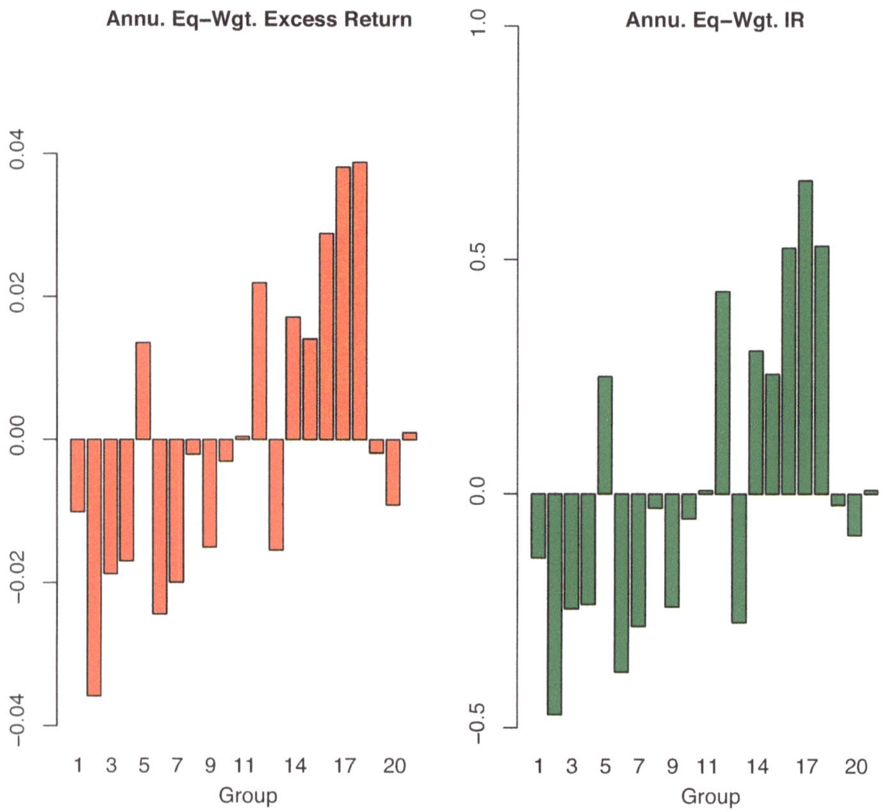

Fig. 5.7 The annual active returns (left plot) and IRs (right plot) by ventile of B/P for the data from January 1995 to December 2004

and for a long-short portfolio, if we long the stocks in the 20th group and short stocks in the 1st group, the portfolio will underperform as well.

The deep value stocks are *cheap for a reason*. For those companies with prices dramatically below the accounting value relative to their peers in the same industry, they may be in financial trouble or have operating issues, or their products may not be competitive in the market, all of which will usually be reflected in their revenue in the first place. For illustration purposes, we use the revenue increase (the change of sales/assets ratio) as a proxy for the cause of nonlinearity. We use a nonparametric approach of kernel density estimation to investigate the interactive effects of this proxy factor and the B/P factor on one-month forward returns. We find that for deep value stocks, it is the companies with decreasing revenue that are associated with low (negative) returns, while the ones with revenue increases have the highest returns. The 3-D plot in Fig. 5.8 shows the nonlinear interactive effects vividly, with both the surface grids and contours.

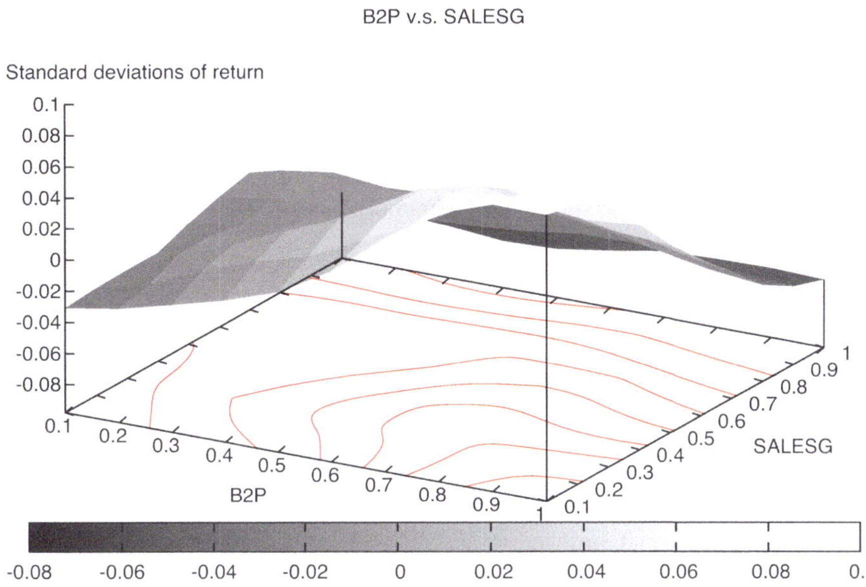

Fig. 5.8 The nonparametric interaction effects of B/P and SALESG (1- sales growth) on forward returns of stocks in the Russell 1000, January 1995 to December 2004

5.6 Industry Insights: A Factor Diagnostics Package

So far in this chapter, we have discussed alpha hunting, risk adjustment, and nonparametric approaches. In this section, we show how industry professionals use nonparametric approaches to analyze a factor—a method called a factor diagnostics package. The diagnostics should cover IPRAE characteristics, as mentioned in the beginning of this chapter, and many other related aspects to assist professionals in making decisions about how to use that factor in an investment. We first present a framework and then give an example of factor diagnostics.

5.6.1 How to Explore the Efficacy of a New Factor

In the industry, there is a general approach to factor diagnostics. It follows the IPRAE guidelines and often includes, but is not limited to, the following parts: factor distribution, forecasting power on returns, risk profiles, information decay, marginal value added, and robustness. We list each part with decomposed items below.

1. Factor distribution
    a. missing values

   b. extreme values
   c. factor distribution
   d. features by industry and country (if global)

2. Forecasting power on returns

   a. factor portfolio and specific risk-adjusted factor portfolio
   b. distribution, long and short, investment horizons
   c. cap and equal weights
   d. industry effects (and country effects if global)
   e. effects across business cycles

3. Risk profiles

   a. correlation with risk factors over time
   b. factor portfolio and factor mimic portfolio exposure to risk factors
   c. sensitivity to macro-level risk variables
   d. volatility of effects

4. Information decay and turnover

   a. information decay of a factor
   b. information decay of factor efficacy
   c. factor portfolio turnover

5. Marginal value and robustness

   a. residual effects
   b. robustness: factor definition change, factor treatment
   c. performance robustness over time and across regimes

Of course, this is just a general framework. Many other characteristics should be added for specific purposes. For example, when we investigate nonlinear effects of a factor, we usually explore the interaction effects with a different factor on forward returns in order to find a solution to treat the nonlinearity.

In the following section, we give an example to show how a factor (which can be a signal, theme, or alpha) is diagnosed by industry professionals.

5.6.2 Factor Diagnostics: An Example

Following the guidelines above, we perform diagnostics for the alpha we built in Chap. 4. The plots for diagnostic results are presented at the end of this chapter. Here we provide brief comments about those plots. Note that the diagnostics for a factor can be simply a brief summary with a few pages or go to very granular levels with hundreds of pages of tables and plots.

A factor diagnostics package usually starts with a title or front page. The title page is very important. It summarizes the information about the factor, investment universe, industry classification, return variable, risk model, data treatment rules, study period, running date/time, and sometimes the person who runs the diagnostics.

Following the title page are the details about diagnostics, for instance, items 2–8 in the list below.

1. Title page
2. Factor: missing, distribution by industry, distribution over time
3. Executable, turnover
4. Predictive: overall, each year, finger plot
5. Robustness, lags of alpha
6. Added value compared with risk factors
7. More on risk exposure
8. Nonlinear effects

The results above help us understand the factor and prepare us for the next stage: whether and how to employ it in a portfolio. For example, the monotonicity and decay information will be considered for the investment horizon of a strategy, and turnover will be taken into account for portfolio construction when we set up the turnover constraint. We discuss portfolio construction in detail in Chap. 7.

An important aspect of diagnostics is that this can become a non-trivial project. Therefore, it requires careful planning and consideration. While we have focused on the IPRAE criteria for a specific factor, other aspects are also very important:

- Data
  - factor components, return data, risk models, raw and treated factor values
- R codes
  - utility functions, structure, run function
- Folder
  - log file, inputs, outputs, results, analysis

We discuss these items in the following chapters, especially Chap. 7.

Based on the diagnostic results, it seems that the alpha we built in Chap. 4 performs well for stock selection purposes. However, we have to remember that our entire construction process for alpha is based on in-sample studies, and the alpha is tested using the same in-sample data. Therefore, we should be very cautious on how the results will repeat in the future.

5.7 R: Refining Plots and Using Parameters

In this section, we demonstrate how to refine plots and use parameters.

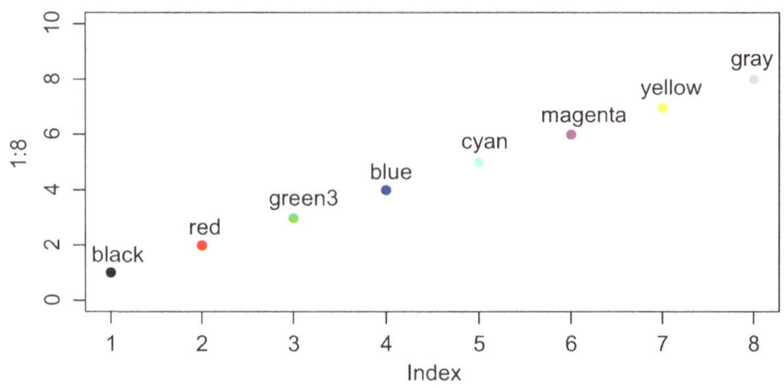

Fig. 5.9 Plot of the default palette in R

5.7.1 *Plot Refinements*

In the R language, plot is a powerful function. We can make many refinements to plots to serve different purposes and requirements of reports and presentations. We list some widely used refinement commands below and present examples using index data for the USA and Chinese stock markets.

- Add color: *col="red"*
- Title: *main ="the title of the plot"*
- Label: *xlab, ylab*
- Add lines/points: *line, points*
- Add new plots: *plot(new=T)*
- Add legends: *legend()*
- Add dates: *xaxt="n"*

Add Color to R Plots R has 657 built-in color names. One can use the command *colors()* to see the list of names. We can also use an integer to specify a color. The default palette contains 8 colors. With a plot of numbers from 1 to 8, the color names are shown as labels in Fig. 5.9.

Plot refinements

```
> palette()
[1] "black"   "red"    "green3"   "blue"    "cyan"    "magenta" "yellow"   "gray"
> plot(1:8,col=palette(),pch=19,ylim=c(0,10))
> text(1:8,c(1:8)+0.5,labels=palette(),cex=0.7)
```

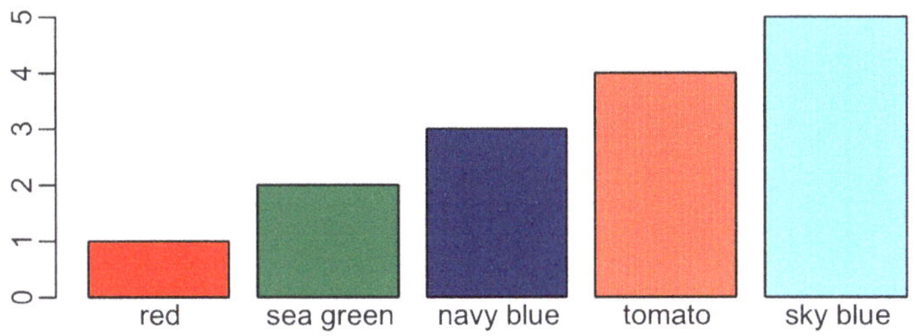

Fig. 5.10 An example of self-defined colors

Fig. 5.11 An example of colors selected from the package *RColorBrewer*

We can also provide our own vector of colors. We present the example R scripts below and the resulting plot in Fig. 5.10.

Plot refinements

```
> my.colors=c("red","sea green","navy blue","tomato","skyblue")
> barplot(1:5,col=my.colors,names.arg=my.colors)
```

There is a very useful package on colors, *RColorBrewer*, designed by Cynthia Brewer, a color specialist (Brewer 1996, 2003). We show an example in Fig. 5.11.

Plot refinements

```
> library(RColorBrewer)
> display.brewer.all(name="qual")
```

For color-blind readers, there is a very helpful R package, *dichromat*. The package specifies 17 color schemes suitable for people with deficient or anomalous red-green vision.

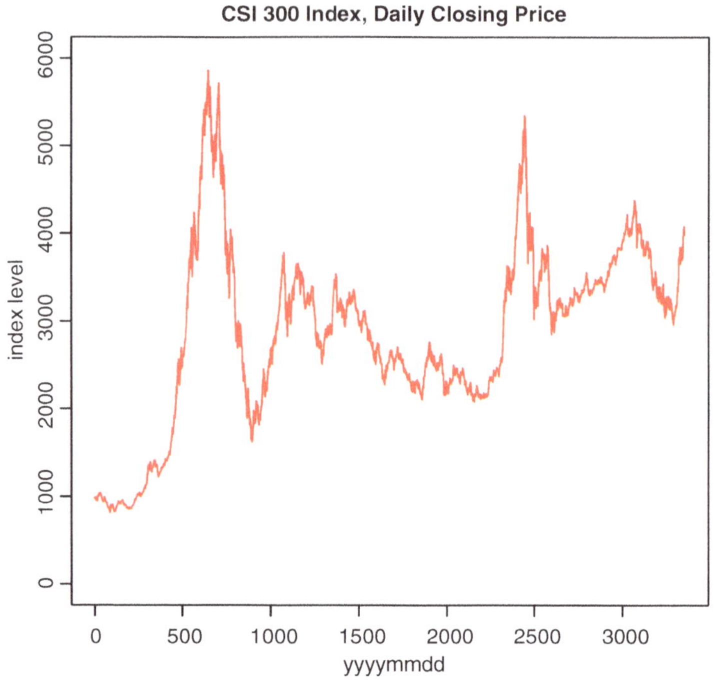

Fig. 5.12 R plot refinement: add title, axis label, and limit

Add Title, Axis Limits, and Labels The commands in R to add a title, axis label, and limit are *main, xlab*, and *xlim*, respectively. We show how to refine the daily price plot for the CSI 300 index from 2005 to 2009 with these commands (Fig. 5.12).

Plot refinements: add title, axis label and limit

```
plot(sp5csi$price.csi,col="tomato",type="l",main="CSI 300 Index, Daily
 Closing Price",xlab="yyyymmdd",ylab="index level",ylim=c(0,6000))
```

Add New Line, Point, and Legend Besides refining an existing plot, we can overlay the current plot with points or lines and add a legend. The R commands to add points and a line are *points, abline*. The command to add a legend to the existing plot is *legend*, with several options. Examples are shown in Fig. 5.13.

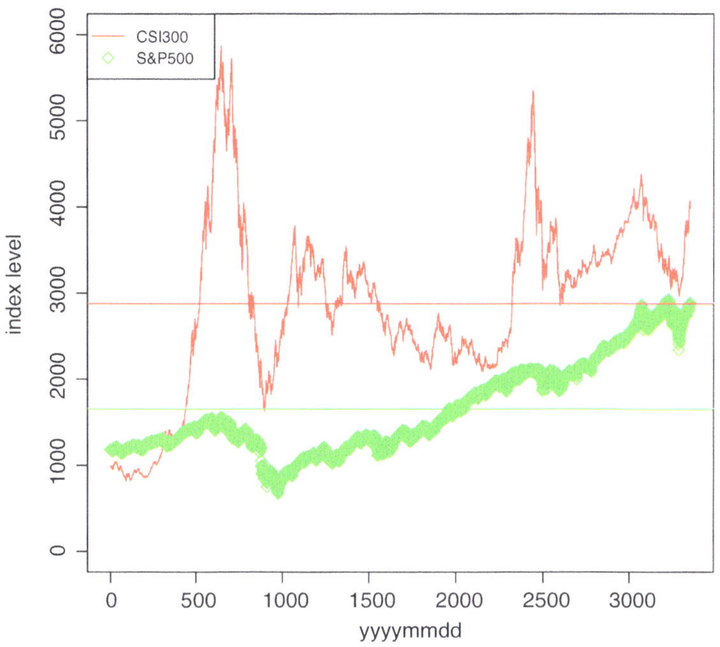

Fig. 5.13 R plot refinement: add line, point, and legend

Plot refinements: add new line, points and legend

```
plot(sp5csi$price.csi,col="tomato",type="l", xlab="yyyymmdd",
       ylab="index level",ylim=c(0,6000))
points(sp5csi$close.sp5, pch=5, col="green")
abline(h=mean(sp5csi$price.csi),col="tomato")
abline(h=mean(sp5csi$close.sp5),col="green")
legend("topleft",legend=c("CSI300","S&P500"), col=c("tomato","green"),
      lty = c(1, -1), pch = c(NA, 5))
```

Add Dates Adding dates is not straightforward. It takes two steps: first, create a plot without the horizontal axis ticks using the command *xaxt="n"*, then add a new axis with positions and ticks. In the example below, we add to the plot the last business day for calendar years in the data. The generated plot is shown in Fig. 5.14.

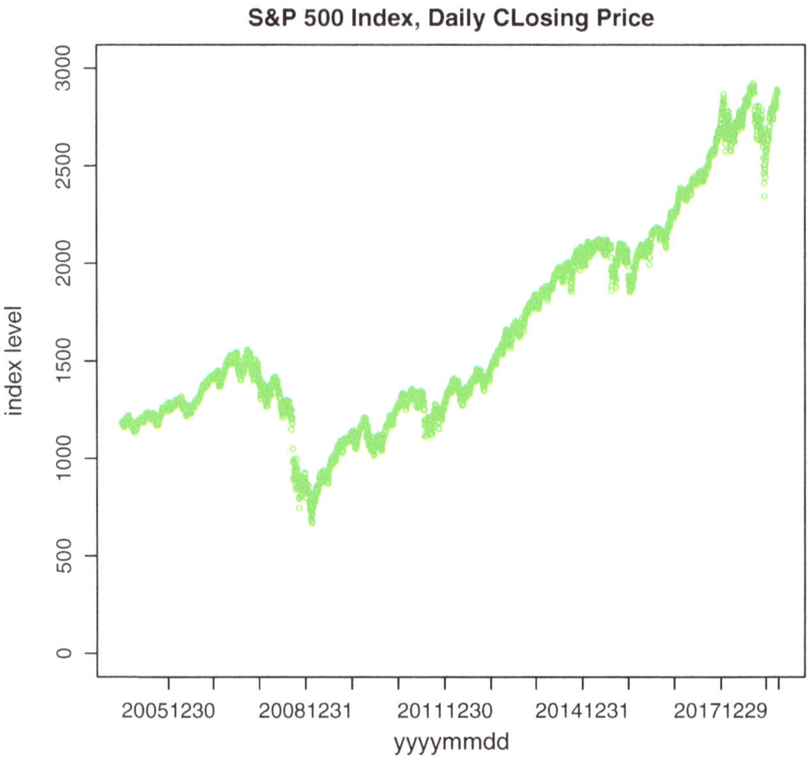

Fig. 5.14 R plot refinement: add dates

Plot refinements: add dates

```
plot(sp5csi$close.sp5,col="green",xaxt="n", main="S&P 500 Index,
 Daily CLosing Price", xlab="yyyymmdd",ylab="index level",ylim=c(0,3000))
sp5csi$yyyy=as.numeric(substr(sp5csi$yyyymmdd,1,4))
year.end=as.vector(tapply(xx$yyyymmdd,xx$yyyy,max))
axis.pos=which(sp5csi$yyyymmdd %in% year.end)
axis(1,axis.pos,year.end)

> xx=sp5csi
> xx$yyyy=as.numeric(substr(xx$yyyymmdd,1,4))
> tapply(xx$yyyymmdd,xx$yyyy,max)
    2005     2006     2007     2008     2009     2010      2011      2012
20051230 20061229 20071228 20081231 20091231 20101231  20111230  20121231
2013     2014     2015     2016     2017     2018         2019
20131231 20141231 20151231 20161230 20171229 20181228 20190411
```

We have introduced 2-D plots. Sometimes, it is convenient to visualize analytical results using 3-D plots, such as the interactional effects of B/P and SALESg on returns. In R, *persp* and *scatter3d* provide advanced plots, such as contour and 3-D plots, which are useful tools for visualizing data and results.

5.7.2 Learning to Use Parameters

We can customize many features of graphs (fonts, colors, axes, titles) by setting graphic parameters. In R programming, there is an important parameter setup command called *par*. Many options we mentioned earlier, such as fonts, colors, and axes, can all be changed through the par( ) function. Note that if you set parameter values here, the changes will be in effect for the rest of the session or until you change them again.

Plot parameters

```
par()               # view current settings
opar <- par()       # copy current settings
hist(mtcars$mpg)    # create a plot with new settings
par(opar)           # restore the original settings
```

Here we list some commonly used parameters for basic plots. They are all used in the plots for a diagnostic package illustrated at the end of this chapter.

- Multiple plots: *mfrow*
- Outer margin space: *oma*
- Inside margin space: *mar*
- Text size: *cex*

When a graph has multiple plots, arranging these plots with proper spacing is important for a nice layout. The R parameter *mfrow* enables you to arrange plots by setting the number of rows and columns. Regarding spacing, there are two margin areas in R plots: the inside and outside margins. One can control their size (number of lines) by calling the par function before your plot and giving the corresponding arguments: *mar()* for inside margin and *oma()* for outer margin area. For both arguments, you must enter four values to set the desired space at the bottom, left, top, and right part of the chart. For example, *par(mar=c(4,2,2,2))* draws a margin of four lines on the bottom and two on the other sides of the chart. One can also use *mai()* and *omi()* to set the areas in inches instead of lines. Other parameters include the control for the font and line types, etc.

The codes below use all the parameters listed above in four plots.

Plot parameters: a 2 x 2 graph

```
par(mfrow=c(2,2), mar=c(3,2,2,2),oma=c(3,2,2,2),mgp=c(1,1,0))
par(cex=0.8) # add four plots
```

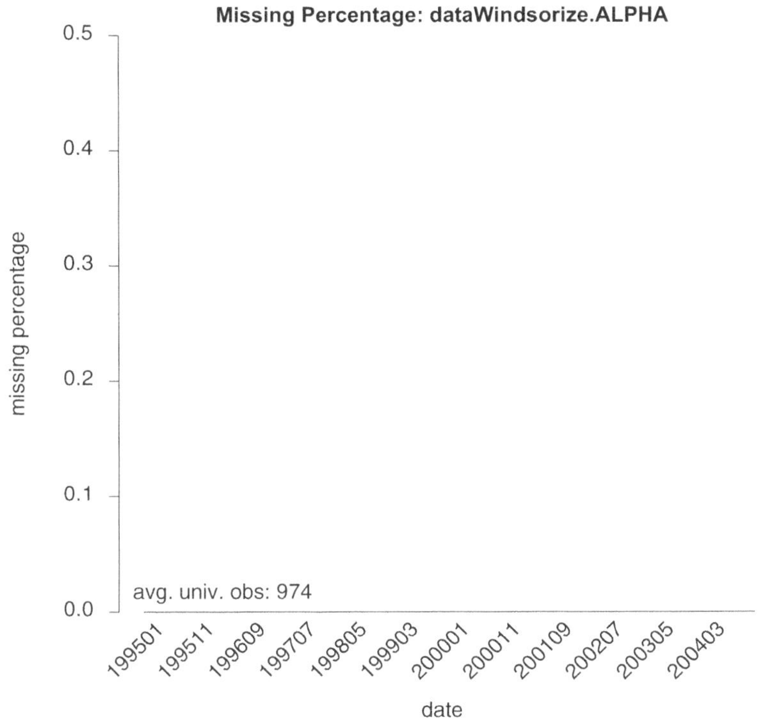
Missing Percentage: dataWindsorize.ALPHA
missing percentage
0.5
0.4
0.3
0.2
0.1
0.0
avg. univ. obs: 974
199501
199511
199609
199707
199805
199903
200001
200011
200109
200207
200305
200403
date

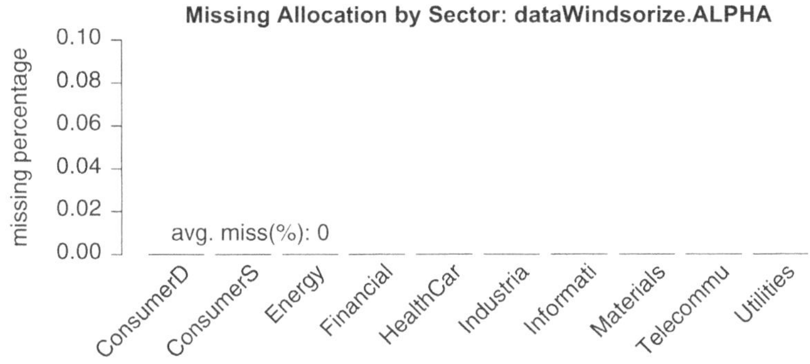
Missing Allocation by Sector: dataWindsorize.ALPHA
missing percentage
0.10
0.08
0.06
0.04
0.02
0.00
avg. miss(%): 0
ConsumerD
ConsumerS
Energy
Financial
HealthCar
Industria
Informati
Materials
Telecommu
Utilities

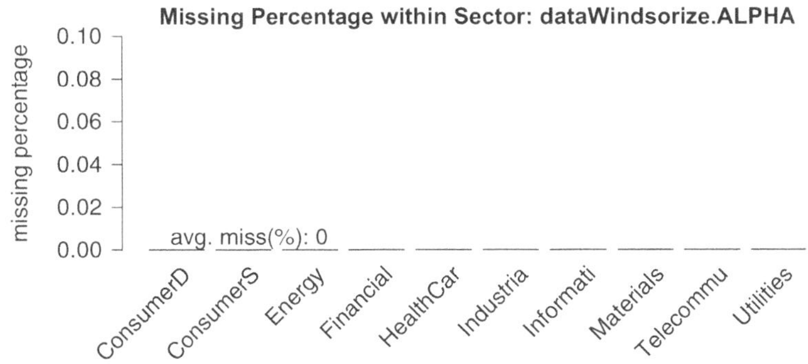
Missing Percentage within Sector: dataWindsorize.ALPHA
missing percentage
0.10
0.08
0.06
0.04
0.02
0.00
avg. miss(%): 0
ConsumerD
ConsumerS
Energy
Financial
HealthCar
Industria
Informati
Materials
Telecommu
Utilities

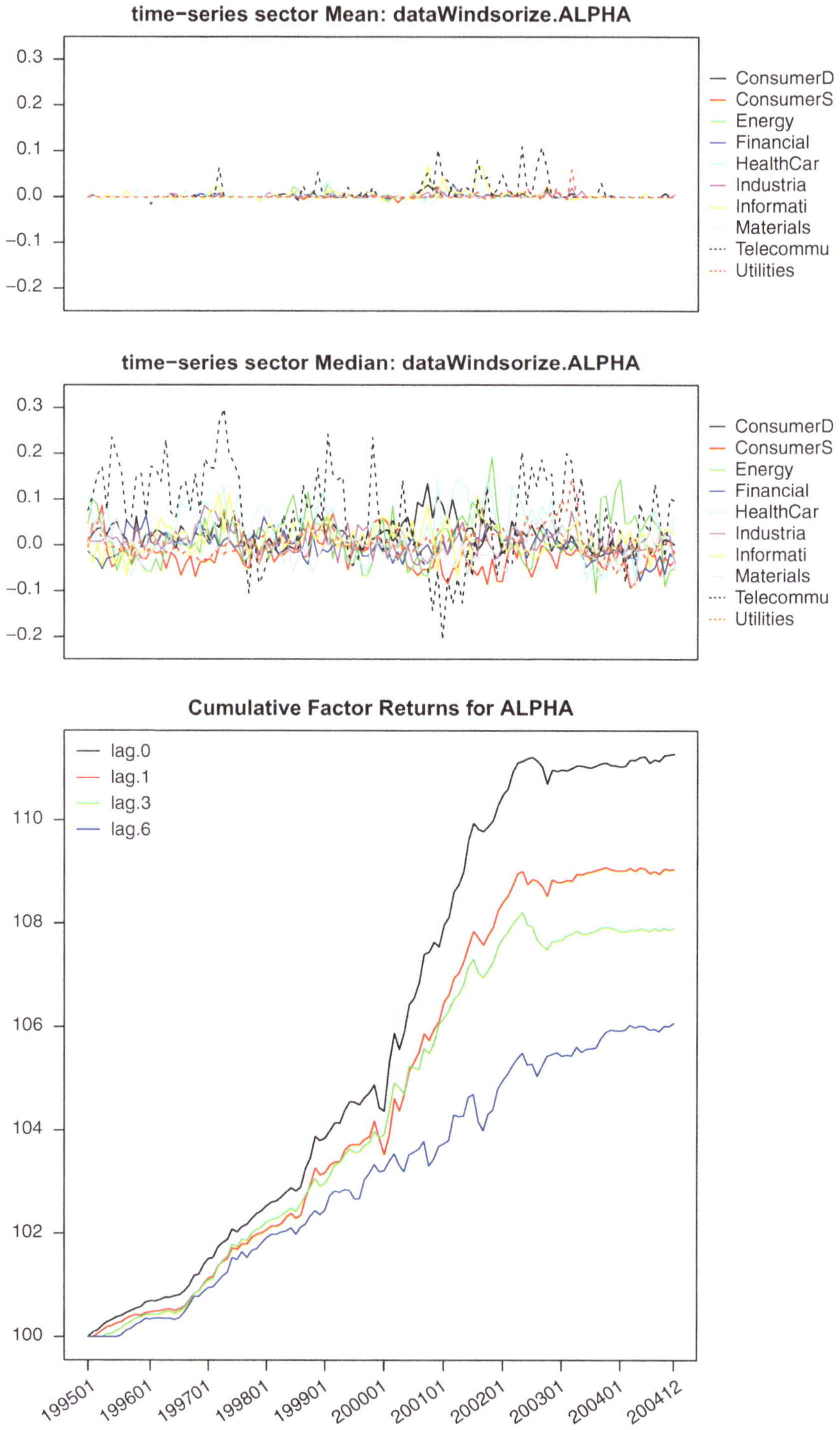
time−series sector Mean: dataWindsorize.ALPHA
0.3
0.2
0.1
0.0
−0.1
−0.2
ConsumerD
ConsumerS
Energy
Financial
HealthCar
Industria
Informati
Materials
Telecommu
Utilities
time−series sector Median: dataWindsorize.ALPHA
0.3
0.2
0.1
0.0
−0.1
−0.2
ConsumerD
ConsumerS
Energy
Financial
HealthCar
Industria
Informati
Materials
Telecommu
Utilities
Cumulative Factor Returns for ALPHA
lag.0
lag.1
lag.3
lag.6
110
108
106
104
102
100
199501
199601
199701
199801
199901
200001
200101
200201
200301
200401
200412

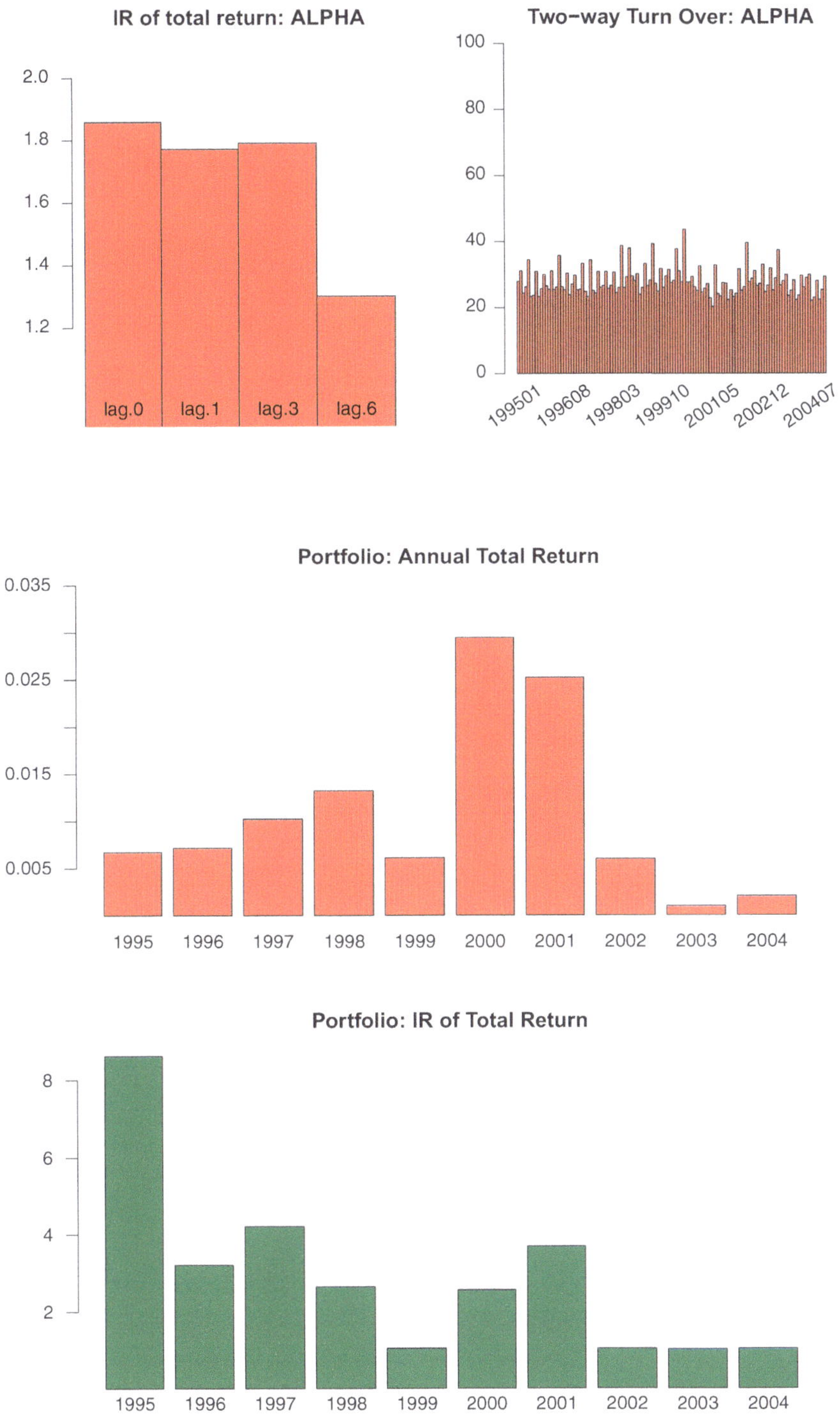
IR of total return: ALPHA
2.0
1.8
1.6
1.4
1.2
lag.0
lag.1
lag.3
lag.6
Two−way Turn Over: ALPHA
100
80
60
40
20
0
199501
199608
199803
199910
200105
200212
200407
Portfolio: Annual Total Return
0.035
0.025
0.015
0.005
1995
1996
1997
1998
1999
2000
2001
2002
2003
2004
Portfolio: IR of Total Return
8
6
4
2
1995
1996
1997
1998
1999
2000
2001
2002
2003
2004

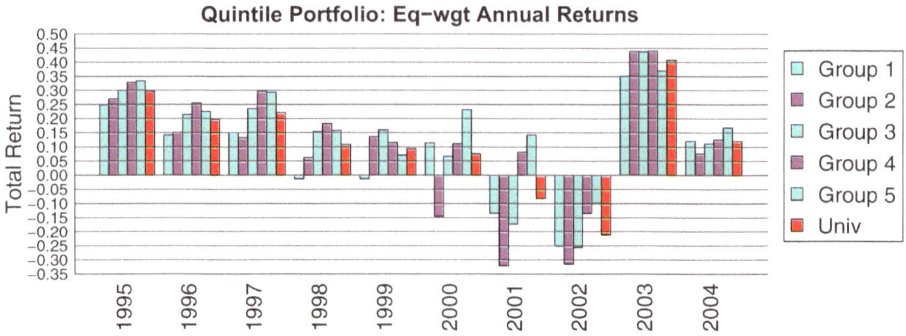
Quintile Portfolio: Eq−wgt Annual Returns
Total Return
0.50
0.45
0.40
0.35
0.30
0.25
0.20
0.15
0.10
0.05
0.00
−0.05
−0.10
−0.15
−0.20
−0.25
−0.30
−0.35
1995
1996
1997
1998
1999
2000
2001
2002
2003
2004
Group 1
Group 2
Group 3
Group 4
Group 5
Univ

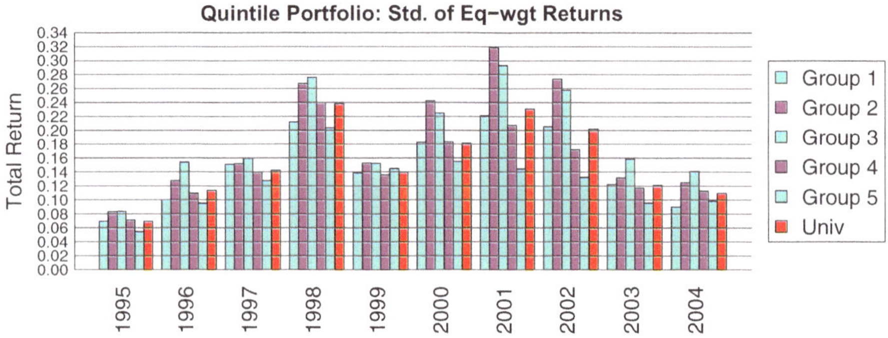
Quintile Portfolio: Std. of Eq−wgt Returns
Total Return
0.34
0.32
0.30
0.28
0.26
0.24
0.22
0.20
0.18
0.16
0.14
0.12
0.10
0.08
0.06
0.04
0.02
0.00
1995
1996
1997
1998
1999
2000
2001
2002
2003
2004
Group 1
Group 2
Group 3
Group 4
Group 5
Univ

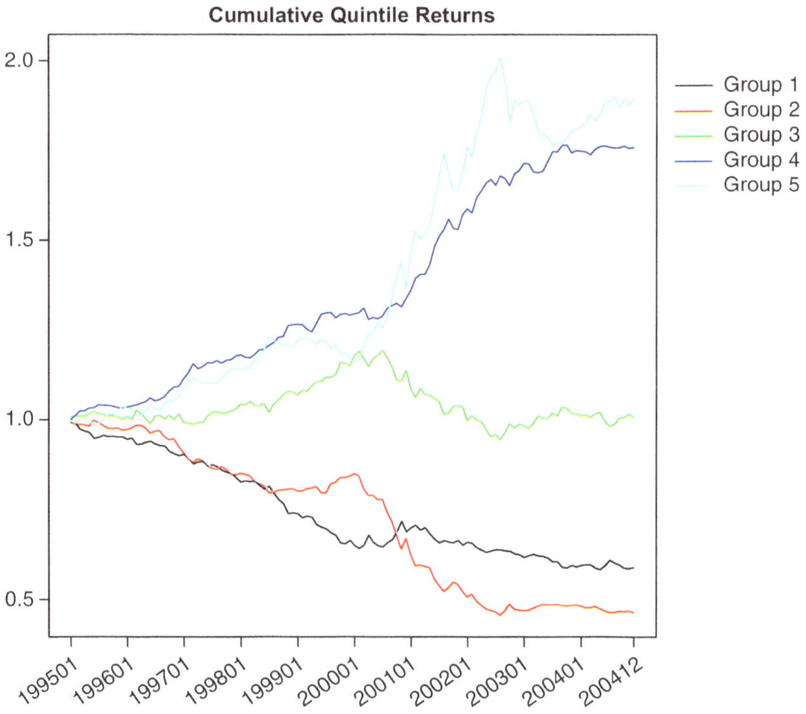
Cumulative Quintile Returns
2.0
1.5
1.0
0.5
199501
199601
199701
199801
199901
200001
200101
200201
200301
200401
200412
Group 1
Group 2
Group 3
Group 4
Group 5

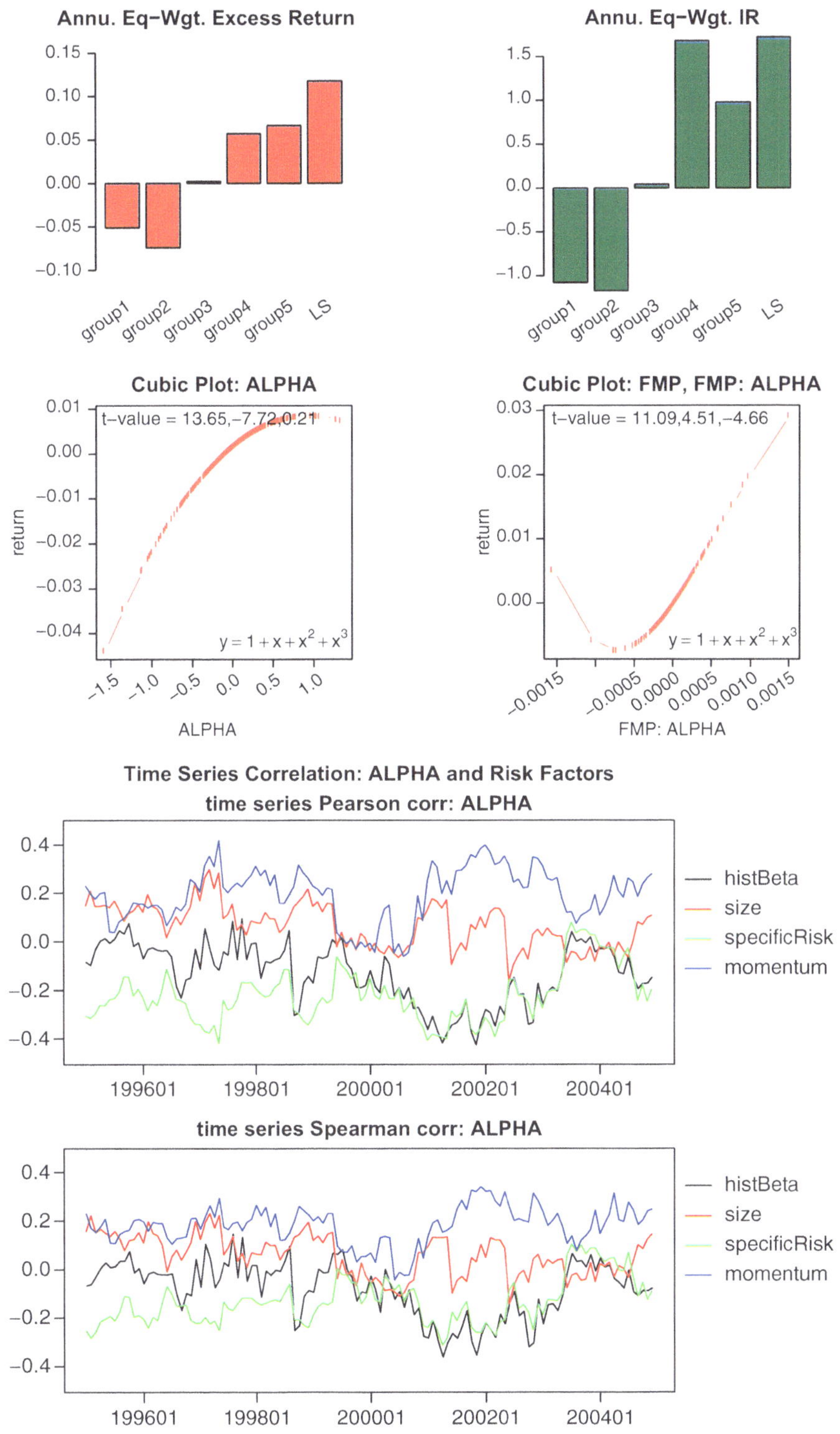
Annu. Eq−Wgt. Excess Return
0.15
0.10
0.05
0.00
−0.05
−0.10
group1
group2
group3
group4
group5
LS
Annu. Eq−Wgt. IR
1.5
1.0
0.5
0.0
−0.5
−1.0
Cubic Plot: ALPHA
t−value = 13.65,−7.72,0.21
return
y = 1 + x + x^2 + x^3
ALPHA
Cubic Plot: FMP, FMP: ALPHA
t−value = 11.09,4.51,−4.66
FMP: ALPHA
Time Series Correlation: ALPHA and Risk Factors
time series Pearson corr: ALPHA
histBeta
size
specificRisk
momentum
199601
199801
200001
200201
200401
time series Spearman corr: ALPHA

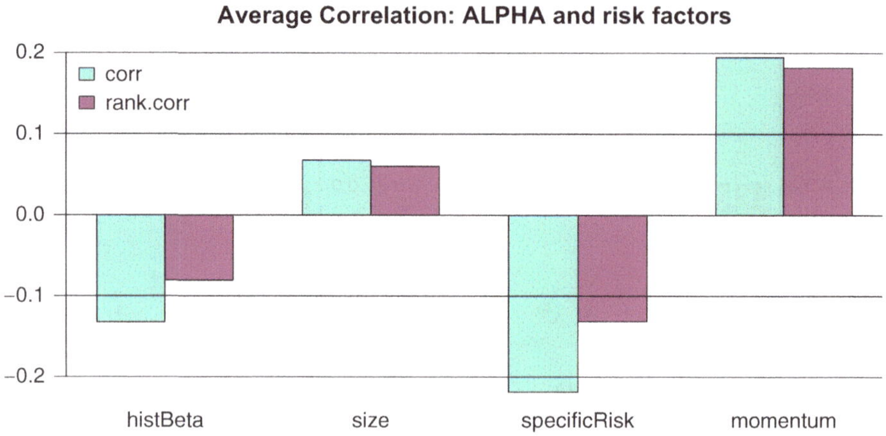
Average Correlation: ALPHA and risk factors
0.2
0.1
0.0
−0.1
−0.2
corr
rank.corr
histBeta
size
specificRisk
momentum

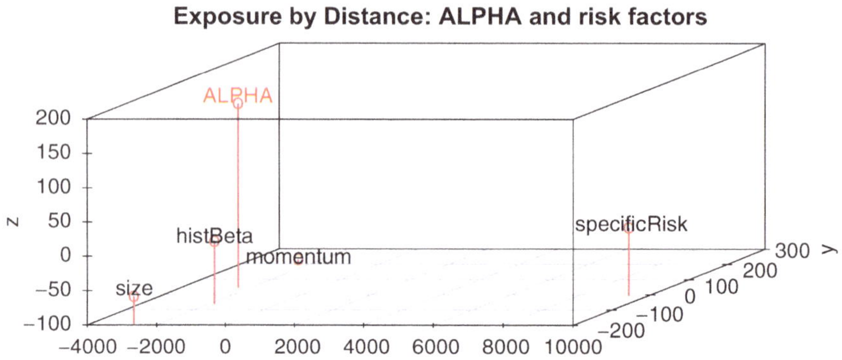
Exposure by Distance: ALPHA and risk factors
ALPHA
histBeta
momentum
size
specificRisk
z
200
150
100
50
0
−50
−100
−4000 −2000 0 2000 4000 6000 8000 10000
y
300
200
100
0
−100
−200

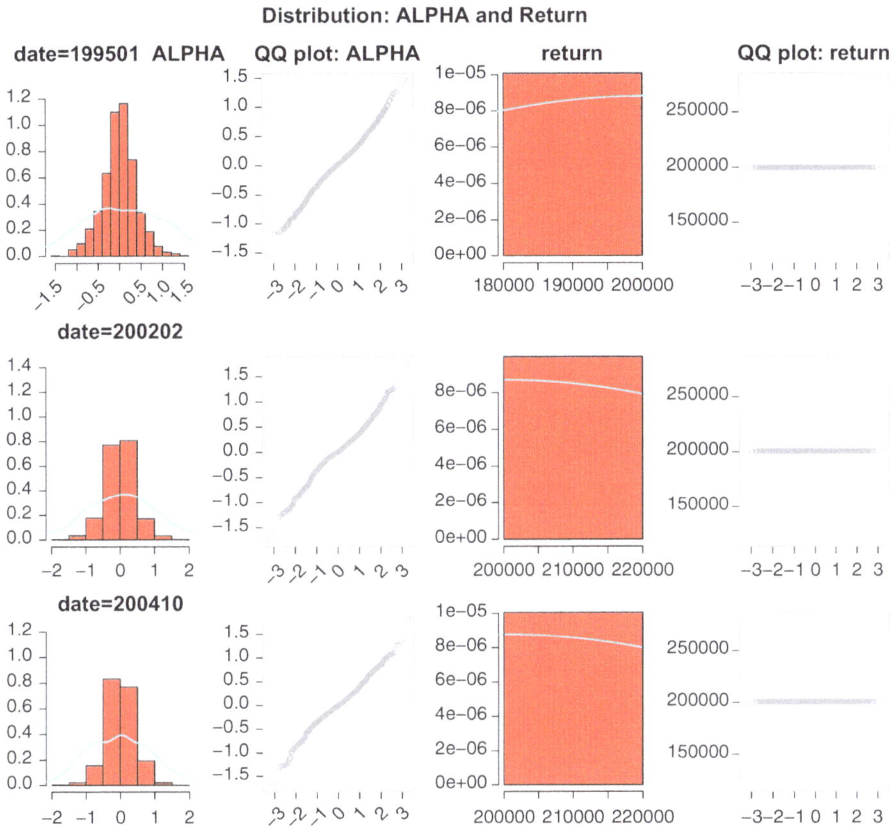

Keywords, Problems, and Group Project

Part I. Keywords

Weighted least squares, endogeneity, exogeneity, heteroscedasticity

Specific risk, risk-adjusted return, APT

IPRAE, decile analysis, factor diagnostics

R plot refinements and parameters

Part II. Problems

Problem 5.1 For the data of stocks from Chap. 4, conduct decile or ventile analysis.

(1) Signals, themes, and alpha
(2) Investigate nonlinear effects of signals on returns, are there any reasons for nonlinearity?

Problem 5.2 Using the same data as you collected from Chap. 4, now build alpha values from themes using WLS.

(1) Run WLS for the multi-factor model.

$$R_F = b_0 + b_1 PROF + b_2 EQ + b_3 VALUE + b_4 PM + b_5 MQ + b_6 MS + \epsilon,$$

where R_F is forward returns of T-month. Run OLS and estimate σ_ϵ for each stock, use $\frac{1}{\hat{\sigma}_i}$ as weight, apply WLS to the model.

(2) Build alpha with weights derived from t-values or coefficients.
(3) Conduct univariate analysis of alpha, correlation between alpha and forward returns, and OLS of returns on alpha. Evaluate the forecasting power of alpha scores.
(4) Plot distribution of return values and alpha scores. Do they follow a Gaussian (normal) distribution?

Problem 5.3 With both $ALPHA_{ols}$ and $ALPHA_{wls}$

(1) For each set of alpha, construct a long-only portfolio with stocks in the top decile, compare the performances of two portfolios.
(2) For each set of alpha, construct a long-short portfolio with stocks in the top decile in the long positions and stocks in the bottom deciles in the short positions, compare the performances of two portfolios.

Problem 5.4 Conduct decile analysis for both VALUE and PROF.

(1) Conduct double sorts of VALUE and PROF.
(2) Investigate interaction effects of VALUE and PROF on forward returns. Identify nonlinear effects.

Part III. Group Project

Problem 5.5 For signals built in Chap. 4, work with a team of 3–5 people to build a factor diagnostic package.

(1) Discuss IPRAE guideline and specify functions for each part of IPRAE.
(2) Build a diagnostic package with the R functions.
(3) Apply the package to a factor (signals, themes, $ALPHA_{ols}$, and $ALPHA_{wls}$).

References

Allison, D. 1999. "Comparing Logit and Probit Coefficients Across Groups." *Sociological Methods and Research* 28: 186–208.

Breusch, T.S., and A.R. Pagan. 1979. "A Simple Test for Heteroscedasticity and Random Coefficient Variation." *Econometrica* 47: 1287–1294.

Brewer, C.A. 1996. "Guidelines for Selecting Colors for Diverging Schemes on Maps." *The Cartographic Journal* 33(2): 79–86.

Brewer, C.A. 2003. "A Transition in Improving Maps: The ColorBrewer Example." *Cartography and Geographic Information Science* 30: 159–162.
Chen, N., R. Roll, and S. Ross. 1986. "Economic Forces and the Stock Market." *Journal of Business* 59(3): 383–403.
Durbin, J. 1954. "Errors in Variables." *Review of the International Statistical Institute* 22: 23–32.
Fan, J. 2005. "A Selective Overview of Nonparametric Methods in Financial Econometrics." Working paper.
Hausman, J. 1978. "Specification Tests in Econometrics." *Econometrica* 46(6): 251–1271.
Heckman, J. 1976. "The Common Structure of Statistical Models of Truncation, Sample Selection and Limited Dependent Variables and a Simple Estimator for Such Models." *Annals of Economic and Social Measurement* 5(4): 475–492.
Heckman, J. 1979. "Sample Selection Bias as a Specification Error." *Econometrica* 47(1): 153–161.
Roll, R., and S. Ross. 1980. "An Empirical Investigation of the Arbitrage Pricing Theory." *Journal of Finance* 35(5): 1073–1103.
Rosenberg, B. 1974. "Extra-Market Components of Covariance in Security Returns." *The Journal of Financial and Quantitative Analysis* 9(2), 263–274.
Ross, S. 1976. "The Arbitrage Theory of Capital Asset Pricing." *Journal of Economic Theory* 13(3): 341–360.
Stock, J., and F. Trebbis. 2003. "Retrospectives, Who Invented Instrumental Variable Regression?" *Journal of Economic Perspectives* 17(3): 177–194.
White, H. 1980. A Heteroskedasticity-consistent covariance matrix estimator and a direct test for Heteroskedasticity. *Econometrica* 48(4): 817–838.
Wright, P.G. 1928. *The Tariff on Animal and Vegetable Oils*. New York: Macmillan.
Wu, D. 1973. "Alternative Tests of Independence Between Stochastic Regressors and Disturbances." *Econometrica* 41(4): 733–750.

Chapter 6
How to Forecast Commodity Price Movements: Time Series Models

Abstract In this chapter, we focus on commodity pricing and investment with time series models. For stock selection strategies, the ability to forecast individual stock returns is critical and usually relies on company-level factors, such as profitability, management quality, etc. In commodity investing, a deep understanding of the geopolitical dynamics and identification of macro-level factors are very important for a successful strategy. A stock selection strategy requires cross-sectional analysis, while commodity investing requires time series analysis. In this chapter, we get into details about the special features of time series models, introduce the concepts of unit root, spurious relationship, and cointegration, and show how they can be employed for quantitative investing in crude oil and pair trading.

6.1 Time Series Data: Three Examples and Common Features

In the previous chapters, we focused on stock selection strategy, that is, how to pick the best stocks from an investment universe to form a portfolio with the purpose of outperforming the market. Hence, cross-sectional analysis is appropriate for this type of investment. Now, we investigate quantitative investing in commodities, where we deal with only one asset, the commodity, such as crude oil. When we forecast the price of a commodity, the historical price of that commodity and factors that impact the price become critical. Commodity pricing factors are usually at the macro level, such as extreme events (e.g., war), supply, and demand. The appropriate model to analyze this type of information is not cross-sectional but rather a time series. In this section, we start with three examples—Chicago daily temperatures, the price of oil, and special items in companies' financial statements to introduce common features of time series data.

L. Ma, *Quantitative Investing*, https://doi.org/10.1007/978-3-030-47202-3_6

6.1.1 Three Examples: Chicago Daily Temperatures, the Price of Oil, and Special Items

We present three examples of time series data in this section: daily temperatures in Chicago, the price of oil, and special items reported by public companies. Through these examples, we illustrate different types of times series data. Moreover, for each example, we show the values of time series data from adjacent periods to challenge readers to consider the reliability of using values from the most recent period to forecast future values.

Example 1: Chicago Temperature Weather is a common example of time series data. In Fig. 6.1, the left plot shows average daily temperatures in Chicago from 2014 to 2018; the right scatter plot compares the average daily temperatures on two

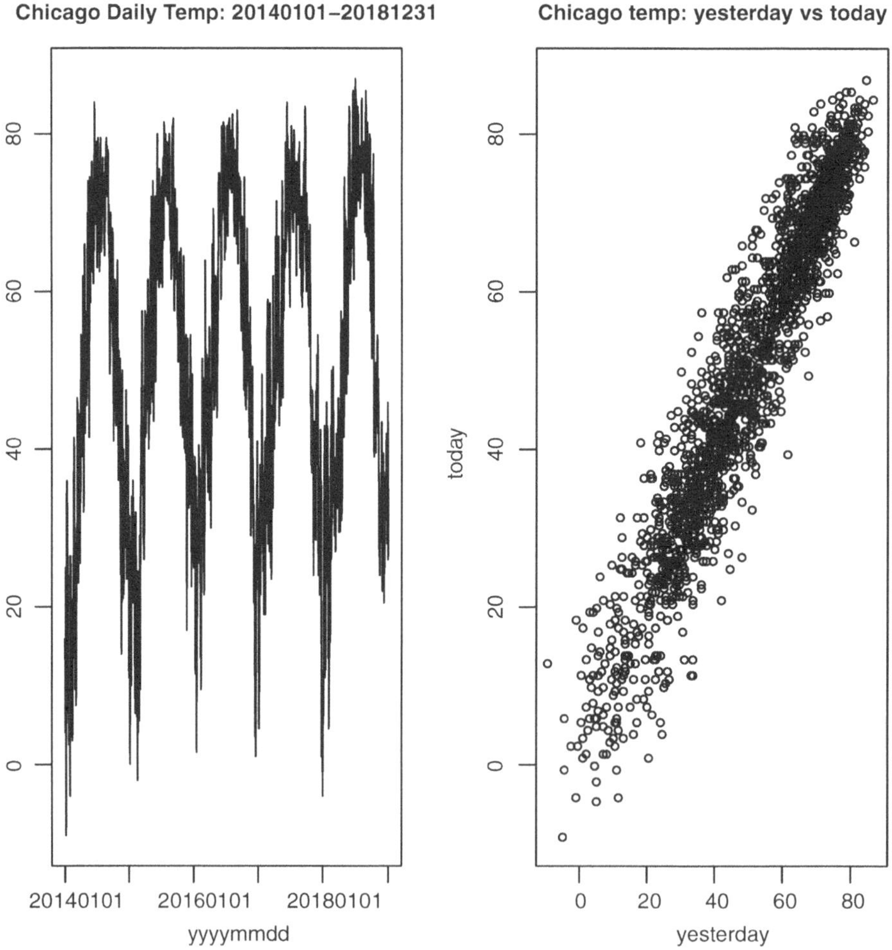

Fig. 6.1 Chicago daily temperatures: annual seasonal cycle, January 1, 2014—December 31, 2018

consecutive days. The left plot shows that there is clearly a seasonal pattern repeated across calendar years. The right plot shows a thick 45-degree line, indicating that while today's temperature is likely to be around yesterday's temperature, there is a wide range of possible outcomes. For example, when 1 day's temperature was 60F, the next day's temperature could be anywhere from 40F to 75F, so local residents should be prepared with either a coat or T-shirt!

We also present a more colorful picture for the year of 2018 in Fig. 6.2, so readers can see more vividly the daily temperature changes over a typical calendar year in Chicago. Indeed, you can see that in Chicago there are dramatic changes of temperature even on a daily basis.

Example 2: The Price of Oil In finance, there are many variables with values in a time series format. For example, the price of oil on each trading day is an example of time series data. Here, we use the Cushing spot price of WTI (West Texas Intermediate)

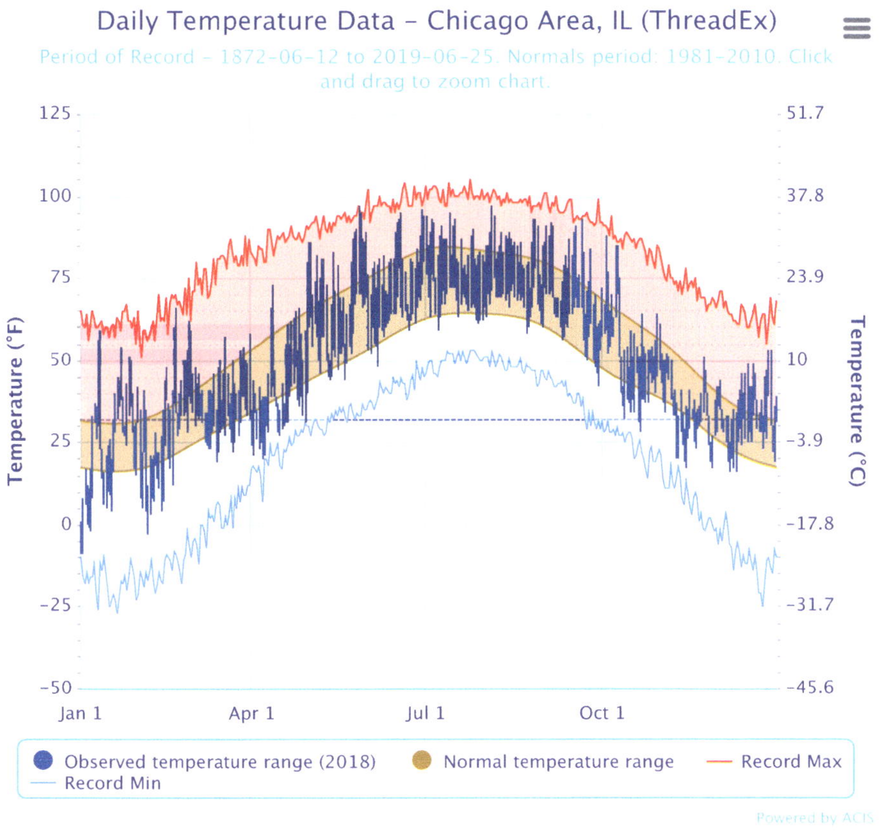

Fig. 6.2 Chicago daily temperatures in 2018: seasonal cycle

as the crude oil price. WTI is the underlying commodity of oil futures contracts, one of the most widely traded derivatives in the world.

In Fig. 6.3, we present the daily prices of WTI from 1986 to 2019 (left plot) and a scatter plot comparing prices between two consecutive days (right plot). We see from the left plot that there is an upward trend in WTI price over different periods with different durations. This non-regular pattern is typical for the prices of commodities.

Regarding daily price changes, we see again that yesterday is potentially a good predictor of today's price. However, again there is a wide range, implying that we have to be very cautious if we use time series data for prediction. If yesterday's price was $80 per barrel, today's price could be anywhere from $70 to $90, which is more than a 10% change in either direction. This is particularly true during periods

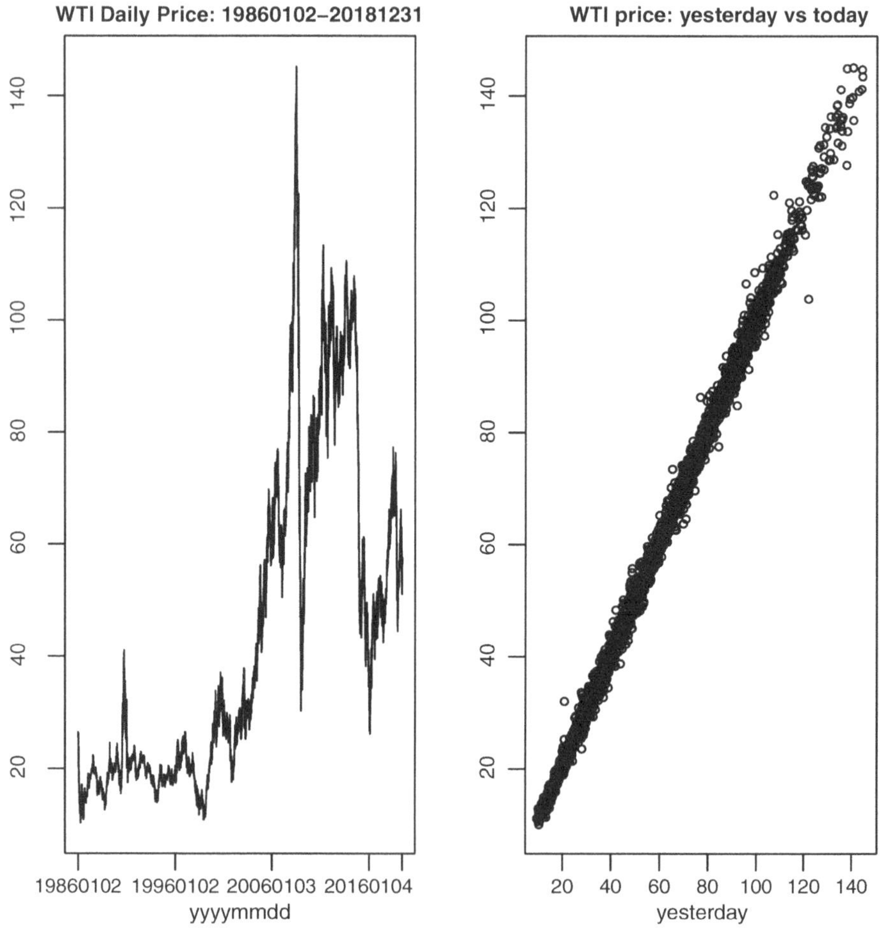

Fig. 6.3 WTI Daily spot price: periodic trend, 1986–2019

of reflection points. For example, the price of oil changed direction dramatically in 2008 and 2015.

Example 3: Special Items Our third example is special items from the financial reports of public companies. Governmental Accounting Standards Board Statement No. 34 defines special items as "significant transactions or other events within the control of management that are either unusual in nature or infrequent in occurrence." A large volume of academic studies document that when firms perform poorly on earnings, there is a tendency to use special items as a way to dilute negative earnings over fiscal periods (Elliott and Shaw 1988; DeAngelo et al. 1994; Carter 2000).

In this example, we center our attention on special items in the North American region. Our universe includes the S&P/TSX Composite for Canada and the MSCI US for the USA. Altogether there are about 800–1300 companies covering over 85% of the equity market capitalization for each country. Our data is month-end spanning from December 1997 to May 2013. The data on special items is obtained from income statements provided by CompuStat. Are there any patterns that characterize special items? Regarding trends over time, Fig. 6.4 shows that from 1997 to 2013, about 30–60% of the firms in the universe reported special items; there was a continuous upward trend from 20% to 45% for negative special items. On the other hand, the percentage of positive special items remained fairly close to 10% over time. This indicates that over time, more and more companies kept reporting special items, particularly *negative* special items. Research has shown that if a public company reports negative special items for several quarters in a row, it will be penalized by the stock market.

6.1.2 Time Series Data: Common Features

Based on the three examples above, we see that a time series data is a series of observations indexed in time order, usually with equally spaced points in time. We list below some common features of times series data.

1. Repeated measurement over time at equally spaced points.
2. Time series data follow a time sequence in which the time order is very critical, differentiating them from cross-sectional data.
3. Time series data can be decomposed into four components: level, trend, seasonality, and noise. We describe each part in detail below.

Level The average value in the series. Sometimes, people use moving average for a fixed period to explore the dynamics of the average over time.

Trend The medium- to long-term direction of the observations over time.

Seasonality Having regularly spaced peaks and troughs with a consistent direction and approximately the same magnitude every period. Seasonal effects are systematic,

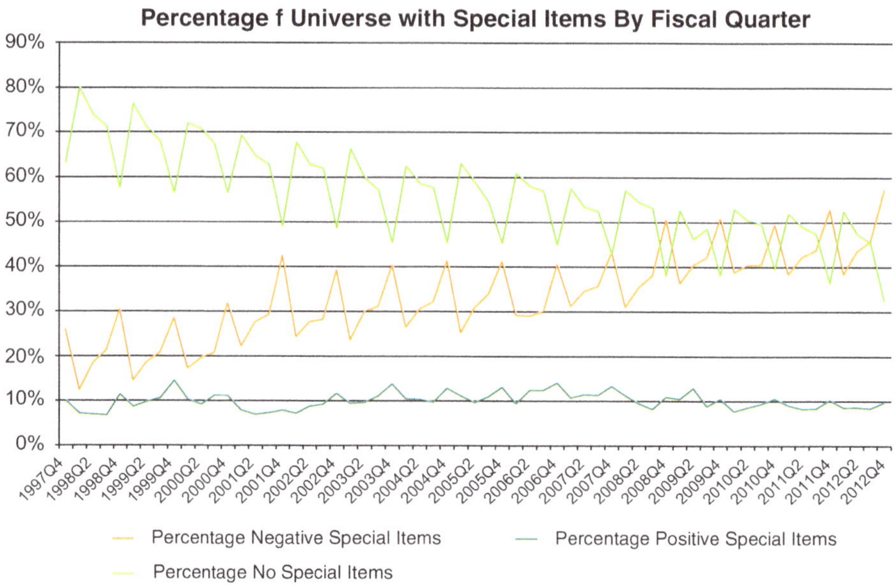

Fig. 6.4 The time series of percentages of positive, negative, and zero special items for companies in North America from December 1997 to May 2013

Table 6.1 Common features shared by three examples

	Level	Trend	Seasonality	Noise
Chicago daily temperature	✓		✓	✓
Price of oil	✓	✓		✓
Special items	✓	✓	✓	✓

calendar-related, and cyclical. A clear example is temperature across seasons. In finance, company earnings announcement effect is an example of seasonal effects.

Noise Random variation in the series. Technically, this is what remains of a times series after the level, trend, and seasonality are subtracted from the model. It is sometimes called shock or the irregular component.

All time series have a level and noise, but the trend and seasonality components are not always present. There are many ways of combining the components, such as additively or multiplicatively. We present common features shared by the three examples in Table 6.1.

While the four-part decomposition outlined above helps to clarify time series data, we need a proper framework for time series analysis to further explore the data. We introduce parametric time series models in the context of quantitative investing in the following sections.

6.2 Time Series Model: Unit Root and Spurious Relationship

In this section, we illustrate characteristics of time series models using the price of oil. We show that for time series data, past values can predict the present value. Therefore, it seems natural to use lagged values to forecast future values. However, there are potentially serious issues if we apply traditional methods, such as ordinary least squares (OLS), directly to such models. We now analyze those issues and introduce unit root in time series analysis. We then demonstrate how to conduct a unit root test and discuss spurious regression for non-stationary time series models.

6.2.1 Unit Root: Definition, Testing, and Treatment

A time series data set is a sequence of observations collected over time and indexed by time. It can be described by a variable Y over time $t = 1, 2, 3, \ldots, T$. For example, daily oil prices at the end of each business day would be a time series data set.

Recall that in Fig. 6.3, the values of Y_t and Y_{t-1} are shown in the right scatter plot. One method we can use to measure such a relationship is autocorrelation between the current and lagged values: $cor(Y_t, Y_{t-1})$. For example, comparing the price of oil between two consecutive days, the Pearson autocorrelation value is 99.93%, and the rank autocorrelation value is 99.89%. Below are the R scripts for autocorrelation computation.

Autocorrelation

```
> dim(oilPrice)
10097    2
> cor(oilPrice$WTI[-1],oilPrice$WTI[-10097])
0.9993236
> cor(oilPrice$WTI[-1],oilPrice$WTI[-10097],method="spearman")
0.9989472
```

Unit Root: Emergence and Consequence As discussed in previous chapters, correlation does not tell us the direction of causality, but we can use a regression framework to detect causality. Autoregressive regression (AR) models use past values to explain the present values.

Example 1: the price of oil on two consecutive days over time.

$$y_t = \beta_0 + \beta_1 y_{t-1} + e_t. \tag{6.1}$$

Example 2: the price of oil on three consecutive days over time.

$$y_t = \beta_0 + \beta_1 y_{t-1} + \beta_2 y_{t-2} + e_t. \tag{6.2}$$

Equation (6.1) is called AR(1) because there is only one lag in the model; Eq. (6.2) is called AR(2) because there are two lags in the model.

How would we estimate parameters in these AR models? For (6.1), we need to estimate $\beta = (\beta_0, \beta_1)$. If we apply OLS, we get the following estimates:

$$\hat{\beta} : \min \sum_{t=1}^{T} (y_t - \beta_0 - \beta_1 y_{t-1})^2.$$

However, we learned in the previous section that time series data may exhibit trends and seasonality. Will this impact the way we estimate the model? Using a simple AR(1) model, we show below that OLS cannot be applied directly to a time series model without further investigation. First we get the expected value,

$$\begin{aligned} E(y_t) &= \beta_0 + \beta_1 E(y_{t-1}) + E(e_t) \\ u_y &= \beta_0 + \beta_1 u_y \\ u_y &= \frac{\beta_0}{1 - \beta_1} \\ \Rightarrow &\quad \beta_1 \neq 1. \end{aligned}$$

Clearly, we cannot have $\beta_1 = 1$!

Next, we calculate the variance.

$$\begin{aligned} Var(y_t) &= \beta_1^2 Var(y_{t-1}) + Var(e_t) \\ \sigma_y^2 &= \beta_1^2 \sigma_y^2 + \sigma_e^2 \\ \sigma_y^2 &= \frac{\sigma_e^2}{1 - \beta_1^2} \\ \Rightarrow &\quad |\beta_1| < 1. \end{aligned}$$

Clearly, we cannot have $|\beta_1| \geq 1$!

What is so distinctive about a time series AR model? For (6.1) to make sense, the absolute value of the coefficient of lagged variables must be less than 1,

$$|\beta_1| < 1.$$

In this case, the time series is called *stationary*. Otherwise, if $|\beta_1| \geq 1$, the time series is called *non-stationary*.

- Case 1: $\beta_1 = 1$, indicating a unit root;
- Case 2: $\beta_1 > 1$, indicating an explosive process over time.

Now consider a simple AR(1) model with a unit root

$$y_t = b_0 + y_{t-1} + e_t. \tag{6.3}$$

In finance, e_t is called white noise or shock, and Eq. (6.3) can be further classified depending on the value of b_0:

$$\begin{aligned} b_0 \neq 0 &: \text{ random walk with a drift} \\ b_0 = 0 &: \text{ a pure random walk} \end{aligned}$$

We learned that a unit root causes a serious non-stationary issue. This is illustrated in the following using the example of Eq. (6.3). Assuming the root observation is y_0, we have

$$\begin{aligned} y_1 &= b_0 + y_0 + e_1 \\ y_2 &= b_0 + y_1 + e_2 \\ &= 2b_0 + y_0 + e_1 + e_2 \\ y_3 &= b_0 + y_2 + e_3 \\ &= 3b_0 + y_0 + \sum_{i=1}^{3} e_i \\ \vdots\ &\quad \vdots \\ y_T &= b_0 + y_{T-1} + e_T \\ &= Tb_0 + y_0 + \sum_{i=1}^{T} e_i. \end{aligned}$$

Assuming we have the first two moments of the error term as following,

$$E(e_i) = 0, \quad Var(e_i) = \sigma^2, \quad Cov(e_i, e_j) = 0, \quad i, j = 1, \ldots, T,$$

we then have

$$\begin{aligned} E(y_T) &= E(Tb_0 + y_0 + \sum_{i=1}^{T} e_i) = Tb_0 + y_0 \\ Var(y_T) &= Var(Tb_0 + y_0 + \sum_{i=1}^{T} e_i) = T^2\sigma^2. \end{aligned}$$

This shows that when T increases, the expected value and variance of the observation of a time series will increase, with the expected mean increasing by b_0 and variance increasing by $(2T - 1)\sigma^2$! The time series y_t is evidently non-stationary. Now, if we have another time series $X_t = \{x_t\}$ with a unit root, then it is not difficult to show that X_t and Y_t can be spuriously correlated because both have the trend T in the data, even when X_t and Y_t are fundamentally unrelated. We discuss spurious relationship in the next section.

Unit Root Test A natural question is, how can we tell if a time series has a unit root? David Dickey and Wayne Fuller proposed a unit root test in 1984, known as the Augmented Dickey–Fuller (ADF) test (Fuller 1976; Said and Dickey 1984). Note that the statistic value from the ADF test is a negative number (Table 6.2). This implies that, assuming H_0 states the presence of a unit root, then the more negative the statistic value, the higher probability of rejecting a unit root presence. We employ an AR(1) model to illustrate the ADF test.

- Method: ADF (Augmented Dickey–Fuller method)

$$
\begin{aligned}
y_t - y_{t-1} &= \beta_0 + \beta_1 y_{t-1} - y_{t-1} + e_t \\
\Delta y_t &= \beta_0 + (\beta_1 - 1) y_{t-1} + e_t \\
\Delta y_t &= \beta_0 + \gamma y_{t-1} + e_t
\end{aligned}
$$

- Unit root test: $H_0 : \gamma = 0, \quad H_1 : \gamma < 0$

$$
ADF = \frac{\hat{\gamma}}{SE_{\hat{\gamma}}}
$$

Note that besides the ADF, there are other unit root tests, for example the Phillips–Perron test proposed by Peter C.B. Phillips and Pierre Perron in 1988 (Perron 1988; Phillips and Perron 1988). The major difference between a PP test and an ADF test is that the former uses asymptotic nonparametric approaches and the latter applies parametric approaches, in cases of autocorrelation between errors.

Table 6.2 Critical values for the Dickey–Fuller t-distribution

Sample size	1%	5%	1%	5%
	Without trend	Without trend	With trend	With trend
T = 25	−3.75	−3.00	−4.38	−3.60
T = 50	−3.58	−2.93	−4.15	−3.50
T = 100	−3.51	−2.89	−4.04	−3.45
T = 250	−3.46	−2.88	−3.99	−3.43
T = 500	−3.44	−2.87	−3.98	−3.42
T = ∞	−3.43	−2.86	−3.96	−3.41

Source: Fuller (1976)

Unit Root Treatment Unfortunately, nonstationarity appears often in commodity investing. What can we do if there are non-stationary issues? We now explore solutions for non-stationary issues in time series analysis.

For non-stationary cases caused by unit roots, we list possible solutions below

- Transformation of variable, such as $\log(y_t)$
- Difference: $\Delta y_t = y_t - y_{t-1}$
- Return type: $R_y = \frac{y_t}{y_{t-1}} - 1$
- De-trend: $y_t = a + bt + e \Rightarrow \hat{e}$

For non-stationary cases caused by time trends, we can simply de-trend by regressing the time series variable against a time variable and then get the residual. The residual values will be stationary.

6.2.2 Spurious Relationship

If there is a unit root in a times series model, we cannot apply OLS to the model for parameter estimation because when a time series is non-stationary, there may be a serious risk of a *spurious relationship*! Spurious relationships were first studied seriously by Yule (1926) and then revisited by Granger and Newbold (1974) where the latter proposed the term spurious regression and caught attention of many applied studies using macroeconomic data series.

Suppose we have two sets of time series data, X_t and Y_t, and both have an upward trend. While the two time series may appear to be highly correlated, they can in fact be totally independent from each other. A famous example of a spurious relationship is between a city's ice cream sales and the rate of drownings. To conclude that ice cream sales cause drowning, of course.

A spurious regression will provide misleading information about the relationship between non-stationary variables. For example, at the macro level, economic variables and commodity prices are likely to be correlated with each other, even when neither has a causal effect on the other. This is because each variable may have a trend or cycle during a given period, and the common presence of the trend or cycle in the two data series causes the spurious relationship.

An Example of a Spurious Relationship In this section, we illustrate spurious relationships through randomly generated samples with a unit root and time trend. We set up four models below and generate a sample of 100 observations for each model:

$$x_t = x_{t-1} + e_t, \qquad w_t = w_{t-1} + T + e_t$$

$$y_t = y_{t-1} + \mu_t; \qquad z_t = z_{t-1} + T + \mu_t.$$

Through the data generation process, both X and Y have a unit root, while $W = \{w_t\}$ and $Z = \{z_t\}$ have a unit root and time trend, e_t and μ_t are independent error terms.

We set the initial value of 5 for X and W and 10 for Y and Z. Thus, by design, X and Y do not have any causal relationship, and the same applies to the relationship between W and Z. However, the OLS estimates and correlation both reveal high levels of relatedness between the pairs. For example, the correlation is 53% between X and Y and 93% between W and Z. Based on the OLS estimates, both factors are significant, and the R^2 values are very high, indicating that the two variables have a strong causal relationship!

$$cor(x, y) = 0.52, \qquad cor(w, z) = 0.93$$
$$lm(y \sim x), R^2 = 28\% \qquad lm(z \sim w), R^2 = 87\%.$$

Below are the R scripts and the results of computation.

Spurious relationship from random data generation

```
> source("/.../quantInvesting/chapter6/spurious.example.R")
> spurious.example()
  ##correlation between x and y
[1] 0.5248156
 ##correlation between w and z
[1] 0.9315895

 ## OLS regression lm(y~x)
Coefficients:
            Estimate Std. Error t value Pr(>|t|)
(Intercept) 11.10231    0.41971  26.452  < 2e-16 ***
x            0.28919    0.04738   6.104 2.08e-08 ***
---
Residual standard error: 2.665 on 98 degrees of freedom
Multiple R-squared:  0.2754,Adjusted R-squared:  0.268
F-statistic: 37.25 on 1 and 98 DF,  p-value: 2.081e-08

## OLS regression lm(z~w)
Coefficients:
            Estimate Std. Error t value Pr(>|t|)
(Intercept) 16.04836    0.62987   25.48   <2e-16 ***
w            0.99210    0.03911   25.37   <2e-16 ***
---
Residual standard error: 4.837 on 98 degrees of freedom
Multiple R-squared:  0.8679,Adjusted R-squared:  0.8665
F-statistic: 643.6 on 1 and 98 DF,  p-value: < 2.2e-16
```

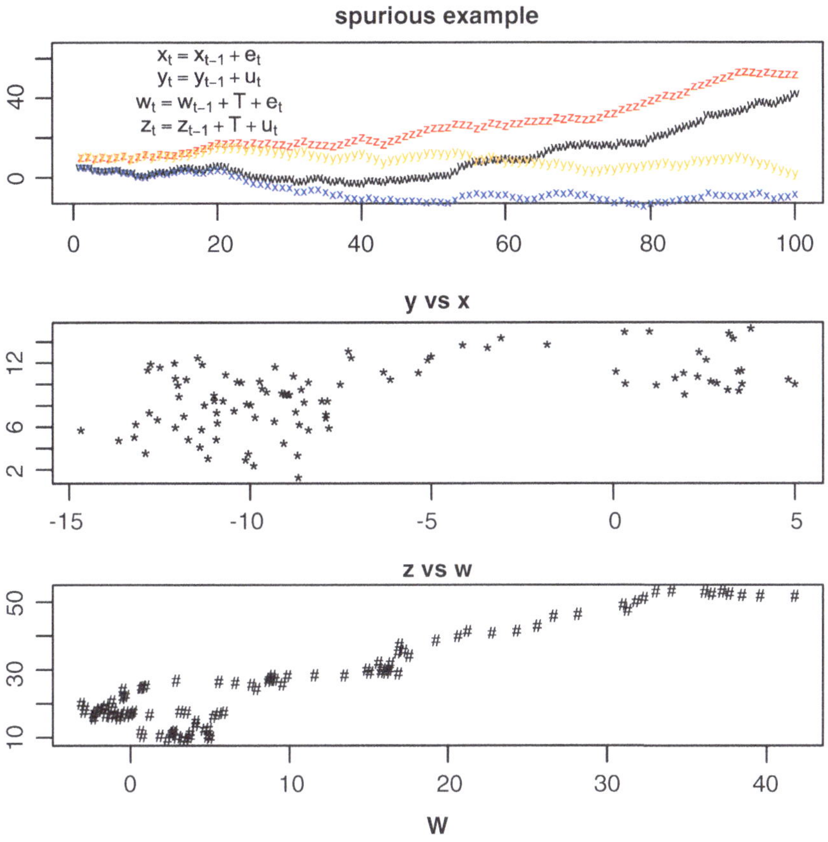

Fig. 6.5 Spurious relationship: randomly generated data with unit root and trend

We present values of the four time series—*X*, *Y*, *W*, *Z*—and the scatter plots between pairs (X, Y) and (W, Z) in Fig. 6.5. The spurious relation between these variables warns us that first, we need to have a good fundamental understanding of time series data and causal relationships, and second, we need to have a proper methodology to deal with time series data.

6.3 The Price of Oil: Is There a Unit Root?

We now apply the concepts and methods we discussed in the previous section to test for the presence of a unit root in oil price data using the daily spot price of WTI from 1986 to 2019. Recall that in Example 2 in Sect. 6.1, we presented the time series data for the price of oil (Fig. 6.3). The price of oil ranged from USD 20–120 per barrel during this period and exhibited trends with different durations. Our design for the unit root test is described below.

- Model: $\text{oilPrice}_t = \beta_0 + \beta_1 \text{oilPrice}_{t-1} + e_t$,
- Unit root test hypothesis: $H_0 : \beta_1 = 1, \quad H_1 : \beta_1 < 1$
- Methods: ADF and PP

We present R scripts for the ADF and PP tests below.[1] The ADF test has a p-value of 27%, and the PP test has a p-value of 39%, both indicating the strong presence of a unit root in the times series data of daily oil prices from 1986 to 2019. This implies that we need to be very cautious when we explore the relationship between the price of oil and other variables because there are potentially spurious results.

Price of oil: unit root test

```
> ### wti is a data frame with a date and daily price of WTI, yyyymmdd and wti
> wti[1:2,]
  yyyymmdd   wti
1 19860102 25.56
2 19860103 26.00

> library(tseries)

> ### ADF test for unit root
> adf.test(wti$wti)
Augmented Dickey-Fuller Test
data:  wti$wti
Dickey-Fuller = -2.7164, Lag order=20, p-value=0.2749
alternative hypothesis: stationary

> ### PP test for unit root
> pp.test(wti$wti)
Phillips-Perron Unit Root Test
data:  wti$wti
Dickey-Fuller Z(alpha) = -12.85, Truncation lag parameter = 12, p-value = 0.3935
alternative hypothesis: stationary
```

There are two ways we can try to avoid a spurious relationship: a quantitative way and a fundamental way. The quantitative way is to transform the price variable by following the suggestions discussed above, such as using log, return, difference, etc. The fundamental way is to identify causal factors or indicators for oil price movements and use those factors to model and forecast the price of oil. In practice, quantitative investing usually entails a combination of the two. We dig into the fundamentals of the price of oil in the next section and then introduce an oil price forecasting model based on both times series and fundamental factors.

[1] For a detailed discussion about R packages, please refer the R section at the end of this chapter.

6.4 Crude Oil: Fundamentals

We now have some basic understanding of time series models and their potential issues. We also learned that there is a unit root in the daily price of WTI from 1986 to 2018. For a successful investment strategy in crude oil, we need to be able to tell the direction of price movement in the future. This is a very challenging task given that the price of a commodity is impacted by many macro-level factors, such as geopolitical power. While it is very difficult or almost impossible to predict those macro-level variables, the good thing is that they will not change overnight, so we just need to be adaptive. As investors, we should focus on what we can do: understand fundamental causes of price movement and incorporate time series features to build a valid time series model for the price of oil.

In this section, we examine the fundamental landscape of crude oil, such as reserves, production, consumption, and transactions. These constitute the big picture of supply and demand for oil, and they all play a very important role in pricing.

6.4.1 Oil Reserves

Oil is a natural resource formed by the decay of organic matter over millions of years. Table 6.3 lists the top ten countries with the most oil reserves in the world based on the data of 2017 proven oil reserves. Proven reserves are measured and approved by technical analysis.[2]

Table 6.3 Top ten countries by oil reserves (1000 MB) at the end of 2017

Country	Proven oil reserves	World share (%)	Daily production (MBD)
Venezuela	303.2	17.9	2.1
Saudi Arabia	266.2	15.7	12.0
Canada	168.9	10.0	4.8
Iran	157.2	9.3	5.0
Iraq	148.8	8.8	4.5
Russian Federation	106.2	6.3	11.3
Kuwait	101.5	6.0	3.0
United Arab Emirates	97.8	5.8	3.9
United States	50.0	2.9	13.1
Libya	48.4	2.9	0.9

MB millions of barrels; *MBD* millions of barrels per day. Data source: BP World Energy Annual Statistic Report

[2]Proven oil reserves data come from the BP Statistical Review of World Energy June 2018 report. Reserves as a share of world total reserves and oil production in barrels also come from BP's report.

In total, the top ten countries make up about 85% of total world oil reserves. Geographically, half of these countries are in the Middle East Gulf region, the others being the USA, Canada, Russia, Venezuela, and Libya. It should be noted that although Venezuela has the most reserves, the reserve quality and hence the cost of oil production are much higher than in Saudi Arabia.

The countries that control the world's oil reserves often have disproportionate geopolitical power, which is a double-edged sword because these countries often become passively involved in war or extreme events due to their strategic positions with oil reserves. The same logic applies to the impacts of oil reserves on economic development. On one hand, oil has been an economic boon for the countries on this list. Often, due in large part to oil production, many of these countries rank among the most prosperous in the world. On the other hand, if mismanaged, oil wealth can also be a curse. Having a diversified economy is always prudent, and many countries on this list are overly dependent on their oil wealth. As a result, many have suffered economically since global oil prices fell precipitously from $115 a barrel in mid-2014 to less than $35 a barrel in early 2016. For example, triggered by an overdependence on oil, Venezuela's economy collapsed in 2018, and there have been social crises from 2015 to 2019.

6.4.2 *Oil Production and Consumption*

Reserves are potentials for production. Some countries have better technology to convert reserves to products, while others have a low transformation ratio. Regardless, most countries in the top ten list of reserves are in the top ten list of producers. In recent years, the top three oil-producing countries are Russia, Saudi Arabia, and the USA; together they produce about 40% of the world's total oil production (Table 6.4). The USA is a swing producer because its production fluctuates with market prices.

Table 6.4 Top ten oil-producing countries in 2018

Country	Daily production (MB)	World share (%)
United States	17.87	18
Saudi Arabia	12.42	12
Russia	11.4	11
Canada	5.27	5
China	4.82	5
Iraq	4.62	5
Iran	4.47	4
United Arab Emirates	3.79	4
Brazil	3.43	3
Kuwait	2.87	3
Total top 10	70.96	70
World total	100.66	

MBD millions of barrels per day. Data source: EIA oil production. Oil includes crude oil, all other petroleum liquids, and biofuels

It is interesting to note that among the top producers, some are top exporters and others are top importers, which differentiates their preferences regarding oil prices. We present more details when we discuss oil transactions later.

We also provide a time series view of crude oil production for the top five countries (Fig. 6.6). It is interesting to note that the top three producers have all been at peak production in recent years. However, about 40 years ago, the USSR was the top producer in the world, with crude oil production of 12 million barrels per day before its collapse in 1991.

There have been production volatilities for the top producers during the last 50 years. For example, it is very interesting to note that the oil production of the USSR (Russia after 1991) was about 9 MBD in the mid-1980s, dropped steadily to only about 5 MBD in 2007, and has increased steadily over the last 5 years. Even more interesting is that while the USA was at peak production in 1985, Saudi Arabia was in a production trough of about 3.5 MBD. Then, while USA production dropped, Saudi oil production increased. The combined oil production of the USA and Saudi

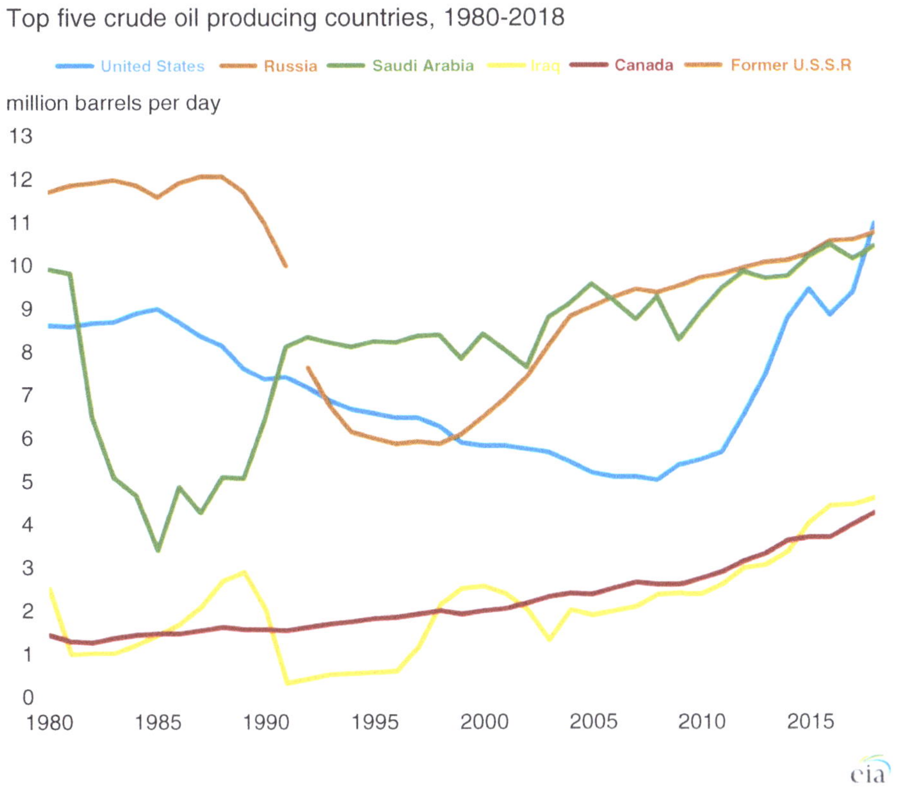

Fig. 6.6 Crude oil production from the top five oil-producing countries, 1981–2018. Note: Includes crude oil and lease condensate. Ranking based on production in 2018. Source: U.S. Energy Information Administration, International Energy Statistics, April 2019

Arabia was about 13–15 MBD, remaining very stable over a long period from 1985 to 2010. After 2010, production from all three countries (the USA, Saudi Arabia, and Russia) increased, with each producing about 9–10 MBD.

Another very important aspect is the ratio of reserves to production. According to the 2018 report of the BP Statistical Review of World Energy (Fig. 6.7), global proven oil reserves in 2017 were 696.6 billion barrels, indicating a 50-year oil production life at the 2017 level. OPEC countries currently hold 71.8% of global proved reserves, of which Venezuela holds 17.9%. If we divide reserves by production for each country for each year from 1987 to 2017, we can see a clear continuous downward trend for Saudi Arabia and a hike for Venezuela in 2007.

Oil Consumption: Who Consumes Oil We list the ten largest oil consumers and their shares of total world oil consumption in 2016 in Table 6.5. Not surprisingly, these countries are among either the top developed economies or the largest emerging economies in the world. For example, the USA, Japan, Canada, South Korea, and Germany belong to the former, while the BRIC nations (Brazil, Russia, India, and China) and Saudi Arabia belong to the latter. It should also be noted that most countries on the list, such as India and Japan, are not on the top producer list, indicating that they have to buy oil from other countries to satisfy their needs. The USA is a both a big oil producer and a big oil consumer.

Fig. 6.7 Crude oil reserve/production ratio by region, 1987–2017

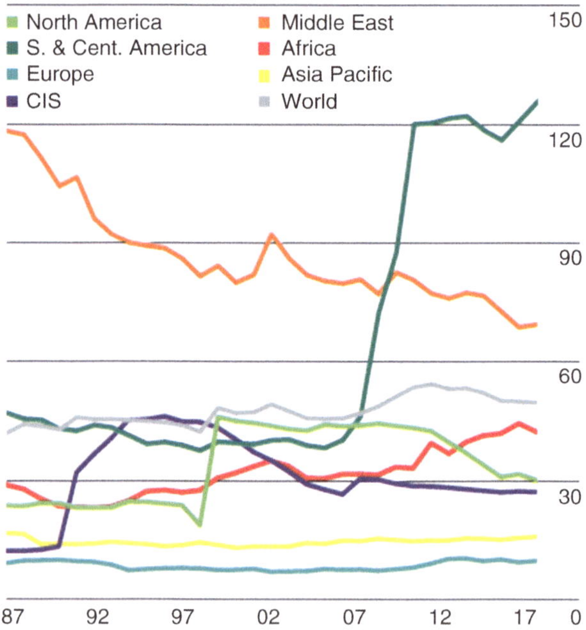

Table 6.5 Top ten oil-consuming countries in 2016

Country	Daily consumption (MB)	World share (%)
United States	19.69	20
China	12.79	13
India	4.44	5
Japan	4.01	4
Russia	3.63	4
Saudi Arabia	3.30	3
Brazil	2.98	3
South Korea	2.61	3
Canada	2.47	3
Germany	2.38	2
Total top 10	58.31	60
World total	96.92	

MBD millions of barrels per day. Data source: EIA oil consumption. Oil includes crude oil, all other petroleum liquids, and biofuels

6.4.3 *Crude Oil Transaction: Petrodollar, Exporters, and Importers*

Transactions involving crude oil are carried out in USD (petrodollar). For example, if a Japanese company, JAP, buys one million barrels of crude oil from a Saudi Arabian company, SAU, the transaction is performed as follows: Using the spot price of WTI on July 4th, 2019 at 4pm EST, the total amount for the deal is USD 56.67 million.

$$\text{WTI transaction amount} = USD\ 56.67$$

$$SAR\ 1 = JPY\ 28.53,$$

$$USD\ 1 = JPY\ 107 = SAR\ 3.75$$

In other words, JAP needs to buy USD 56.67 million with Japanese currency, 6.06369 billion Japanese Yen (JPY), and then pay SAU with USD. Once SAU receives the payment in USD, it can exchange for 212.5125 million Riyal (SAR), the Saudi currency.

The USD became the designated currency for oil transactions after the collapse of the Bretton Woods gold standard in the early 1970s. In June 1974, the United States-Saudi Arabian Joint Commission on Economic Cooperation was established, under which Saudi Arabia would invoice oil in USD. Those US dollars would be recycled back to the USA through contracts with US companies. Hence the birth of the petrodollar. The effect of this tacit agreement is conditioned on the massive production of oil in GCC countries and the economic power of the USA in the world.

Table 6.6 Top ten crude oil export (bottom) and import (top) countries in 2018

Country	Import (USD billion)	World import share (%)
China	239.20	20.20
United States	163.10	13.80
India	114.50	9.70
Japan	80.60	6.80
South Korea	80.40	6.80
Netherlands	48.80	4.10
Germany	45.10	3.80
Spain	34.20	2.90
Italy	32.60	2.80
France	28.50	2.40
Country	Export (USD billion)	World Export share (%)
Saudi Arabia	182.5	15.90
Russia	129	11.30
Iraq	91.1	7.90
Canada	66.9	5.80
United Arab Emirates	66.8	5.80
Kuwait	49.8	4.30
United States	47.2	4.10
Iran	45.7	4
Nigeria	43.6	3.80
Angola	38.4	3.40

Now let us take a look at the parties on the two sides of crude oil transactions: crude oil exporters and importers. Table 6.6 lists the top ten exporting and importing countries, by total amount and percentage share in the world in 2018. Note that the USA appears in both lists, indicating that the USA is a trade dealer in the market. This is another way that the USA plays a critical role in the crude oil market in addition to the business rule that crude oil transactions are required to use USD. In terms of geography, the countries that import oil are from Asia and Europe, while the countries that export oil are the Gulf countries, Russia, and Canada. In terms of their interests, oil-exporting countries hope there is strong demand for crude oil and the price of oil stays high, while oil-importing countries hope for the opposite. Eventually, in a free market setting, there should be an equilibrium price, though it may have a wide range.

Usually, crude oil transactions are based on long-term contracts with prices and amounts specified. Nowadays, price is settled in the derivatives market, where geopolitical power, speculation, and technology all play significant roles.

Oil Price is Crucial for Some Countries While crude oil is crucial for modern industry and households, it has different implications for different countries. For a superpower like the USA, accessibility and control of oil reserves are an organic part of the country's long-term strategy. For developed and emerging economies depending on oil as an input, accessibility and a steady oil supply are important for normal operation. For oil-producing and -exporting countries, their economies and people's lives depend heavily on oil revenue, that is, both oil production and the price of oil are very important!

We analyze countries' economic dependence using the ratio of crude oil profit (revenue minus production cost) to GDP in each country and list the twenty most dependent countries (Table 6.7). Since oil-rich countries depend on oil revenue, which is very sensitive to the price of oil, we list three columns of ratios in the table: the ratios for 2012 and 2017 based on actual oil revenue and GDP and the hypothetical ratios for 2017. The hypothetical ratio is calculated using the oil price difference between 2012 and 2017 (assuming everything else stayed the same between 2012 and 2017). The average oil price (WTI) was USD 94 per barrel in 2012 and USD 51 per barrel in 2017. Assuming the ratio of oil revenue to GDP is x in 2012, then the new ratio in 2017 based on the oil price change is

$$\text{hypothetical ratio for 2017} = \frac{x \times \frac{51}{94}}{100 - x \times \frac{43}{94}}.$$

Table 6.7 Top twenty oil-dependent countries in 2017

Country	Oil profit/GDP, 2017	Oil profit/GDP, 2012	Pure oil price driven, 2017
Iraq	37.78	48.43	33.31
Libya	37.29	60.79	45.19
R. of Congo	36.73	51.79	36.36
Kuwait	36.61	61.07	45.48
Saudi Arabia	23.1	47.22	32.24
Oman	21.8	42.76	28.43
Eq. Guinea	19.23	36.76	23.61
Azerbaijan	17.87	27.8	16.99
Angola	15.75	35.25	22.45
Gabon	15.34	35.46	22.61
Iran	15.34	20.2	11.86
Chad	15.25	24.33	14.60
Qatar	14.23	29.06	17.89
UA Emirates	13.13	28.5	17.49
Algeria	12.31	26.17	15.86
Kazakhstan	10.19	17.26	9.98
Brunei	8.843	21.05	12.42
Russia	6.43	10.33	5.77
Nigeria	6.12	13.92	7.92
Suriname	5.24	7.93	4.38

The ratio = (Crude oil revenue - production cost)/ GDP. The theoretical ratio is based on the average price of WTI in 2017 (USD 50 per barrel) and 2012 (USD 90 barrel)

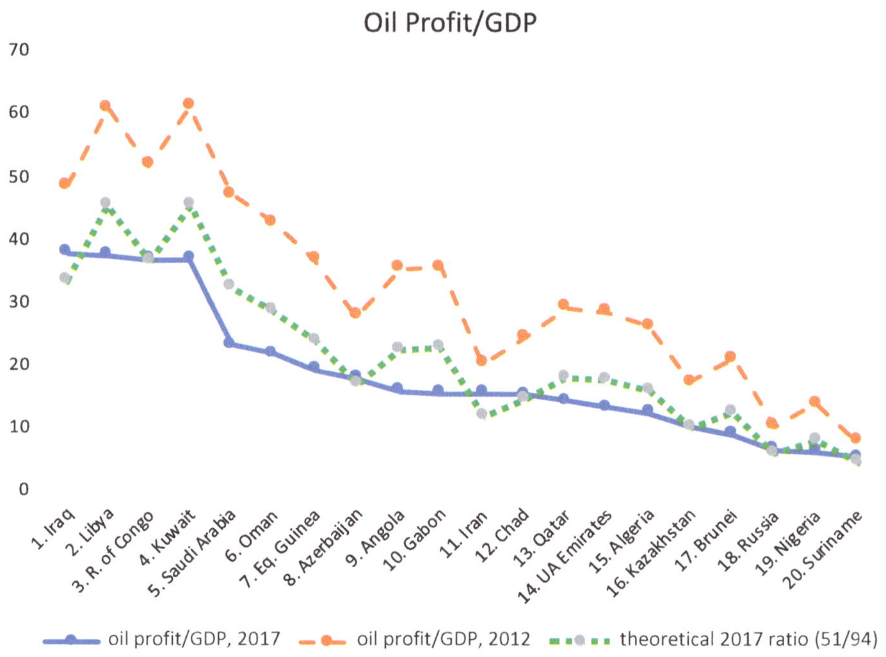

Fig. 6.8 Oil profit as a share of GDP: 2012, 2017, and the theoretical ratio for 2017

The rationale for the validity of using oil prices to calculate the ratio is that the economic structure of these oil-dependent countries would not change much over a short time. Second, their oil production will not change much because these countries face a dilemma when the price of oil drops dramatically. The results are shown in Fig. 6.8, where we see that the predicted oil profit to GDP ratios (dotted line) and the actual ratios (solid line) for the year of 2017 are very close to each other, indicating high accuracy of the hypothetical ratios.

6.5 Crude Oil: 100 Years of Price Change

Oil, through energy and its derivative products, plays a significant role in modern industry and people's everyday lives. During the past 100 years, the price of oil has undergone dramatic ups and downs. To build a successful investment strategy for crude oil, it is important to understand not only the economic landscape and the time series of prices, but also the causes of price changes.

We present two pictures of the price of oil in this section: a long history starting in 1861 (Fig. 6.9) and a recent history since 1970 (Fig. 6.12). Based on the two plots, we describe the pricing regimes and the events that impacted the price of oil throughout these periods.

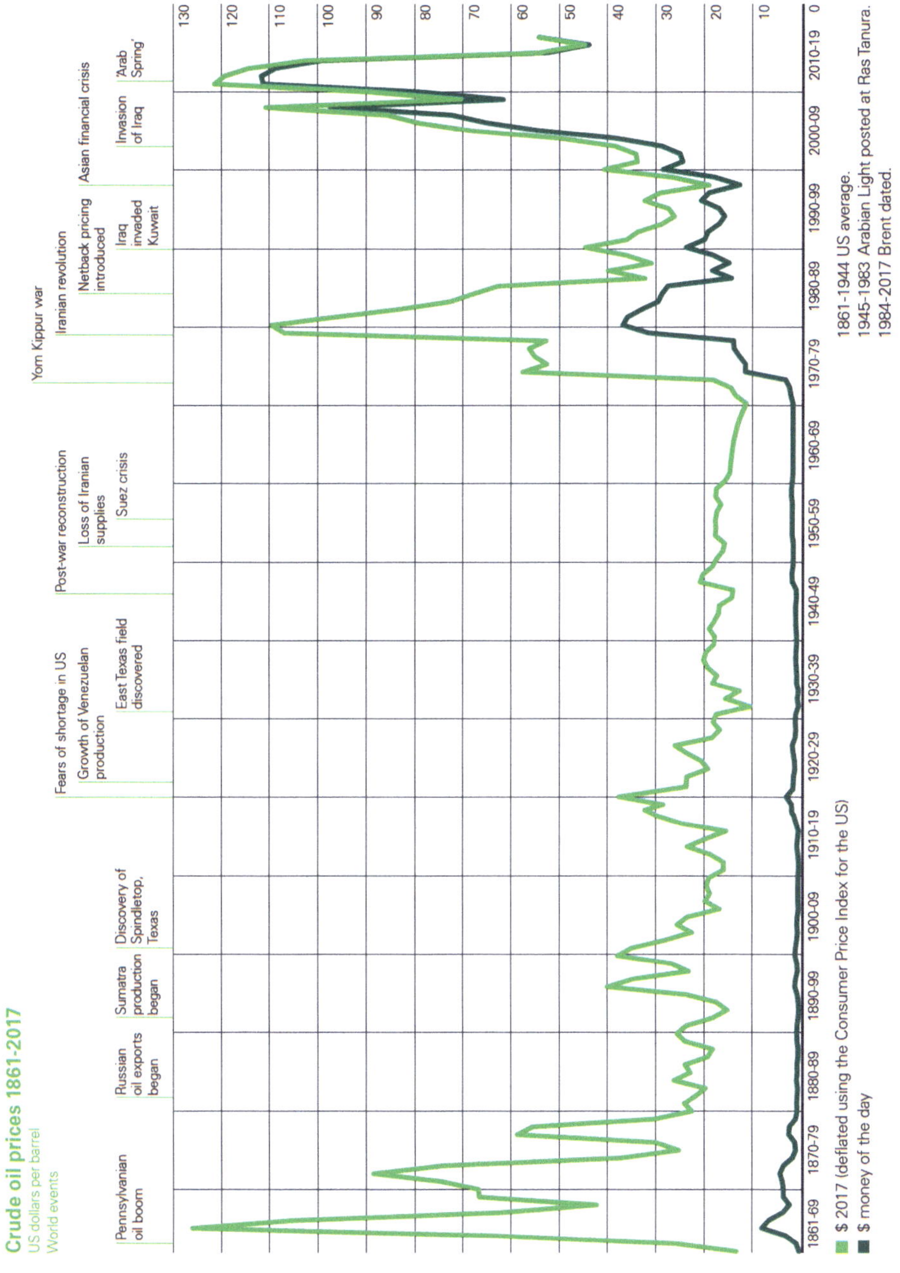

Fig. 6.9 The price of oil from 1861 to 2017 with events. Data Source: BP Statistical Review of World Energy, 2018 report

6.5.1 Historical Pricing Regimes: Seven Sisters, OPEC, and the Oil Market

There have been several pricing regimes for crude oil in modern history. From 1920 to 1972, a few Western oil companies had control over both world oil reserves and price. From 1972 to 1985, a cartel of oil-producing countries in the Middle East headed by Saudi Arabia had control over oil production and price. From 1985 onward, the three markets—Brent, WTI, and Dubai—have served as major references for settlement prices. We briefly discuss each pricing regime below.

Pricing Regime 1: Seven Sisters, 1920–1972 Empire Country and the seven sisters cartel controlled both reserves and the price of oil during this period. The "Seven Sisters," coined by Enrico Mattei, include seven oil companies (Ammann 2009): "Anglo-Iranian (initially Anglo-Persian) Oil Company (now BP), Gulf Oil (later part of Chevron), Royal Dutch Shell, Standard Oil Company of California (SoCal, now Chevron), Standard Oil Company of New Jersey (Esso, later Exxon, now part of ExxonMobil), Standard Oil Company of New York (Socony, later Mobil, also now part of ExxonMobil), and Texaco (later merged into Chevron)."

Long before and even after the WWII, the seven sisters controlled about 85% of the world oil reserves and the Middle East's oil production. They also had considerable power over energy production in emerging countries. Because of the cartel's monopoly power, the seven sisters made billions of dollars each year during this period, which strengthened their political influence and in turn made the cartel even more powerful. The seven sisters were the largest and most profitable companies in the world back then. During the period from 1920 to 1972, the price of oil was around USD 20.

Pricing Regime 2: OPEC, 1973–1985 The Organization of the Petroleum Exporting Countries (OPEC) was established in 1960 with the purpose to fight against the control over oil reserves and production of seven sisters. There were only five members, Iran, Iraq, Kuwait, Saudi Arabia, and Venezuela, at the beginning, but many other oil-producing countries at the Gulf region joined in the 1970s.

In the previous section, we learned that OPEC held 75% of global oil reserves. Moreover, OPEC has the world's lowest barrel production costs. These two factors give OPEC tremendous influence over crude oil prices. A typical practice by OPEC is to control price through production: cut production when there seems a glut and increase production if there is less demand.

After the 1980s, non-OPEC oil production increased to more than 50% of the global oil supply. OPEC's ability to control oil prices through production has been weakened in the sense that it cannot set prices directly. However, given its significant oil production share (40%) and oil export share (60%), OPEC continues to influence oil prices in the market.

The pricing power of OPEC is decided by its reserve and production shares in the world. We present the reserve share of the Middle East in Fig. 6.10 and the

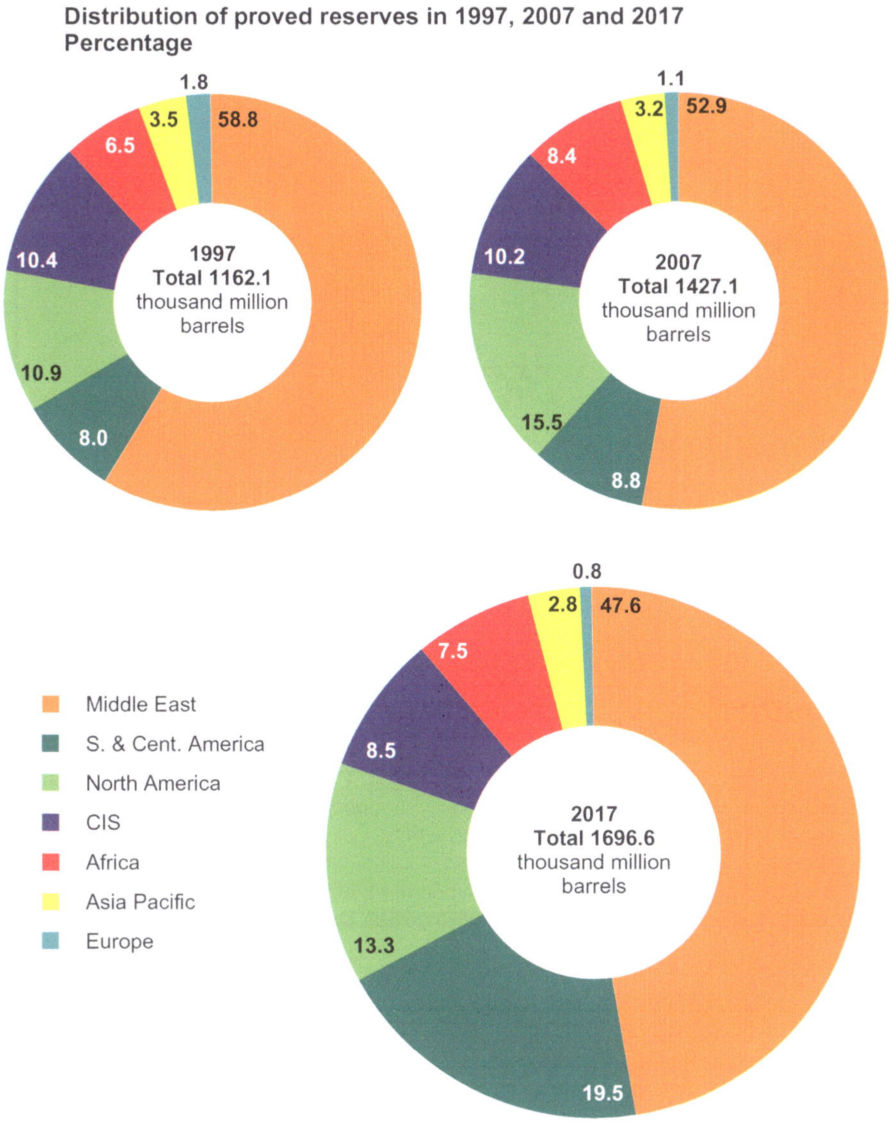

Fig. 6.10 The world share of crude oil reserves of the Middle East and other regions in 1997, 2007, and 2017. Data source: BP Statistical Review of World Energy 2018

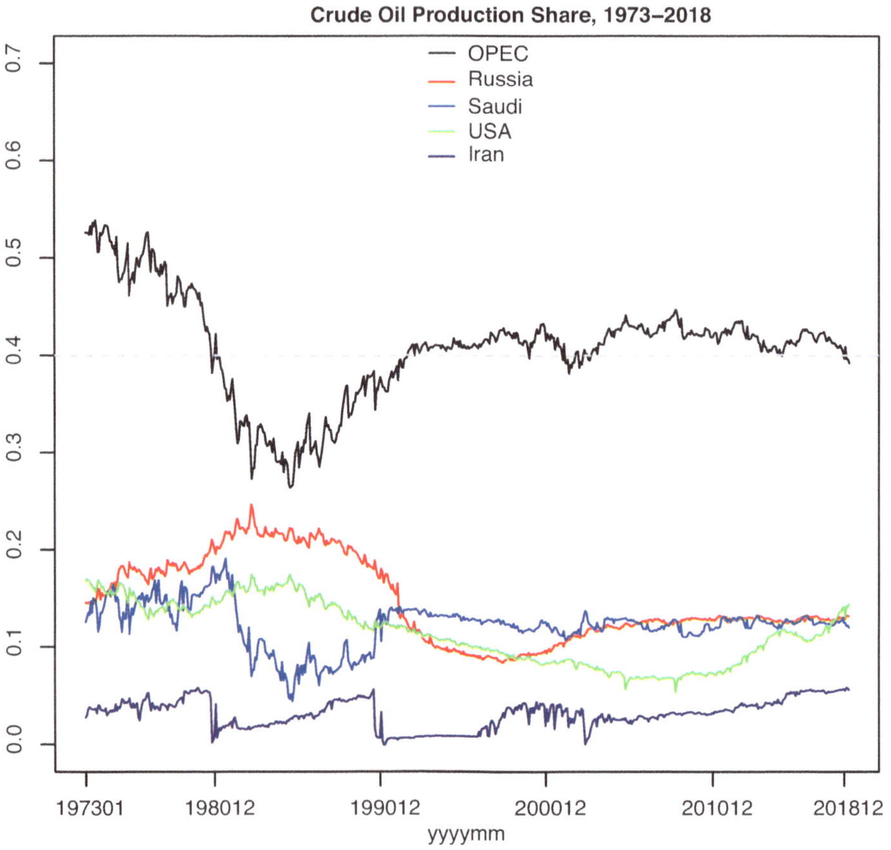

Fig. 6.11 The world share of crude oil production of OPEC and single country (Russia, USA, Saudi Arabia, and Iran) from January 1973 to March 2019. Data source: EIA energy international

production share of OPEC and Saudi Arabia together with other important oil-producing countries in Fig. 6.11.

We see a clear decrease in both the reserve share and production share of Gulf countries. The reserve share of the Middle East region dropped from 56% in 1997 to 48% in 2017. The production share of OPEC dropped from 50% in the early 1970s to 40% after the 1990s. With the increasing oil production of the USA and Russia, this share is decreasing even further. In recent years, Saudi production is at best on par with Russia and the US.

During this period (1973–1985), oil prices were very volatile, ranging from 10 to 35 US dollars per barrel.

Pricing Regime 3: Market, 1985–Present The price of oil is mostly settled by oil markets, through lense of many factors such as supply and demand, speculation, and geopolitical forces. Since the 1980s, there has been a gradual shift from OPEC to the oil market, in light of factors that impact the price of crude oil. Needless to say, given the size of the US economy and geopolitical power, it has one of the biggest influence on the price of crude oil.

6.5.2 *Events Impacting the Price of Oil*

We turn now to major events in history that have impacted the price of oil from 1970 to 2019. Given the critical position of crude oil in the modern world, nations sought to secure access to oil and wave impacts on reserves, supply, and demand. Understanding these events and their impacts on the price of oil will help us to model and forecast the price of oil from a quantitative investing perspective. We outline major events associated with significant oil price changes during this period.

- Oil Embargo, October 16, 1973–March 17, 1974.
  During this period, Arab oil producers cut production by about 5–25% and instituted an oil embargo against Israel's allies including the USA. The price of oil increased fourfold to nearly $12. This was later called the "first oil shock" and was followed by the 1979 oil crisis, termed the "second oil shock."
- Iranian Revolution, November 1978–December 1979.
  During the period of regime change in Iran in November 1978, Iran's oil production reduced from 6 MBD to about 1.5. The shock caused panic and drove the price of oil up from $15 in November 1978 to more than double at $38 per barrel at the end of 1979.
- Non-OPEC oil production.
  Oil production by non-OPEC countries has increased from 40% of world oil production to about 60% in recent years. This mainly refers to the production of the North Sea region, the USA, and Russia (the USSR before 1991).
- Collapse of the USSR, 1986–1991.
  During the period from 1986 to 1991, the Saudis used their oil capacity to raise and lower world prices. For example, the Saudis increased oil supply dramatically in the 1980s resulting in a sharp oil price drop when the USA launched an economic war against the Soviet Union.
- Iraq–Kuwait War: 2 August 1990–28 February 1991.
  This was also called the Gulf War. Iraq invaded Kuwait and the USA led a coalition force to fight and defeat Iraq. During the war, Iraq dumped 400 million gallons of crude oil into the Persian Gulf and set fire to oil wells causing about six million barrels loss each day. Oil supply was reduced by about 4.3 million barrels a day and the price of oil increased from $21 per barrel at the end of July to $46 in mid-October of 1991.
- Strong oil demand from China since 2000
  China accounted for approximately 40% of global oil demand growth between 2009 and 2019. To put China's oil demand growth during this period into perspective, China's crude oil demand in 2009 grew by 0.28 MBD—roughly

the daily consumption of the Philippines and by 1.16 MBD in 2010—roughly the daily consumption of Taiwan, one of the world's most industrialized economies.

- The US financial crisis: 2008
  The financial crisis triggered fear during the economic slowdown and hence less demand for crude oil. The price of oil dropped temporarily. On December 23, 2008, the WTI crude oil spot price fell to USD 30 a barrel from above USD 100 a few months before. After the crisis, the price of oil recovered and rose to USD 80 a barrel in 2009.
- Arab Spring: 7 December 2010–mid-2012.
  The Arab Spring refers to the protests and uprisings in the Middle East and North Africa region from 2010 to 2012. In Libya, authoritarian dictator Colonel Muammar Gaddafi was overthrown and the country has remained in a state of civil war. The price of oil (Brent) rose from USD 95 per barrel in January 2011 to USD 120 a barrel in February 2011.
- Crimean Crisis, 2014, and sanctions on Russia, 2014–present
  The Russian Federation annexed the Crimean Peninsula in March 2014, which the USA and its allies in Europe protested strongly. The USA and EU countries imposed sanctions, and there was a dramatic oil price drop, from about $100 in early 2014 to about $25–60 per barrel thereafter. Oil revenue is about 10% of Russia's GDP. The sanctions, together with the oil price drop, contributed to the Russian ruble collapse and the Russian financial crisis in 2015–2016.
- Shale boom: 2015–2019
  The shale oil industry boomed in 2015, with USA oil production increasing by 20–30% on a daily basis. This supply glut put consistent downward pressure on oil prices. US crude oil production reached ten million barrels a day, which had not happened since the 1980s.

6.6 Crude Oil: A Pricing Model

Based on our analysis of oil prices over the past 150 years, especially the last 40 years, we identify the following factors that impact the price of oil:

1. Demand: measured by the GDP growth rate of China and the USA
2. Supply disruption: measured by agreed cut or increase in production
3. US dollar strength: measured by the US Dollar Index
4. War and social crisis: measured by a score from -3 to 3
5. Time series trend: measured by a 3-month trend

Using P as the price of oil, we specify a linear oil pricing model as follows:

$$P = b_0 + b_1 GDPg + b_2 SUD + b_3 USDX + b_4 WSR + b_5 TRD + \epsilon \quad (6.4)$$

where GDPg is the combined real GDP growth rate of China and the USA, SUD is oil supply disruption, USDX is the US dollar index, WSR is war and social crisis, and TRD is the time series trend. We briefly discuss the data for each variable. While some variables have a daily frequency, such as US dollar strength and the price of oil, most variables have a much lower frequency, such as oil production and GDP data. In this study, we use monthly frequency data. We outline rationales for each factor below.

GDP: We use the combined real GDP growth rate of China and the USA to approximate demand. The quarterly GDP data for China is only available after 1992. We transform the data from quarterly to monthly and add a 2-month lag to ensure information availability. For example, the GDP growth rate for 2019 Q1 will be known as of the end of May 2019. Before 1992, we use only the US GDP growth rate and then both the USA and China after 1992. It is expected that a higher GDP growth rate will increase the price of oil in the long run.

USDX: We use the future data of DX (DX-Y). We use the average of the daily data for each month for this variable. Because oil transactions are in US dollars, lower values of USDX will trigger higher demand, so it is expected that USDX has positive impacts on the price of oil.

SUD: Supply disruption. We construct the oil production disruption factor with major oil-producing countries: the USA, Russia, Arab Saudi.

WSR: depends on three considerations: seriousness of the event, whether the entity imports or exports oil, and relationship with the USA. Depending on the latter two, the impact of an event on oil price can be negative or positive. If the entity exports oil and has a negative relationship with the USA, the score is negative; if the entity imports oil, the score is positive. We assign a score from -3 to 3 to indicate the direction and magnitude of impacts.

TRD: this is measured by the 3-month moving average,

$$TRD = \frac{2P_t}{P_{t-1} + Pt - 2},$$

where P_t is the average of daily prices for the month t.

Given that crude oil prices stayed constant before the 1970s, the above pricing model is more suitable for the period from 1973 to 2018 for identification reasons. It is fairly straightforward to collect data for each variable and construct a month-end data set from 1973 to 2018.

We leave further exploration to readers. We suggest readers run an in-sample study first and then explore the out-of-sample forecasting power of this model. We remind readers to pay special attention to unit root and spurious relationship.

6.7 Price of Oil and Price of Gold: Cointegration

In the previous sections, we have explored issues for single time series data. What if there are two time series? How can we characterize the relationship between them? In this section, we use the examples of the price of oil and the price of gold to explore the relationship between two time series. We first introduce the cointegration concept through the famous example of the drunkard and her dog, then analyze whether the price of gold and the price of oil are cointegrated.

6.7.1 Two Time Series: The Price of Oil and the Price of Gold

In the commodity space, gold and oil are two very important assets with very different characteristics: while oil is used everywhere, gold has little use in the real world. The price of gold is also strongly correlated with the price of oil throughout modern history. In fact, the prices of these two commodities are so intertwined that they seem almost incapable of heading in separate directions over long time frames. Their close relationship arises from both economic and structural linkages. Most importantly, higher oil prices tend to slow down the US economy as a whole and reduce disposable income for most Americans, which adversely affects the financial markets and the economies of both the USA and other economic entities. Gold shines brilliantly at such times. In addition, the price of oil is a significant cost factor for gold production. Higher oil prices increase the cost of extracting gold and negatively impact the long-term gold supply.

We investigate the relationship between two time series—the price of oil and the price of gold—from 1968 to 2018. We use the starting year of 1968 because before 1968, there was no free market for the price of gold in the USA as the result of the Bretton Woods Agreement that linked gold to the USD directly. However, in May 1968, the free market was allowed to establish its own price. We collect monthly average data for prices of gold from 1968 to 2018.[3] We present time series plots in Fig. 6.12 and a scatter plot in Fig. 6.13. We see clearly in Fig. 6.12 that the price of oil and the price of gold move together most of the time from 1968 to 2018. Focusing on the price of gold first, we see that gold price moved quickly away from the fixed price of $35 per ounce to approximately $40 per ounce by the end of May 1968 and reached over $100 in 1973. The price of gold moved up rapidly over 1978 and 1979, culminating in a 48% increase from December 1979 to January 1980 during the Iran hostage crisis. The price of gold peaked at $675.30 per ounce in January 1980, and this was to remain the highest price ever observed until April 2006. Following the surge in the late 1970s and early 1980s, the price of gold declined and fluctuated in the range of $200–$500 per ounce until 2001. Since April 2001, the price of gold has been on an upward trend and has set new record highs in recent years, which has garnered renewed attention from the media. The price of gold reached $800 in 2007, $1000 in 2009, and $1600 by the end of 2011, then dropped to $1300 in 2014 and stayed in the range of $1200–1600 thereafter.

The pair's relationship is shown more clearly in the scatter plot in Fig. 6.12, with the order of magnitude versus the order of time in Fig. 6.13. The scatter plot indicates that while overall there is a strong positive relationship between price movements of

[3] The price of gold is based on the London PM price, measured by USD per ounce.

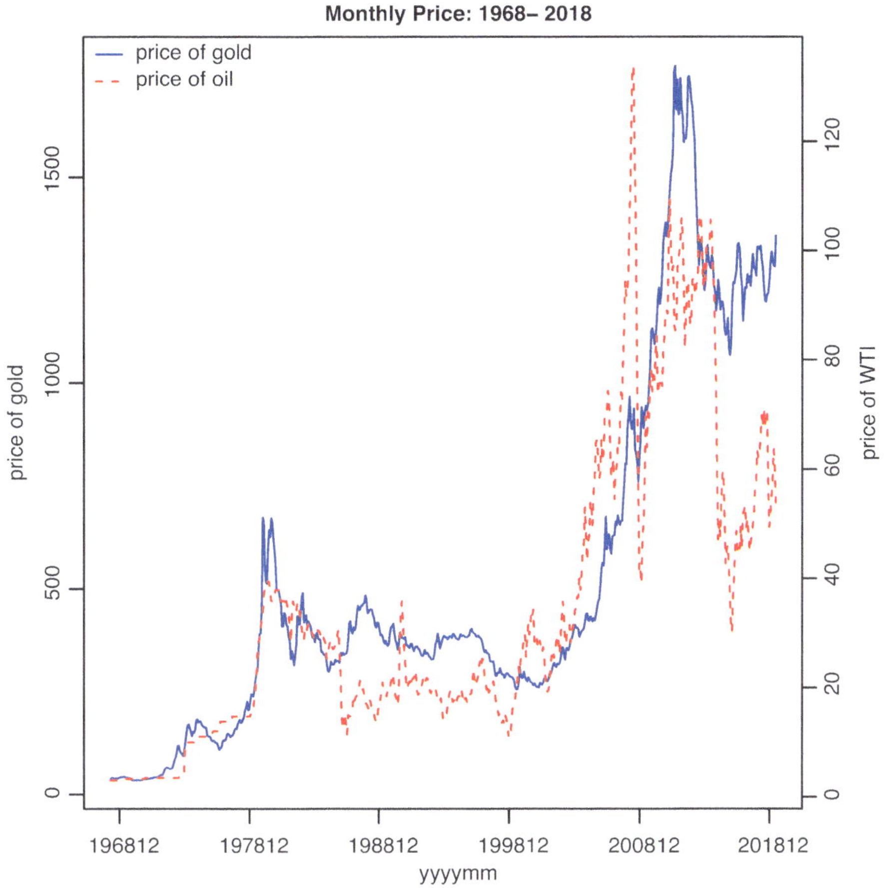

Fig. 6.12 The price of gold and WTI from 1968 to 2018. Data source: price of gold, Kitco; price of WTI, EIA

the two commodities, the relationship becomes weaker when the price of oil is high, indicating heterogeneity.

For any two time series data sets, their relationship will fall into one of the following three cases:

1. both are stationary.
2. one is stationary and the other is non-stationary, and
3. both are non-stationary.

The first case is the easiest, and we can apply OLS directly to derive the relationship. The second and third cases involve non-stationary series, so we need to pay special attention for spurious relationships. For the second case, to avoid spurious relationships, we can transform the non-stationary variable to make it stationary,

Fig. 6.13 The scatter plot of the price of gold and WTI from 1968 to 2018. Data source: price of gold, Kitco; price of WTI, EIA

then apply OLS to estimate the relationship. In the third case, the relationship can be very complicated. Two non-stationary random variables can move together or move purely independently from each other. To find out which case the paired data series belong to, we first need to know whether each single time series is stationary.

To explore the quantitative relationship between the prices of gold and oil, we first conduct univariate and bivariate analysis for each time series. We calculated the four moments and correlations (both Pearson and rank) as shown below using R scripts. We see that both time series are highly skewed to the right, which is supported by the fact that the means are much higher than the medians. Matching the results from the scatter plots, the correlations between the price of oil and the price of gold are very high: 86% for Pearson and 87% for rank.

Price of oil and gold: four moments and correlations

```
> ## univariate: summary and four moments
> summary(gold$gold.price)
   Min. 1st Qu.  Median    Mean 3rd Qu.    Max.
  34.94  276.50  380.70  527.60  649.60 1772.00
>  summary(gold$WTI)
   Min. 1st Qu.  Median    Mean 3rd Qu.    Max.
   3.07   15.01   25.74   34.63   47.90  133.90

> sd(gold$gold.price)
[1] 435.8839
> sd(gold$WTI)
[1] 27.96233

> skewness(gold$gold.price)
[1] 1.181013
> skewness(gold$WTI)
[1] 1.202785

> kurtosis(gold$gold.price)
[1] 3.273274
> kurtosis(gold$WTI)
[1] 3.656487

> ##bi-variate: Pearson and rank correlation
>  cor(gold$gold.price,gold$WTI)
[1] 0.8592699
>  cor(gold$gold.price,gold$WTI,method="spearman")
[1] 0.8728937
```

To test whether there is a unit root in the prices of gold and oil, we next carry out an ADF test for each time series. The ADF test produces a p-value of 0.7432 for the price of gold and 0.3102 for the price of oil, indicating the strong presence of a unit root in both time series. Below are the R scripts for the tests.

Unit root test: price of gold and oil

```
> ## unit root test for the price of gold
> adf.test(commodity$gold.price)
Augmented Dickey-Fuller Test
data:  commodity$gold.price
Dickey-Fuller = -1.6103, Lag order = 8, p-value = 0.7432
```

```
alternative hypothesis: stationary

> ## unit root test for the price of oil
> adf.test(commodity$WTI)
Augmented Dickey-Fuller Test
data:  gold$WTI
Dickey-Fuller = -2.6333, Lag order = 8, p-value = 0.3102
alternative hypothesis: stationary
```

Now it seems that we are facing a puzzle: intuitively, there do exist fundamental links between the price movements of the two commodities, but both time series are non-stationary and thus any quantitative relationship can be spurious. Therefore, there is a possibility that the spuriousness exaggerates the fundamental relationship. The question is, how do we know the proportions of the relationship that are fundamental and spurious if we have an overall high correlation between the two price data sets? We explore answers to this question in more detail in the next section.

6.7.2 A Drunk Man and His Dog: Cointegration

In the previous section, we found that both the price of oil and the price of gold have a unit root. How do we define and analyze the relationship between two non-stationary data sets? We can explore the answer using the classic example of a drunk lady and her dog in Murray (1994): "Suppose we follow a drunk out of the bar after a night of drinking. We observe that her path looks much like a random walk. Also, if we have ever watched a dog freely exploring, its path also looks much like a random walk. The dog will go this way and that, wherever its nose leads it. If, additionally, the dog belongs to the drunk, and the drunk calls periodically for the dog, the dog will stay pretty close to the drunk. So, the drunk and her dog go forth wandering aimlessly, together."

This is exactly the concept of cointegration: two (or more) series wandering aimlessly, together! The term "cointegration" was coined by Engle and Granger (1987) as a continuity of early studies on spurious relationship by Granger and Newbold (1974) and Granger (1981).

Definition 6.1 Suppose we have two time series of random variables, X_t and Y_t, and over time $t = 1, 2, 3, \ldots, T$, both have integration of one, I(1). If the combined movement of $aX_t + bY_t$ is stationary, where a and b are constant numbers, then X_t and Y_t are said to be cointegrated, or there is cointegration for X_t and Y_t.

Cointegration occurs very often in time series macro variables, such as household income and consumption, currency values, and commodity prices. Cointegration can also occur at the company level, such as peer companies' stock prices.

We present below the procedure to test for cointegration of two time series.

H_0: X_t and Y_t are cointegrated, H_1: X_t and Y_t are NOT cointegrated.

- Step 1. Test if X_t and Y_t both follow a unit root process, integration one, I(1).
- Step 2. If both X_t and Y_t are I(1), run OLS to get residuals, $y_t = \beta_0 + \beta_1 x_t + \epsilon_t$.
- Step 3. Run a unit root test to investigate if $\hat{\epsilon}_t$ is stationary.
- Step 4. If $\hat{\epsilon}_t$ is stationary, we say that X_t and Y_t are *cointegrated*; otherwise, there is no cointegration.

The above test is called the Granger–Engle test (Engle and Granger 1987); another similar test for cointegration for multiple time series is the Johansen test (Johansen 1991).

When two non-stationary variables are not cointegrated, they may move independently from each other, or the combination of the two may be non-stationary. Recall the third case specified in the previous section where both time series are non-stationary. We can further classify the relationship into three types, extending the classic example of a drunk and her dog:

- two drunk men who are strangers walk out of the bar:
  random walks with independent paths
- two drunk men who are friends walk out of the bar:
  random walks with the same path for a while then deviate (to their own homes)
- a drunk man walks out of the bar with his dog:
  random walks that stray together

Clearly, cointegration is only a special case for two non-stationary time series.

6.7.3 The Prices of Gold and Oil: Are They Cointegrated?

We know from the previous section that both gold and oil prices have a unit root. To find out if the two data series are cointegrated, we carry out the Granger–Engle test. We first run OLS for the model

$$G_t = b_0 + b_1 O_t + \epsilon$$

where G_t is the price of gold and O_t is the price of oil. We obtain $\hat{\epsilon}$ and then use the ADF test to determine whether the residuals are stationary. The OLS results show that $b = (b_0, b_1)$ are very significant and the $R^2 = 0.74$ is a very high value, implying that the two series have a very significant relationship: the price of oil can explain about 74% of the price of gold! However, we have to remember that the two series are non-stationary, so the high significance could arise from both fundamental links and spurious phenomena. The unit root test of OLS residuals indicates that the residuals are not stationary (p-value is 0.41), so the price of gold and the price of oil are not cointegrated. We present the R codes below for the test.

Test of cointegration between the price of gold and the price of oil

```
> coin=lm(gold.price~WTI,data=gold)
> print(summary(coin))
Coefficients:
            Estimate Std. Error t value Pr(>|t|)
(Intercept)  63.7879    14.3295   4.452 1.01e-05 ***
WTI          13.3945     0.3221  41.591  < 2e-16 ***
---
Residual standard error: 223.1 on 613 degrees of freedom
Multiple R-squared:  0.7383,Adjusted R-squared:  0.7379
F-statistic:  1730 on 1 and 613 DF,  p-value: < 2.2e-16

> coin.i1=adf.test(coin$resid)
> print(coin.i1)
Augmented Dickey-Fuller Test
data:  coin$resid
Dickey-Fuller = -2.3877, Lag order = 8, p-value = 0.4142
alternative hypothesis: stationary
```

Cointegration test can help to detect spurious relationship. For non-stationary time series, another significant benefit brought by cointegration is that there is no need to transform non-stationary data into stationary.

Now that we know the prices of gold and oil are not cointegrated, we need to transform both variables to minimize or avoid spurious relationships. We use returns of prices for both series.

$$R_{g,t} = \frac{P_{g,t}}{P_{g,t-1}} - 1, \quad R_{o,t} = \frac{P_{o,t}}{P_{o,t-1}} - 1,$$

where R_g is the return for gold and R_o is the return for oil. If there is indeed a strong fundamental relationship between the prices of these two commodities, it will be reflected in the relationship between R_g and R_o, though with some information loss.

After the transformation, we conduct unit root tests for both time series. The ADF test has a p-value of 0.01 for both return series, rejecting the hypothesis that the return series have a unit root. Regarding the bivariate analysis, the Pearson correlation between returns of oil and returns of gold is 18%, while the rank correlation is 13%, indicating some outliers in the returns data set. The R scripts below carry out the unit root tests and compute correlation values between the two returns.

Test of unit root for returns of gold and returns of oil

```
> ## test the unit root for Returns of GOLD
> return.gold.i1=adf.test(gold$return.gold)
> print(return.gold.i1)
Augmented Dickey-Fuller Test
data:  gold$return.gold
Dickey-Fuller = -6.4087, Lag order = 8, p-value = 0.01
alternative hypothesis: stationary

> ## test unit root for Returns of OIL
> return.wti.i1=adf.test(gold$return.wti)
> print(return.wti.i1)
Augmented Dickey-Fuller Test
data:  gold$return.wti
Dickey-Fuller = -8.3753, Lag order = 8, p-value = 0.01
alternative hypothesis: stationary

> ## Run correlations between returns: Pearson and rank
> print(cor(gold$return.gold,gold$return.wti))
[1] 0.1815562
> print(cor(gold$return.gold,gold$return.wti, method=
  "spearman"))
[1] 0.1395845
```

Given that both returns time series are stationary, we run OLS for the returns model

$$R_{g,t} = \gamma_0 + \gamma_1 R_{o,t} + \mu_t,$$

and present the R codes and results below. We see that the returns of oil are a significant factor (t-value = 4.57) to explain contemporaneous returns of gold, with $R^2 = 0.03$.

Contemporaneous relationship between returns of oil and returns of gold

```
> return.coin=lm(return.gold~return.wti, data=gold)
> print(summary(return.coin))
Residuals:
     Min       1Q   Median       3Q      Max
-0.17681 -0.02492 -0.00537  0.02031  0.47771

Coefficients:
            Estimate Std. Error t value Pr(>|t|)
```

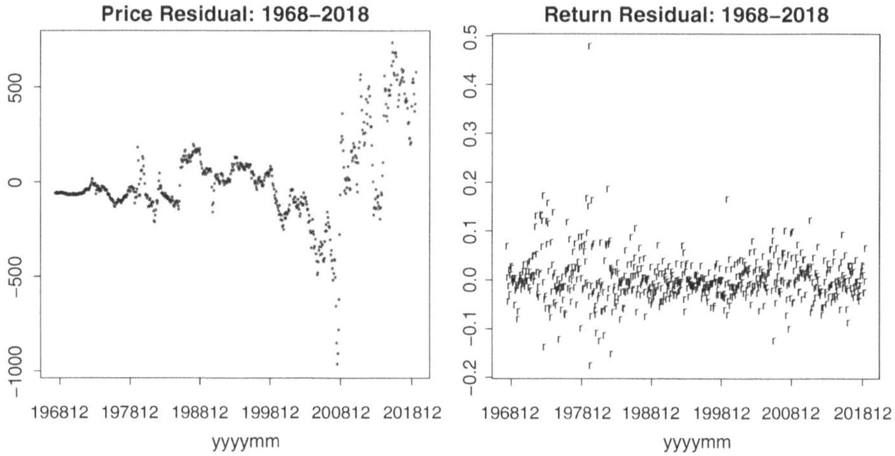

Fig. 6.14 The residuals from price data (non-stationary, left plot) and the residuals from returns data (stationary, right plot)

```
(Intercept)  0.90777    0.02182  41.605  < 2e-16 ***
return.wti   0.09844    0.02155   4.567 5.97e-06 ***
---
Residual standard error:0.0489 on 612 degrees of freedom
Multiple R-squared: 0.03296,Adjusted R-squared: 0.03138
F-statistic: 20.86 on 1 and 612 DF,  p-value: 5.974e-06
```

The above result is supported by fundamental links between the price movements of the two commodities at both the macro and micro levels, as we discussed early in this chapter. Recall that we produced residuals for the price data, and those residuals are non-stationary. To illustrate the difference between a non-stationary price residual and a stationary returns residual, we present the plots of both residuals in Fig. 6.14.

6.8 Industry Insights: Pair Trading Strategy

In quantitative investing, one immediate application of cointegration is pair trading. Suppose you have two stocks, X_1 and X_2. Both prices have unit roots, but they are cointegrated. A linear combination, $aX_1 + bX_2$, is stationary, that is, the two series never stray very far from one another. If 1 day this "spread" between them is unusually large, we can profit from the spread as we know that the large spread would not last long given the characteristics of the cointegration. This serves as the basis for pair trading.

We use the example of stock prices for American Airlines and United Airlines to illustrate the industry approach to a pair trading strategy.

6.8.1 *Cointegration Application: A Pair Trading Strategy*

We select two peer companies in the airline industry: United Airlines (UA) and American Airlines (AA). In addition to firm-specific factors, anything impacting American Airlines will impact United Airlines as well. For example, the price of oil is a significant part of the costs, and hence profits, for both companies. We expect that (1) the stock prices of both companies have a unit root and (2) the two stock prices are cointegrated. If this is confirmed, we can then proceed to design a pair trading strategy.

We collect daily closing price data for both stocks from January 31, 2008 to July 24, 2019. To visualize stock price movements of the two companies, we have a time series plot for both AA and UA in Fig. 6.15. We see that the two prices move together

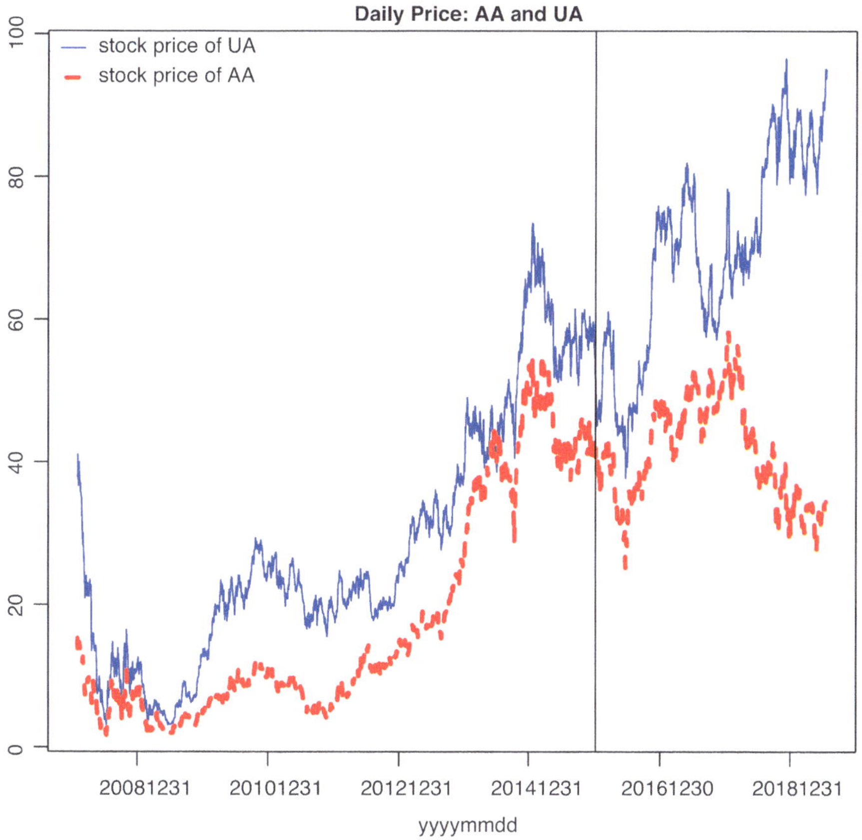

Fig. 6.15 The stock prices of AA and UA from January 31, 2018 to July 24, 2019. Data source: finance.yahoo.com

in a very similar pattern, and the price of UA is higher than the price of AA most of the time. Moreover, we observe that the two prices move closer to or farther from each other over the study period, which provides an opportunity for a pair trading strategy.

To define the quantitative relationship, we carry out a cointegration test for the stock prices of AA and UA. We divide all the data into two parts: the first part is from January 31, 2008 to January 11, 2016, and the second part is from January 12, 2016 to July 24, 2019. We use the first part of the data as an in-sample to test for cointegration and estimate the quantitative relationship between the two stock prices. Once we confirm their cointegration, we specify details of a pair trading strategy between the two prices and implement the strategy with the out-sample, the second part of the data. Thus, the division of in-sample and out-sample helps us when we derive the in-sample results and obtain out-sample portfolio performance, thus validating the strategy in the real world. The R scripts for the cointegration test are presented below.

Cointegration test of stock prices of AA and UA

```
>## Step 1, test unit root for stock prices of AA and UA
> aa.i1=adf.test(airlines$Close.AA)
> print(aa.i1)
Augmented Dickey-Fuller Test
data:  airlines$Close.AA
Dickey-Fuller = -1.9931, Lag order=13, p-value=0.5813
alternative hypothesis: stationary

> ua.i1=adf.test(airlines$Close.UA)
> print(ua.i1)
Augmented Dickey-Fuller Test
data:  airlines$Close.UA
Dickey-Fuller = -1.6127, Lag order = 13, p-value = 0.7423
alternative hypothesis: stationary

> ##   Step2, run OLS regression
> AAregUA=lm(Close.AA~Close.UA, data=airlines)

> ## Step 3, run ADF test for residuals
> AAandUA.coin=adf.test(AAregUA$resid)
> print(AAandUA.coin)
Augmented Dickey-Fuller Test
data:  AAregUA$resid
Dickey-Fuller = -3.1576, Lag order=13, p-value=0.0952
alternative hypothesis: stationary

> ## Step 4, make reference on cointegration: 90\% confidence level.
```

We follow the four-step cointegration test procedure to study whether the prices of AA and UA are cointegrated. The results from Step 1 indicate that both data series have a unit root. We then apply OLS regression to the following model,

$$P_{AA} = b_0 + b_1 P_{UA} + \epsilon,$$

where P_{AA} is the stock price for AA and P_{UA} is the stock price for UA. The residuals from the OLS method were used for an ADF test, which shows that at the 10% level, the residuals are stationary, implying that the two companies' stock prices are cointegrated.

Using the cointegration relationship, and following the industry approach, we specify a simple trading strategy for AA stock and UA stock in the next section.

6.8.2 *Stock Pairs: Entering and Exiting*

A pair trading strategy requires following specifications:

1. There exists a cointegration relationship between the two price series.
2. Define the quantitative relationship between the two series.
3. Conditions for entering and exiting the trade.
4. Entering the pair trades.
5. Exiting the pair trades.
6. Risk control: buffer zone.

Before using the real-world data, we use simulated data to show how a pair trading strategy works. Suppose we have two stocks X and Y, their price movements are cointegrated (i.e., each time series is a I(1)), and the linear combination of the two series is I(0), stationary.

$$y = 5 + x + \epsilon$$

$$y - x = 5 + \epsilon$$

The distance of the two prices will fluctuate around 5; the average of ϵ is zero.

We have simulated values for Y and X and present their prices for 14 business days in Fig. 6.16. We employ this simple example to show the principles and mechanism of a pair trading strategy. When the price of *Y* increases on Day 4, the distance between X and Y is 7, larger than 5; on Day 5, the price of Y increases even further to 25, making the distance between Y and X 10. We know that this will not be sustainable because X and Y are cointegrated. Any large deviation will revert to the long-term average range, 5. This will occur through either a decrease in Y or an increase in X. That is, the expected price of Y will decrease and/or the expected price of X will increase.

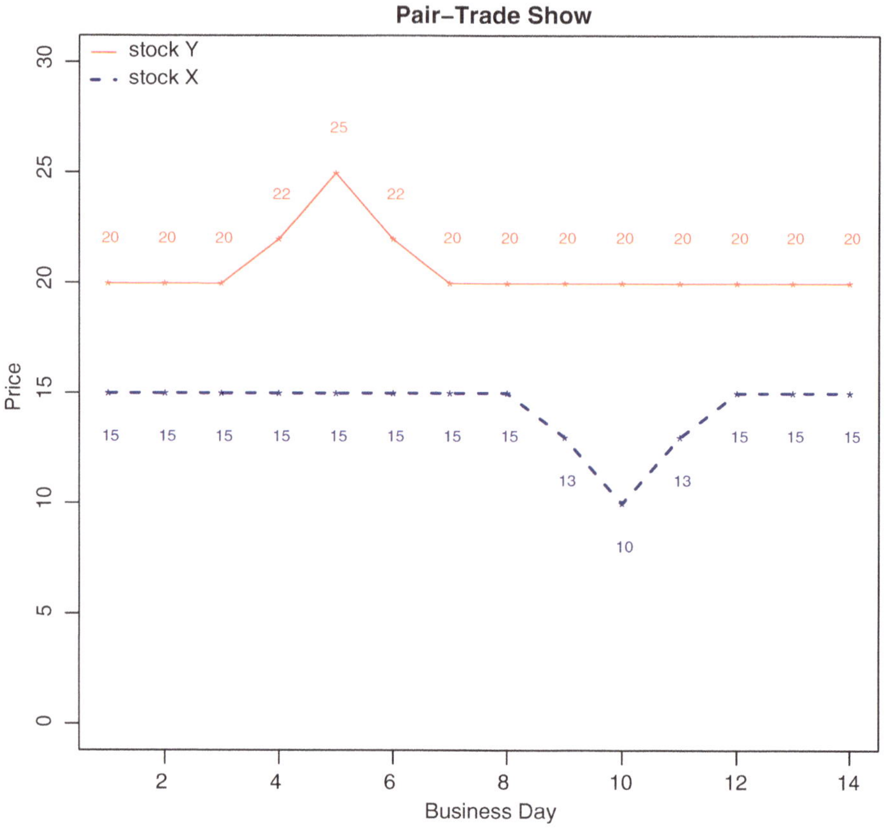

Fig. 6.16 Illustration of a pair trading strategy using simulated data

In terms of investment strategy, we could short sell stock Y and buy long stock X at the same time. So, on Day 5, we short sell 100 shares of Y at a price of \$25 and buy 100 shares of X at a price of \$15. When the distance reverts to 5, we can close the positions. On Day 6, the price of Y drops to 22 and the distance is 7. On Day 7, the price of Y drops to \$20, and the distance is 5, the long-term norm. Now, we can buy 100 shares of Y at a price of \$20 to cover the short position and sell X at \$15 to close the long position. We made $100 \times (25 - 20) = \$500$ on stock Y and $100 \times (15 - 15) = \$0$ on stock X, thus the return is 5/25=20% on stock Y and 0% on stock X. Life continues. However, on Day 9, the price of X drops from \$15 to \$13, then \$10 on Day 10, causing the distance between Y and X to be 10 now, far from the long-term norm of 5. We think this will not last long, expecting the price of X to increase and/or the price of Y to decrease to revert to the normal distance. Taking the opportunity on Day 10, we short 100 shares of Y at a price of \$20 and buy 100 shares of X at a price of \$10. Then we wait. On day 12, stock X's price increases to \$15, and the distance between X and Y is back to the long-term norm.

We close the positions by buying 100 shares of Y at \$20 and selling 100 shares of X at \$15. From this pair trading strategy, we made $100 \times (20 - 20) = 0$ on stock Y but $100 \times (15 - 10) = \$1500$ on stock X.

Now we focus on a pair trading strategy for a real-world example: stock prices of AA and UA. We know from the previous section that the stock prices of AA and UA are cointegrated. Now we need to discover the quantitative relationship between the two time series using the in-sample data. We present the OLS results computed by R codes below.

Cointegration test of stock prices of AA and UA

```
> ## OLS regression: stock price of AA ~ stock price of UA
> AAregUA=lm(Close.AA~Close.UA, data=airlines)
> print(summary(AAregUA))
Residuals:
    Min      1Q  Median      3Q     Max
-15.684  -5.530  -1.931   2.701  33.306

Coefficients:
            Estimate Std. Error t value Pr(>|t|)
(Intercept) -5.25117    0.36411  -14.42   <2e-16 ***
Close.UA     0.88295    0.01065   82.94   <2e-16 ***
---
Residual standard error: 8.593 on 2399 degrees of freedom
Multiple R-squared:  0.7414,Adjusted R-squared:  0.7413
F-statistic:  6878 on 1 and 2399 DF,  p-value: < 2.2e-16
```

We run OLS regression for the model and get the following estimates:

$$P_{AA} = \underset{(0.36)}{-5.25} + \underset{(0.01)}{0.88} P_{UA} + \hat{\epsilon}$$

which implies that

$$P_{AA} - 0.88 P_{UA} = -5.25 + \hat{\epsilon}, \quad E(\hat{\epsilon}) = 0.$$

Note that the estimates for coefficients have very high t-values, indicating that they are very significant from zero. The R^2 is 74%, indicating a strong relationship between the two time series.

If $P_{AA} - 0.88 P_{UA} = -5.25$, then the distance is constant and there would be no investment opportunities. Thus, the value of $\hat{\sigma}$ decides the investment strategy: if $\hat{\epsilon}$ is positive and large, we expect that the distance will be smaller. Either a decrease in P_{AA} or an increase in P_{UA} will cause the distance to revert to -5.25 at some point because the two data series are cointegrated. From the OLS results above, we see that the residuals vary widely from −15.684 to 33.306, with a median of −1.931 (the mean is zero). The standard deviation of the residuals is 4.23.

Based on the quantitative relationship from the in-sample data, we design the following simple pair trading strategy for AA and UA:

Enter condition: $P_{AA} - 0.88P_{UA} >= a + 0.5\sigma = -5.25 + 0.5 \times 4.23 = -3.135$

Pair trades: *short sell AA and long buy UA;*

Exit condition: $P_{AA} - 0.88P_{UA} <= -5.25$

Pair trades: *cover AA and sell UA.*

We execute the pair trading strategy both in-sample and out-sample. The in-sample results confirm the profitability of the strategy. We present the results of the entry and exit dates, price, and returns in Table 6.8.

While the in-sample portfolio results simply confirm that the strategy may work, the out-sample results are a real test of the profitability of the strategy. Of course, this will be based on the assumption that the cointegration and quantitative relationship defined from the in-sample period will more or less continue into the out-sample period. We execute the pair trading strategy for the out-sample period and present the results in Table 6.9. There are 6 pair trades, with each trade including entry and exit, during this out-sample period from January 12, 2016 to July 24, 2019. The performance is positive for each trade, with returns ranging from 6.38% to 13.04%.

We summarize the portfolio's performance in Table 6.10 with annualized returns, risk, and Sharpe ratios for in-sample and out-sample periods. For comparison purposes, we also include the performance of the S&P 500 index for the same period. We see that first, the in-sample portfolio has much higher returns than the returns of both the index and out-sample because of the in-sample attribution, as expected. For the out-sample period, the pair trading strategy delivers a 6.32% annualized return, almost double the S&P 500 index return of 3.36%, but with the same level of risk at 12.52%, resulting in a Sharpe ratio of 0.50. This simple example shows that the pair trading can be a profitable strategy (Figs. 6.17 and 6.18).

Table 6.8 The in-sample results of pair trading strategy for daily stock prices of AA and UA from January 31, 2008 to January 11, 2016

Enter date	Exit date	Enter price, AA	Enter price, UA	Exit price, AA	Exit price, UA	Return, pair trade
20080428	20100105	8.62	14.81	5.31	13.91	0.5626
20121031	20121213	12.18	19.21	12.97	23.06	0.1395
20131024	20131029	22.67	31.30	22.37	33.91.	0.0968
20140203	20150126	33.96	43.82	55.45	73.62	0.2925
20150318	20150721	54.13	68.10	40.90	55.97	0.1454
20151016	20151106	43.71	55.97	45.34	61.49	0.0627

Table 6.9 The out-sample results of pair trading strategy for daily stock prices of AA and UA from January 12, 2016 to July 24, 2019

Enter date	Exit date	Enter price, AA	Enter price, UA	Exit price, AA	Exit price, UA	Return, pair trade
20160112	20160223	42.00	50.86	40.38	55.45	0.1304
20160429	20160527	34.69	45.81	31.65	45.18	0.0823
20160722	20160805	36.36	47.57	34.44	48.41	0.0734
20160816	20160831	36.75	47.99	36.30	50.41	0.0628
20170918	20180110	45.31	58.11	53.78	73.08	0.1001
20180124	20180327	54.79	69.05	50.90	68.18	0.0638

Table 6.10 The in-sample and out-sample portfolio performance and S&P 500 index performance (return and risk in %)

	In-sample period S&P 500 index	In-sample period Pair-trade strategy	Out-sample period S&P 500 index	Out-sample period Pair-trade strategy
Annual return	4.46	15.05	3.36	6.32
Annual risk	22.12	23.85	12.85	12.52
Sharpe ratio	0.20	0.63	0.26	0.50

While pair trading can be a profitable strategy, it has risks just like any other investment strategies. One of the biggest challenges is the possibility that the spread can be wider and last longer. For example, when one of the paired companies shifts its core business to a new area, the stock prices of these two companies would not be cointegrated any more. This can cause dramatic loss for investments. Therefore, it is wise to have some risk management built into the pair trading strategy.

6.9 R Packages, Database Connection, and Time Series Commands

We discussed how to write an R function in Chap. 5. A function can be made to deal with each subject area. Regardless of differences across these subject areas, there are many tasks that are common to almost all users, such as data import, data wrangling, and data visualization. This is the base for R packages.

In this section, we discuss R packages, connections to external databases, and analysis of time series data with R. We start with R packages first.

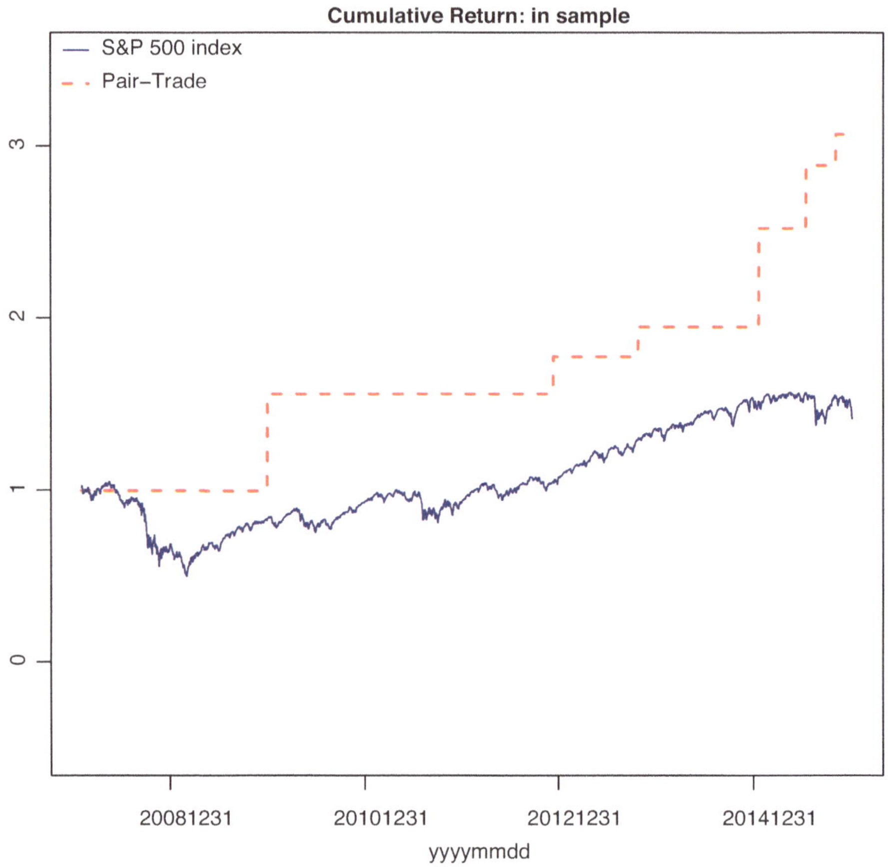

Fig. 6.17 The cumulative returns of the S&P 500 index and pair trade (in-sample) from January 31, 2008 to January 11, 2016

6.9.1 R Packages

R has thousands of available packages. There is a standard set of packages when R is installed. These default packages are available when we open an R session. Others are available for download and installation when needed.

Typically, a package will include documentation, R functions, and data sets. The basic information about a package is provided in the *description* file, which describes the main function, the author and other information such as dates and license of the package.

Download and Install a Package We only need to do this once. The most common way is to use the CRAN repository, where we just need the name of the package. For

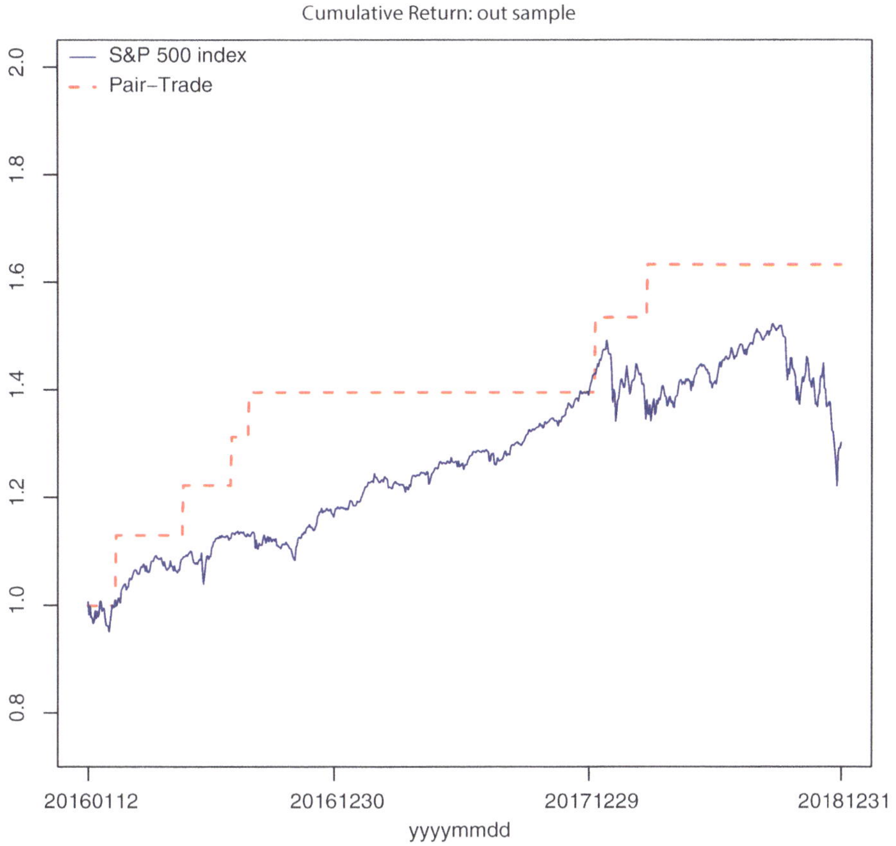

Fig. 6.18 The cumulative returns of the S&P 500 index and pair trade (out-sample) from January 12, 2016 to July 24, 2019

example, we can use the command *install.packages("vioplot")*, where "vioplot" is the name of the first R package written by Daniel Adler.

You can also install packages manually. Here is a brief description of the steps to install packages for Windows and Mac computers.

Windows:

1. Go to the Packages menu, click on "Install Packages".
2. Select a mirror and a package (e.g., *moments*), complete installation.
3. Use the R command *library* in an R session, e.g., library(moments).

Mac:

1. Choose Package Installer from the Packages and Data menu.
2. Select a package repository, such as CRAN (binaries).
3. Type the package name in the box "package search".

4. Click on "Get List" and select the package.
5. Select a proper location and click on "Install Selected" to install the package.

After the installation, there is information on the R console about the status of the installation. Depending on what operating system you are using, the information usually includes dependencies and whether the package was successfully installed. For example, we install the "vioplot" package in R, and the following shows information about the installation.

Install an R package

```
> install.packages("vioplot")
  There is a binary version available but the source version is later:
        binary source needs_compilation
vioplot    0.2  0.3.2             FALSE

installing the source package ?vioplot?
...
* DONE (vioplot)

The downloaded source packages are in
.../downloaded_packages
```

To Use the Package To use the package, we need to use the command *library* to load the package into the session. For example, if we want to have the *moments* package, we need to type *library(moments)*. We present some basic information about a package by using the command *library(help=package name)*.

How to use an R package

```
 > library(moments)
 > library(help=moments)
 Package:           moments
Type:              Package
Title:             Moments, cumulants, skewness, kurtosis and related tests
Version:           0.14
Date:              2015-01-05
Author:            Lukasz Komsta <lukasz.komsta@umlub.pl>,
                   Frederick Novomestky <fnovomes@poly.edu>
Maintainer:        Lukasz Komsta <lukasz.komsta@umlub.pl>
Description:       Functions to calculate: moments, Pearson's kurtosis, Geary's kurtosis
                   and skewness; tests related to them
                   (Anscombe-Glynn, D'Agostino, Bonett-Seier).
License:           GPL (>= 2)
URL:               http://www.r-project.org, http://www.komsta.net/
Packaged:          2015-01-05 11:30:01 UTC;
                   Administrator
NeedsCompilation:  no
Repository:        CRAN
```

```
Date/Publication:   2015-01-05 12:58:47
Built:              R 3.3.0; ; 2016-05-04 14:45:31 UTC; unix

Index:

agostino.test          D'Agostino test of skewness
...
jarque.test            Jarque-Bera test for normality
kurtosis               Pearson's measure of kurtosis
skewness               Skewness of the sample
```

More on Packages

1. Updating an existing package: Many package authors continually update the contents in their packages. We can download updates for a package already installed in R.
2. Unload the package: remove.packages(package, lib).
3. Select a package if there are multiple packages for the same task. This will depend on the quality of the functions you are to use from that package, so this is a case-by-case selection. Of course, the best way is to use both and then see which is more suitable for your purpose. Some factors that can be of help to judge the function are opinions from other users, the reputation of the author, and the frequency of package updates. For example, for the same subject, if one package is written by an expert in that field while another is written by a lay person, it is usually better to use the package from the expert.

6.9.2 Database Connection

For quant investing, many data sets are stored in a database, such as Oracle or MS Access. Access to a data set in a database management system usually requires the path, folder name, and password. In this section, we briefly describe how to import data to and export data from a database.

Get Data from a Database One way to access an Oracle database from R is to use the package, *RODBC*. ODBC stands for Open DataBase Connectivity, an open standard application programming interface (API) for databases. We provide example scripts on how to connect to a data set within a database. Note that we need to first create a data source name (dsn) which includes the connection information.

RODBC: connection to a database

```
> install.packages("RODBC")
> RShowDoc("RODBC", package="RODBC").)
> # Open a connection to an ODBC database
> the.path = odbcConnect(dsn, uid=''id", pwd=''password")
> # terminate the connection
> odbcClose(the.path)
```

Once we connect to the database, we can then work on the data sets stored in the database. In the RODBC package, *sqlQuery* is the workhorse function around data. Writing SQL queries is beyond the scope of this book. Readers can refer to many excellent resources on the SQL language for relational databases.

6.9.3 Time Series Analysis with R

In this subsection, we introduce some R commands related to time series analysis. We first introduce a set of R commands on dates, then discuss R commands for a unit root test and a cointegration test.

Time Series Dates First of all, the R command *format* is very useful to set a desired date format.

Dates: format

```
> today <- Sys.Date()
> format(today, "%Y%m%d")
[1] "20191213"
> format(today, "%Y%b%d")
[1] "2019Dec13"
```

In R, there are many date category commands, such as weekdays, months, and quarters. Each gets a date attribute which is very useful in time series analysis. There are also useful commands for the lag of dates and differences between days.

Dates: format

```
> weekdays(today)
[1] "Friday"
> months(today)
[1] "December"
> quarters(today)
[1] "Q4"
> ceiling(as.numeric(difftime(Sys.time(),"2010-12-27",units="days")))
[1] 3288
```

In addition, there are many R packages, such as *timeDate*, *seasonal*, and *RQuantLib*, for times and dates, especially holidays and trading dates. These packages are very helpful for global portfolios as each country has its own holidays and trading days.

Unit Root Tests Unit root tests are critical for time series models. There are two widely used unit root tests: the augmented Dicky–Fuller (ADF) test and Phillips–Perron (PP) test. We introduce R command and usage for each and then briefly discuss their differences.

ADF and PP tests templates

```
##ADF
 adf.test(x, alternative = c("stationary", "explosive"),
            k = trunc((length(x)-1)^(1/3)))
##PP
 pp.test(x, alternative = c("stationary", "explosive"),
           type = c("Z(alpha)", "Z(t_alpha)"), lshort = TRUE)
```

While both test for the null that x has a unit root, there are some differences between the ADF and PP tests. The PP test is based on the ADF model but uses a nonparametric approach to correct for autocorrelation and heteroscedasticity with asymptotic properties. On one side, if there is a large number of observations, the PP test can be more robust than the ADF test; but on the other side, for finite sample size, the PP test may underperform the ADF test since the PP test is based on asymptotics.

We display the differences by applying the two tests to the same data set. The function below displays R scripts for the ADF and PP tests of the monthly average prices of both gold and oil from April 1968 to June 2019.

ADF and PP unit root tests for prices of gold and oil

```
golddata.dir=c("/.../book/quantInvesting/springerLatex/bookQI/chapter6/")
golddata.name="gold.price.196804.csv"

oil.gold.UnitRootTest<-function(data.dir=golddata.dir,data.name=golddata.name)
{
     ## time series package, tseries
     library(tseries)

     gold=read.csv(paste(data.dir,data.name,sep=""), sep=",",header=T)

     ### tests of unit root for price of oil and price of gold
     cat("\n Unit root test for gold prices: ADF and PP \n")
     gold.adf= adf.test(gold$gold.price)
     gold.pp = pp.test(gold$gold.price)
     print(gold.adf)
     print(gold.pp)

    cat("\n unit root test for WTI prices: ADF and PP \n")
    wti.adf =adf.test(gold$WTI)
    wti.pp  = pp.test(gold$WTI)
    print(wti.adf)
    print(wti.pp)
}
```

Running the R function above produces the following results. Note that while ADF and PP tests agree that there is a unit root in the price of gold, the two tests disagree on whether a unit root is present in the price of oil. The list below presents the p-values. Clearly, for the price of WTI, the test results differ dramatically: the ADF test implies a unit root while the PP test implies no unit root. Note that there are about 500 data points, so the number of observations is not small but not large either. For a challenging case like this, we can seek help from visualization and other tests such as *kpss* to check if there are breaks. Now, we see that quant analysis is not a pure science, it can be an art! So does investment based on quantitative modeling, it needs fundamental support to validate statistical results.[4] This example also shows that it is important not only to know how to apply quantitative analysis to investment, but also to understand the assumptions and conditions required for the validity of these quantitative models, as we have stressed in the preface and throughout the book.

	Price of Gold	Price of WTI
ADF test	0.7432	0.3102
PP test	0.8418	0.0685

[4]We will discuss the combination of fundamental and quantitative analysis in detail in Chap. 9.

ADF and PP unit root tests for prices of gold and oil

```
> source("/.../book/quantInvesting/chapter6/oil.gold.R")
> oil.gold.Rsection()

# Unit root test for gold prices: ADF and PP
Augmented Dickey-Fuller Test
data:  gold$gold.price
Dickey-Fuller = -1.6103, Lag order = 8, p-value=0.7432
alternative hypothesis: stationary

Phillips-Perron Unit Root Test
data:  gold$gold.price
Dickey-Fuller Z(alpha) = -4.8027, Truncation lag parameter = 6, p-value=0.8418
alternative hypothesis: stationary

# unit root test for WTI prices: ADF and PP
Augmented Dickey-Fuller Test
data:  gold$WTI
Dickey-Fuller = -2.6333, Lag order = 8, p-value = 0.3102
alternative hypothesis: stationary

Phillips-Perron Unit Root Test
data:  gold$WTI
Dickey-Fuller Z(alpha) = -20.24, Truncation lag parameter = 6, p-value=0.06853
alternative hypothesis: stationary
```

R scripts for a cointegration test of two time series data (both have unit roots) are straightforward. It is simply another unit root test but now on the residuals obtained by running one variable against the other.

Keywords, Problems, and Group Project

Part I. Keywords

Unit root, ADF test, cointegration, Granger–Engle test, spurious relationship

Data mining, price of oil, price of gold, Bretton Woods System, USDX

Seven Sisters, OPEC, pair trading, geopolitical power

R packages, database connection, time series analysis

Part II. Problems

Problem 6.1 Get 10-year daily temperature data for your hometown (or the nearest city).

(1) Plot time series data and identify any patterns in the data.

(2) Scatter plot the temperature of today versus yesterday, 5 days ago and 10 days ago, and run regression

$$y_t = b_0 + b_1 y_{t-1} + b_2 y_{t-5} + b_3 y_{t-10} + \epsilon$$

(3) Conduct unit root tests for the model in (2) using both ADF and PP methods.
(4) Discuss pros and cons of using the model in (2) to forecast today's temperature.

Problem 6.2 Using daily oil (WTI) price data, repeat the analysis in 6.1.

Problem 6.3 Using daily price data for oil and gold.

(1) Test for unit roots for each variable using both ADF and PP methods.
(2) Test for cointegration between the price of gold and the price of oil.

Problem 6.4 Recall that you picked a stock in Problem 2.3, now pick another stock in the same industry, making sure they are the closest competitors in the industry. Collect data on daily prices of both stocks for as long as possible.

(1) Run a unit root test for each time series of stock prices.
(2) Run a cointegration test between the two stock prices.
(3) Design a pair trading strategy, specifying the entering and exiting conditions.
(4) Perform both in-sample out-of-sample (using the second half data) trades, compare portfolio performance between in-sample and out-of-sample methods, and also against the benchmark S& P 500.

Part III. Group Project

Problem 6.5 Continuing from the last problem, now with a group of 3–5 people. Now with 3–5 pairs, build a portfolio to allocate money between these 3–5 pairs, investigate portfolio performance and compare with single-pair portfolios.

(1) Does the multi-pair portfolio outperform single-pair portfolios in terms of returns? Are there any benefits in terms of risk reduction?
(2) Discuss challenges of pair trading strategies.

References

Ammann, D. 2009. *The King of Oil: The Secret Lives of Marc Rich*. New York: St. Martin's Press. ISBN 0312570740.

Carter, M. 2000. "*Operating performance following corporate restructurings*." Working paper: Columbia University.

DeAngelo, H., L. DeAngelo, and D. Skinner. 1994. "Accounting choice in troubled companies." *The Journal of Accounting and Economics* 17: 113–143.

Elliott, J., and W. Shaw. 1988. "Write-offs as accounting procedures to manage perceptions." *Journal of Accounting Research* 26(3): 91–119.

Engle, R., and C. Granger. 1987. "Co-integration and Error Correction: Representation, Estimation and Testing." *Econometrica* 55(2): 251–276.

Fuller, W.A. 1976. *Introduction to Statistical Time Series*. New York: Wiley. ISBN 0471287156.

Granger, C. 1981. "Some Properties of Time Series Data and Their Use in Econometric Model Specification." *Journal of Econometrics* 16(1): 121–130.

Granger, C., and P. Newbold. 1974. "Spurious Regressions in Econometrics." *Journal of Econometrics* 2(2): 111–120.

Johansen, S. 1991. "Estimation and Hypothesis Testing of Cointegration Vectors in Gaussian Vector Autoregressive Models." *Econometrica* 59(6): 1551–1580.

Murray, M.P. 1994. "A Drunk and Her Dog: An Illustration of Cointegration and Error Correction." *The American Statistician* 48(1): 37–39.

Perron, P. 1988. "Trends and Random Walks in Macroeconomic Time Series." *Journal of Economic Dynamics and Control* 12: 297–332.

Phillips, P.C., P. Perron. 1988. "Testing for a Unit Root in Time Series Regression." *Biometrika* 75(2): 335–346.

Said, S.E., and D.A. Dickey. 1984. "Testing for Unit Roots in Autoregressive-Moving Average Models of Unknown Order." *Biometrika* 71: 599–607.

Yule, U. 1926. "Why Do We Sometimes Get Nonsense-Correlations Between Time Series?–A Study in Sampling and the Nature of Time Series." *Journal of the Royal Statistical Society* 89(1): 11–63.

Chapter 7
Portfolio Construction: From Alpha/Risk to Portfolio Weights

Abstract In this chapter, we focus on portfolio construction. In particular, we present details of the classical mean–variance approach, including principles, algorithms, and examples of a long only and a market neutral long-short portfolio. We also discuss backtesting and portfolio performance attribution. We introduce Harry Markowitz who made important contributions to modern portfolio theory. Regarding industry insights, we show how industry practitioners build MV portfolios with practical constraints. For R programming, we discuss the structure of R codes and functions.

7.1 Aspects of a Portfolio

In quantitative investing, all brilliant investment ideas and methods ultimately have to boil down to a portfolio: what to buy/sell and for how much. A portfolio is simply a grouping of financial assets. The assets can be publicly traded securities, such as stocks and commodities, or non-publicly tradable securities, such as real estate and art. A portfolio can also include exchange-traded funds (ETFs) or mutual funds.

In this chapter, we focus on the public equity market. In this context, a portfolio is a collection of stocks. What stocks, and how many shares of each, should be included in a portfolio? This is the central question of portfolio construction. To generate a portfolio, we have to consider many factors:

- What is the investment goal of the portfolio? Is it to avoid loss or achieve capital growth?
- What is the investment universe? Does it include only DM, or will we consider EM as well?
- Is the portfolio long-only or long-short?
- What are the risk appetite and expected return?
- What other constraints exist, such as a stipulation not to invest in certain industries?

Before we get into details about portfolio construction, we introduce basic concepts about portfolios. For the public equity markets, two of the most common quantitative strategies in the industry are long-only and long-short market neutral. We discussed these two portfolios in Chaps. 4 and 5 from an alpha perspective. We now revisit them from a portfolio construction perspective.

L. Ma, *Quantitative Investing*, https://doi.org/10.1007/978-3-030-47202-3_7

7.1.1 A Long-Only Portfolio

Consider a long-only stock selection portfolio with an investment universe of the S&P 500. We present the constituents' market capitalization weights on August 1, 2019 (505 stocks) in Fig. 7.1. The left plot shows the weights from highest to lowest. There are four large stocks with weights over 2% and many stocks with weights less than 0.10%. Thus, the price changes of large stocks will have huge impacts on the index's performance. The right plot shows the density of the weights, with positive skewness and a fat right tail. We present the ten largest and ten smallest stocks by weight in Table 7.1. The top ten stocks represent more than 21.74% of the total weight, while the bottom ten represent only 0.15%.

We introduced benchmark and portfolio weights in Chap. 1. We present them again here for the readers' convenience. For a long-only strategy, suppose we have a portfolio with K stocks,

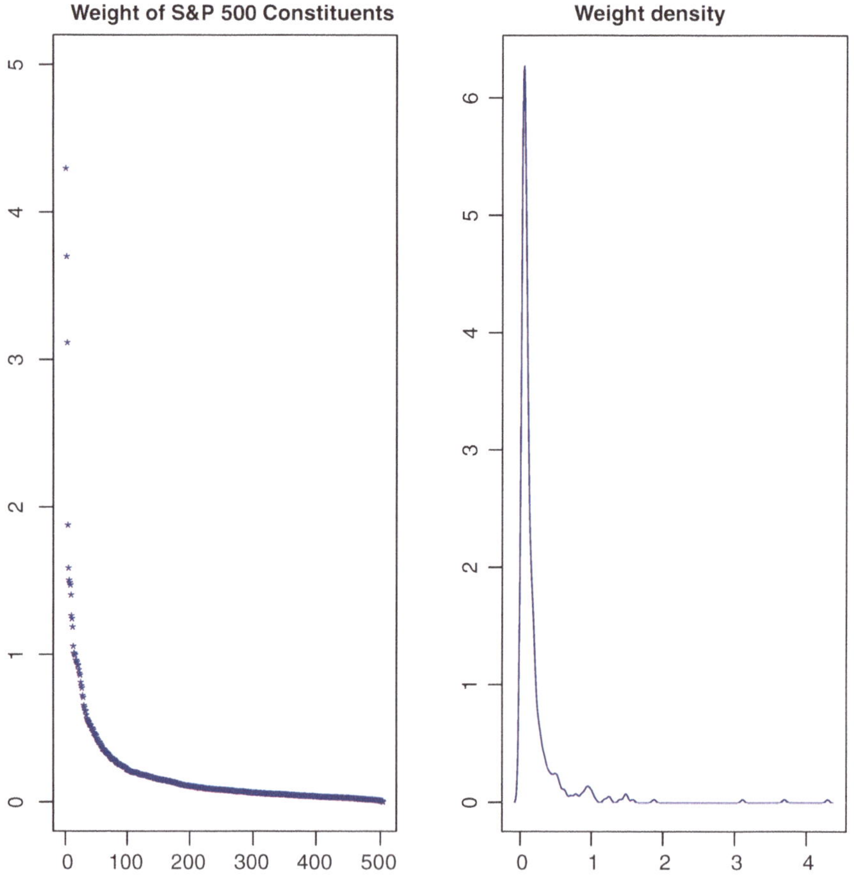

Fig. 7.1 The weights of 505 stocks in the S&P 500 index on August 1, 2019

Table 7.1 Weights of the top ten and bottom ten stocks in the S&P 500 index on August 1, 2019

Top ten companies	Index weight (%)	Bottom ten companies	Index weight (%)
Microsoft Corporation	4.301	IPG Photonics Corporation	0.018
Apple Inc.	3.704	TripAdvisor Inc.	0.018
Amazon.com Inc.	3.119	Affiliated Managers Group Inc.	0.017
Facebook Inc. Class A	1.882	Under Armour Inc. Class A	0.016
Berkshire Hathaway Inc. Class B	1.589	500 Gap Inc.	0.016
Alphabet Inc. Class C	1.506	Under Armour Inc. Class C	0.014
JPMorgan Chase & Co.	1.489	Macerich Company	0.014
Alphabet Inc. Class A	1.475	Coty Inc. Class A	0.013
Johnson & Johnson	1.406	Nordstrom Inc.	0.013
Visa Inc. Class A	1.267	News Corporation Class B	0.006

$$\text{Security return: } R = (r_1, r_2, \ldots, r_N)$$

$$\text{Index weight: } W^b = (w_1^b, w_2^b, \ldots, w_N^b), \quad \sum_{i=1}^{N} w_i^b = 1 \text{ and } w_i^b > 0,$$

$$\text{Portfolio weight: } W_p = (w_1^p, w_2^p, \ldots, w_k^p), \quad \sum_{i=1}^{k} w_i^p = 1 \text{ and } w_i^p > 0,$$

$$\text{Active weight: } W_a = (w_i^p - w_i^b).$$

Note that for a long-only portfolio, there is an inherent bound for underweights: only up to the index weight, $\max w_i^a = -w_i^b$. The benchmark and portfolio returns are defined as

$$R^b = W^b R = \sum_{i=1}^{N} w_i^b r_i, \quad R^p = W^p R^\top = \sum_{i=1}^{N} w_i^p r_i.$$

Tracking Error Tracking error (TE) is the deviation of a portfolio from its benchmark. TE is measured by the standard deviation of active returns, the differences in return between a portfolio and a benchmark.

$$TE = std(R^p - R^b).$$

To demonstrate the TE calculation, we use the annual returns of Berkshire Hathaway during the period of 2008–2018 (Table 7.2). The standard deviation of active returns is 11.10%, indicating that Buffett's fund performance deviated significantly from the S&P 500 index.

Table 7.2 Annual returns (%) of Berkshire Hathaway and the S&P 500 index, 2008–2018

Year	Buffett fund return	S&P 500 return	Active return
2008	−31.8	−37.0	5.2
2009	2.7	26.5	−23.8
2010	21.4	15.1	6.3
2011	−4.7	2.1	−6.8
2012	16.8	16.0	0.8
2013	32.7	32.4	0.3
2014	27.0	13.7	13.3
2015	−12.5	1.4	−13.9
2016	23.4	12.0	11.4
2017	21.9	21.8	0.1
2018	2.8	−4.4	7.2

Data source: Warren Buffett's annual letter to the shareholders of Berkshire Hathaway in February 2019

Table 7.3 Products and management fees based on TE

TE	Product	Mgmt fee
0–1%	Index	5–25 bps
1–2%	Index plus	15–50 bps
2–5%	Active	25–100 bps
5–10%	Highly active	50–150 bps
>=10%	Concentrated portfolio	50–200 bps

Tracking error is an important concept in the industry because it measures how active a fund is relative to the benchmark. If a portfolio has a high tracking error, it is expected that the returns will be higher than the benchmark. Thus, one significant use of tracking error is to evaluate portfolio performance. A portfolio with low returns and a high tracking error signals a lack of skill and expertise needed to handle the investment. On the other hand, high tracking error products are, in general, associated with high management fees. Usually, the industry uses the annualized version of TE so that it can be compared apples-to-apples across different portfolios versus a benchmark. In the context of quantitative investing, products are categorized into different groups based on their TE. See Table 7.3 for reference.

Of course, we need to be very cautious about interpreting tracking error. For example, one large outlier can cause the tracking error to be very high. A related concept is the information ratio, the annualized return difference between a portfolio and its benchmark divided by TE,

$$IR = \frac{R^p - R^b}{TE}.$$

Using the values in Table 7.2, we calculate the annualized returns and TE. During the period of 2008–2018, Buffett's fund had an annualized return of 7.24%, while the S&P 500's was 7.26%. The similar returns result in an information ratio of nearly zero:

$$IR = \frac{7.24 - 7.26}{11.10} = -0.0018.$$

We need to be cautious, though, because Buffet's fund returns are after tax.

7.1.2 A Long-Short Portfolio

A long-only portfolio prohibits short sales. Even if we know that the price of a stock will go down, the best decision for a long-only portfolio is not to hold it. However, if we allow for short sales, we can harvest returns by selling shares of a stock and buying back the same number of shares later.

Now consider a long-short strategy in which short sales are allowed. From an accounting perspective, a long-short portfolio has two arms, the long part and the short part. Note that the long and short parts are not generated separately; rather, they are generated simultaneously during the portfolio construction process. The following specifies a portfolio with K stocks in the long position and S stocks in the short position,

$$Portfolio = (w_1, w_2, \ldots, w_k, w_{k+1}, \ldots, w_{k+s}),$$
$$\sum_{i=1}^{k} w_i = 1 \text{ and } w_i > 0; \quad \sum_{j=k+1}^{k+s} w_j = -1 \text{ and } w_j < 0.$$

Of course, we cannot long and short the same stock. So a stock in an investment universe will be either in a long or short or not-held position in a portfolio.

For a long-short portfolio, a security's portfolio weight is its active weight, as the benchmark is usually the cash. In the following, we describe a general mechanism for the short sales arm.

1. Get a primary broker and negotiate the borrowing rates.
2. Borrow shares from the primary broker for sale in the secondary stock market.
3. The proceedings from short sales will be used as collateral and earn some interest (usually very small compared with the borrowing rate).
4. After buying shares (cover) for a security in the short position, the shares are "returned" to the primary broker.

Note that borrowed shares are usually supplied by index fund (or index plus or structured products) holders who have tremendous amounts of assets and very low turnover. By lending shares out, they can make extra profits. The primary broker can make the arrangement because its trading desk can earn profits by carrying out trades for investors.

One important term for measuring the borrowing capital in a long-short strategy is the leverage ratio, which is defined as follows:

$$\text{Leverage ratio} = \frac{\text{values of short positions} + \text{value of long positions}}{\text{capital employed}}.$$

Note that this should be under the condition of full equitization, meaning that all the investment money is spent on stocks and no cash is left in the account. For a long-short market neutral strategy, the leverage ratio is 200%.

Portfolio Turnover Portfolio turnover measures the changes in weights required by transactions.

$$\text{Portfolio Turnover at t} = \sum_{i=1}^{N} |w_{i,t} - w_{i,t-1}|,$$

where $w_{i,t}$ is the weight of security i in the position at time t. For a long-short market neutral portfolio, the maximum turnover is to convert positions to cash and then buy a different set of stocks, in which case the portfolio turnover is 400%. Sometimes, people use a two-way definition of turnover (buy and sell), which is the definition above divided by two.

In reality, portfolio turnover is usually measured by annual turnover, the overall portfolio turnover in a calendar year, or the time it takes to have 200% portfolio turnover. Portfolio turnover is very important because it is related to transaction costs. All else being equal, the higher the turnover, the higher the transaction costs. Transaction costs may not be a big issue for large cap stocks, such as the S&P 500 universe, but it is definitely important to consider for small-cap stocks and stocks in emerging markets.

Note that simple methods of portfolio construction, such as equal weight, do not take into account "forecasts" of stock returns and/or risk. In the next section, we introduce modern portfolio theory, which shows how to combine risk and returns to generate a desired portfolio.

7.2 Modern Portfolio Theory: Mean and Variance

Modern portfolio theory (MPT) describes how to construct a portfolio that maximizes expected returns with a given risk level. Economist Harry Markowitz (Fig. 7.2) proposed MPT in his seminal work on portfolio selection (Markowitz 1952).[1]

[1] Photo credit: The CME Group Melamed-Arditti Innovation Award.

Fig. 7.2 Harry Markowitz, 1927–

For quantitative investing, Markowitz's most significant contributions are the efficient frontier and portfolio construction with mean and variance. The mean–variance optimization determines not only the theoretical value of an optimal portfolio, but also shows how to achieve that optimal portfolio. In this section, we introduce the efficient frontier concept and mean–variance optimality.

7.2.1 Efficient Frontier

According to MPT, the efficient frontier refers to a set of optimal portfolios with the highest returns for a given series of risk levels.

Suppose we have an investment universe of 500 stocks, and we want to select some stocks to form a portfolio. What should we do? Well, the first question we need to consider is, how high will the return from this portfolio be? That is the highest achievable return given a risk level, which gives us the boundary or the best theoretical value we can obtain. If we can find such a boundary for each risk level, then for a series of risk levels in that market, we can derive a curve that describes the whole range of optimal returns we could obtain from any combination of securities in that investment universe. In Markowitz's words, this is called the *efficient frontier*.

In the real world, we could define risk as the standard deviation of returns. Assume we know the expected returns and risk levels of three portfolios, A, B, and C.

Portfolio A: std=3%, expected return = 1%,

Portfolio B: std=3%, expected return = 2%,

Portfolio C: std=3%, expected return = 3%, OPTIMAL

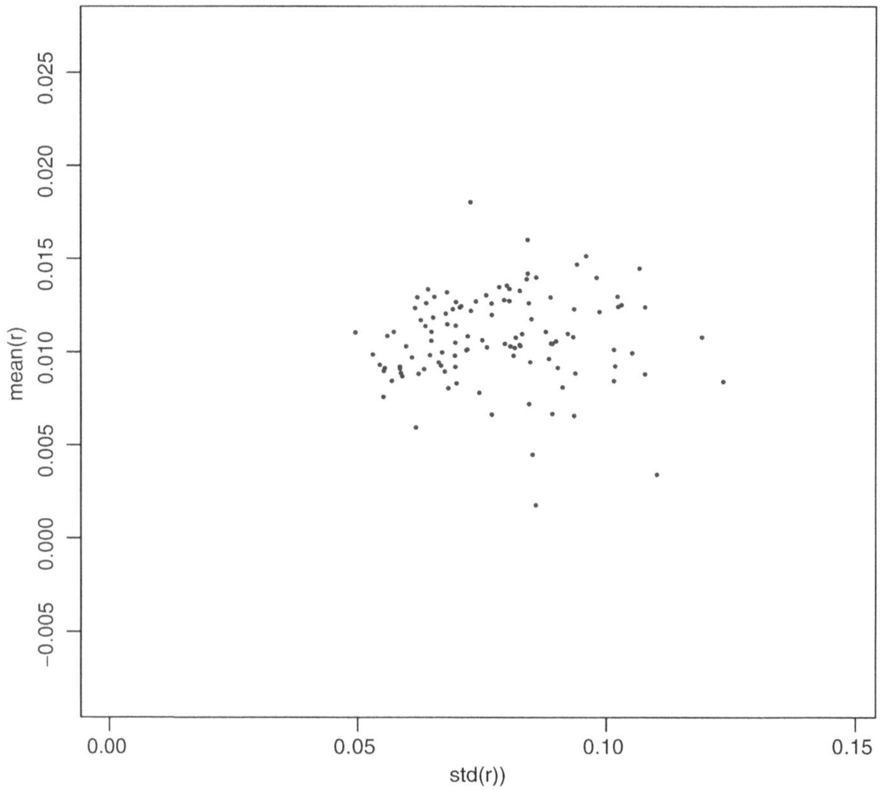

Fig. 7.3 The means and standard deviations of monthly returns for stocks in the S&P 500 from 1965 to 2014

It is easy to tell that portfolio C is the optimal portfolio because it has the highest return given the same level of risk. Or we could have the set below, where portfolio D is the optimal portfolio.

Portfolio D:	std=4%, expected return = 5%, OPTIMAL
Portfolio E:	std=5%, expected return = 5%,
Portfolio F:	std=6%, expected return = 5%,

The above examples demonstrate the principle of the efficient frontier. We provide a more realistic scenario in Fig. 7.3 with the S&P 500 index constituents using the average returns as expected returns and standard deviation of risk levels. The risk level is between 5–13%, while the average returns are in the range of 0–2%.

One concept derived from the efficient frontier is diversification: If we combine securities that are very different, it is possible to lower risk levels. As we see from the plot in Fig. 7.3, if two stocks have the same risk and return but are negatively correlated, the combination of the two stocks will yield the same return but lower risk, thus being more efficient due to diversification. Consider an example. Suppose the correlation between stock 1 and stock 2 is −0.5, and the expected return and risk are specified as follows:

$$r_1 = r_2 = 5\%, \qquad \sigma_1 = \sigma_2 = 10\%,$$

then the variance of the optimal portfolio return is

$$Var(0.5r_1 + 0.5r_2) = 0.25 * 100 + 0.25 * 100 - 2 * 0.5 * 100 * 0.25 = 25\%.$$

The risk of the portfolio is only 5%—the combined investments have lower volatility than any individual component. This is the benefit of diversification, hence the rationale of the efficient frontier.

How can we achieve this efficient frontier if we have 500 stocks? What procedure should we follow to derive the solution? Is there a unique solution? The efficient frontier can be obtained using mean–variance optimization. We discuss MPT in detail in the following section.

7.2.2 Modern Portfolio Theory

Modern portfolio theory was first proposed by Markowitz in 1952 and then expanded by many other economists, such as Sharpe with CAPM. MPT is an analytical framework for portfolio construction: for a risk aversion investor, how to achieve optimality with the information of the first two moments of a return distribution.

In a portfolio with K securities, each security has an expected return of $E(r_i)$, variance $V(r_i)$, and weight w_i.

Portfolio expected return: It is calculated as the weighted sum of the individual assets' expected returns: $E(R^p) = \sum_i w_i E(r_i)$.

Portfolio return variance: The portfolio's risk is a function of the variance of each asset and the correlation of each pair of assets.

$$\sigma_p^2 = \sum_i w_i^2 \sigma_i^2 + \sum_i \sum_{j \neq i} w_i w_j \sigma_i \sigma_j \rho_{ij},$$

where σ is the (sample) standard deviation of the periodic returns on an asset, and ρ_{ij} is the correlation coefficient between the returns on assets i and j. Thus, MPT shows that investment is not only about picking stocks but also building the right combination of stocks. The "right" combination means diversification, such as adding a security that maintains expected return but reduces the risk level. Mathematically, this is a stock that has a negative correlation with the existing stocks in the portfolio. In a fundamental sense, what are those stocks, and what kind of risk can be diversified away? This is related to the concepts introduced by William Sharpe—systematic risk and nonsystematic risk—where only the latter is diversifiable. Moreover, Sharpe sought to resolve the issue of expected return in the CAPM. However, it should be noted here that practical and large-scale usage of forecast for expected return or alpha has been employed in quantitative investing only since the emergence of multi-factor models such as APT and the FF three-factor model.

MPT paved a significant foundation for quantitative investing. It has been studied extensively since its birth in the 1950s and used widely in the finance industry, particularly since the emergence and boom of quantitative investing since the 1990s. MPT offers a systematic tool for investors to analyze a portfolio in a return-risk framework.

On the other hand, MPT has been criticized for many reasons. We list below the main arguments against MPT.

- The framework assumes linear risk aversion.
- It depends on expected values.
- It assumes that returns follow a normal distribution.
- Risk is measured by variance.
- The portfolio is built on a point estimation.

More advanced portfolio construction methodologies have been studied recently, especially since the 2008 financial crisis. We focus on the mean–variance approach in this chapter and explore advanced methodologies in Chap. 8.

7.3 Mean–Variance Portfolio: Optimization, Covariance Estimation, and Least Squares

Assuming we know the risk and expected return for each security in an investment universe, how should we select securities to form an optimal portfolio? In other words, how should we allocate funds to each security so that their weights will be optimal? One way to achieve this goal is through mean–variance optimization, a practical application of MPT. In this section, we lay out the general framework of mean–variance optimization and two important related issues: equivalence to GLS and covariance estimation. We then illustrate the MV approach by a long-only and a long-short portfolio with the investment universe of the S&P 500.

7.3.1 Mean–Variance Portfolio: Optimization with the First Two Moments

According to modern portfolio theory, we can formulate an optimal portfolio using the mean–variance framework:

$$\textbf{max}\ \text{expected return}, \qquad s.t.\ \text{risk level} = S.$$

If we let α be the vector of expected returns for securities in a defined investment universe, Ω is the variance defining risk levels, and W is a vector of portfolio weights.

$$\alpha = (E(r_1), E(r_2), \ldots, E(r_n))$$
$$\Omega = Var((E(r_1), E(r_2), \ldots, E(r_n))$$
$$W = (w_1, w_2, \ldots, w_n)$$

then according to the mean–variance approach, we have

$$\max_W W^\top \alpha - \lambda \left(W^\top \Omega W - S \right), \tag{7.1}$$

where $\lambda > 0$ is the risk aversion parameter, usually defined by the mandate of an investment strategy. To find the optimal allocation, we solve the Lagrangian:

$$\mathcal{L} = W^\top \alpha - \lambda \left(W^\top \Omega W - S \right)$$

The first order conditions are

$$\frac{\partial \mathcal{L}}{\partial W} = \alpha - 2\lambda \Omega W = 0 \tag{7.2}$$
$$\frac{\partial \mathcal{L}}{\partial \gamma} = W^\top \Omega W - S = 0. \tag{7.3}$$

Solving $\frac{\partial \mathcal{L}}{\partial W}$ for W, we get

$$W = \frac{1}{2\lambda} \Omega^{-1} \alpha. \tag{7.4}$$

Equation (7.4) states that optimal portfolio weight is proportional to risk: the higher the risk level, the smaller the weight; the higher the alpha, the larger the weight. The risk aversion parameter has a negative relationship with portfolio weights.

At this stage, it is helpful to illustrate the mean–variance portfolio with an example. Suppose we have only two stocks, A and B, that are not correlated. Their expected returns and risk, measured by standard deviations, are

$$\text{stock A: } \alpha_1, \sigma_1; \qquad \text{stock B: } \alpha_2, \sigma_2.$$

Following the procedure above, we have

$$w_1 = \frac{1}{2\lambda} \frac{\alpha_1}{\sigma_1^2}, \quad w_2 = \frac{1}{2\lambda} \frac{\alpha_2}{\sigma_2^2}, \quad w1 + w2 = 1.$$

Hence, the optimal solution is as follows:

$$\lambda^\star = \frac{\alpha_1}{2\sigma_1^2} + \frac{\alpha_2}{2\sigma_2^2}, \quad w_1\star = \frac{\alpha_1}{\alpha_1 + \frac{\sigma_1^2}{\sigma_2^2}\alpha_2}, \quad w_2\star = \frac{\alpha_2}{\alpha_2 + \frac{\sigma_2^2}{\sigma_1^2}\alpha_1}.$$

We see that the optimal weights from the mean–variance approach represent a tradeoff between the alpha and risk of both securities.

7.3.2 MV Optimization is Equivalent to GLS

In a naked version without any constraints, the mean–variance optimization expressed in (7.2) is equivalent to the GLS solution to the alpha model below:

$$R = F\gamma + e, \quad W = \Omega^{-1}, \tag{7.5}$$

where F is a vector of factors. As discussed in Chap. 5, the GLS estimate solves

$$\min_{\gamma} e^\top \Omega^{-1} e \tag{7.6}$$

with the solution

$$\gamma_{gls} = F\Omega^{-1}(F^\top \Omega - 1F)^{-1} F^\top \Omega^{-1} R.$$

Now, consider MVP with $\alpha = E(R) = F\gamma$, the optimal portfolio return is

$$\begin{aligned} R^p &= \max_W W^\top R \\ &= \max_W \alpha^\top \Omega^{-1} R \\ &= \max_W (R - e)^\top \Omega^{-1} R \\ &\equiv R^\top \Omega^{-1} R + \min_\gamma e^\top \Omega^{-1} e, \end{aligned} \tag{7.7}$$

where the second term in (7.7) is exactly the same as GLS.

7.3.3 Covariance Matrix Estimation

In the analysis above, we assume that both α and Ω are known. We explored how to build α in Chaps. 4, 5, and 6. To estimate Ω, one way is to use historical returns. For example, assuming we have t periods for a security i,

$$r_i = \left(r_{i1}, r_{i2}, \ldots, r_{i,t-1}, r_{it}\right), \quad i = 1, \ldots, n,$$

we can proceed to estimate Ω,

$$\hat{\Omega} = Var((r_1, r_2, \ldots, r_n)\,.$$

However, there are three related issues we need to address when estimating Ω,

1. $T \gg N$, the number of periods should be much larger than the number of stocks.
2. Decomposition of Ω into systematic and nonsystematic risk.
3. Forecasting power of estimation from historical returns.

We discuss each issue in detail below.

Condition $T \gg N$ Note that Ω is symmetric and always semi-positive definite. However, in order for the covariance to have the full rank of N, $T > N$ is required, where T is the number of periods and N is the number of securities. That is, we need Ω to be positive definite. Otherwise, there is perfect collinearity, a problem for obtaining the inverse of the covariance, Ω^{-1}. Moreover, given that the prices of stocks within the same industry usually move together, the very high correlation of those pairs requires T to be much larger than N. However, quantitative strategies usually encompass a large investment universe to achieve breadth, the condition $T \gg N$ can be difficult to meet. For example, in emerging markets, one of the most widely used investment universes is the MSCI Emerging Markets Index, which includes large- and mid-cap stocks across emerging stock markets. As of July 31, 2019, the index covers 26 EM countries with 1193 constituents. The covariance matrix for stocks in this index requires a period that is a multiple of 1193. Given the relatively short history of the emerging market index–the MSCI Emerging Markets Index was launched on January 1, 2001– this can be very challenging. If we use daily returns, there are 4500 trading days, a fairly small T relative to N.

Decomposition of Ω Ω represents total risk. We know from MPT that the benefit of diversification is to lower overall risk by reducing the diversifiable risk. Consider the following decomposition:

$$\Omega = \Omega_1 + \Omega_2,$$

where Ω_1 is systematic risk and Ω_2 is nonsystematic risk. The systematic risk is the risk at the whole market level and cannot be diversified away. The nonsystematic risk is the risk at the individual company or group of companies level. Nonsystematic risk is also called individual risk or specific risk. An example of specific risk is that a pharmaceutical company has a lawsuit on patents, which had negative impacts on its stock price. But this has no impact on the healthcare industry, let alone the overall stock market. An investor could invest in other companies to reduce this particular risk. Nonsystematic risk is diversifiable.

Because nonsystematic risk is specific to each company and changes overtime, it is impossible to identify and include all specific risks. However, since systematic risk applies to the whole market, impacts all securities in the investment universe, and is not expected to change often over time, it is easy to identify and quantify from a practical standpoint. Following this logic, we can first identify Ω_1 and then obtain $\Omega_2 = \Omega - \Omega_1$.

- Identify systematic risk
  - Definition: how to define the systematic risk
  - Factors: what factors can be used to approximate systematic risk
  - Quantify: how to quantify these factors
- Separate systematic risk from total risk; the remainder is unsystematic risk
  - Multi-factor model provides an analytical framework
  - Residuals stand for specific risk
  - More information can be obtained with moments of residual and tails

Suppose we identify k systematic factors F. We specify a linear model:

$$r = F\gamma + \epsilon. \tag{7.8}$$

Of course, in the real world, systematic risk will have a far more complicated impact on returns than a linear model represents. For simplicity, we assume a linear relationship. Systematic and nonsystematic risk do not overlap and are not correlated. Therefore, we have the following variance decomposition from (7.8),

$$\begin{aligned}\Omega &= Var(F\gamma) + Var(\epsilon) \\ &= \Omega_1 + \Omega_2,\end{aligned} \tag{7.9}$$

where $\Omega_1 = Var(F)[\gamma^\top \gamma]$ is systematic risk, and $\Omega_2 = Var(\epsilon)$ is nonsystematic risk.

Note that one advantage of the return spanning multiple risk factors is dimension reduction for the computation. Because the number of systematic factors is far less than the number of securities ($K \ll N$), the dimension of securities is reduced to the number of factors. The $N \times N$ variance matrix for returns is calculated through the variance of the F factors and residuals. Since the condition $T \gg K$ can be met easily, the $T \not\gg N$ issue is resolved by common systematic factors faced by all securities.

Now, a natural question is, how do we determine those systematic factors F? For a factor to qualify as systematic, it needs to satisfy two conditions: (1) the scope of impact on price movement is the whole market, not just a company or group of companies and (2) the level of impact on price movement is significant. In short, systematic factors should be the significant common drivers of returns in the investment universe. The exposure of securities to systematic risk takes the form of factor values. As we discussed in Chaps. 4 and 5, there are many systematic factors identified and studied in academia and employed in quantitative investment, such as the three factors identified by the FF three-factor model, the momentum, profitability, and market sentiment themes. Of course, systematic factors may differ over time and across different markets. For example, in general, momentum is a systematic factor for the US stock market but not for the Japanese stock market.

Once we identify systematic factors, we can proceed to estimate Ω_1 and Ω_2. For illustration purposes, suppose we have beta and profitability as systematic factors for the US stock market. In the investment universe of the S&P 500, for 10 years of daily data (N=500, T=2500, K=3), we have

$$r = \gamma_0 + \gamma_1 BETA + \gamma_2 PROF + \epsilon. \tag{7.10}$$

Applying OLS to (7.10) yields $\hat{\gamma} = (\hat{\gamma}_0, \hat{\gamma}_1, \hat{\gamma}_2)$,

$$r = \hat{\gamma}_0 + \hat{\gamma}_1 BETA + \hat{\gamma}_2 PROF + \hat{\epsilon}.$$

Thus, we obtain both systematic and nonsystematic risk:

$$\hat{\Omega}_1 = Var(F)[\hat{\gamma}^\top \hat{\gamma}], \quad \hat{\Omega}_2 = Var(\hat{\epsilon}).$$

Examples: A Long-Only and a Market-Neutral Portfolio We implement the MV approach to build a long-only and a market neutral long-short stock selection portfolio. We use month-end data for the returns of selected constituents of S&P 500 from January 1989 to June 2014. For the sake of simplicity, we use the average of historical returns as expected return and standard deviation of historical returns as risk. The mean–variance optimization uses the classical quadratic form,

$$\max -d^\top b + b^\top Db, \qquad s.t. A^\top b \le b_0, \tag{7.11}$$

where b_0 is the target of the linear constraints.[2] We present sample R codes below for the long-only portfolio.

MV portfolio: long-only strategy

```
## call for the quadprog package
library(quadprog)
## set up A and b_0
A1=rep(1,nstock)
b1=1
A2=matrix(0,nstock,nstock)
diag(A2)=1
```

[2]In R, the package *quadprog* implements the dual method of Goldfarb and Idnani (1982, 1983). The package has a function, *solve.QP*, to solve quadratic programming problems (7.11).

```
b2=rep(0,nstock)
Amat=t(rbind(A1,A2))
bvec=c(b1,b2)
## set up covariance matrix (risk), D and  expected return, d
dmat=var(xx.in)*lambda ## xx.in is the in sample data with historical returns
dvec=colMeans(xx.in)
## run quadratic optimization
opt=solve.QP(dmat, dvec, Amat, bvec, meq=1)
```

We run a series of values for λ to obtain different risk levels. Labeling the optimal weights as $W^\star$, we obtain portfolio performance:

$$\text{risk: } W^{\star T}\Omega W^\star, \text{ in-sample return: } W^{\star T}\bar{R}, \text{ out-sample return: } W^{\star T}R_t,$$

where $\bar{R}$ is for in-sample, the average of historical stock returns up to time $T-1$ and R_t is for out-of-sample, the stock returns of the period T. We present portfolio performance for a series of λ values in Table 7.4.

The pair values of λ and portfolio risk are plotted in Fig. 7.4. We see that when λ increases, portfolio risk decreases, but the relationship is not linear. For example, for the long-only portfolio, the risk drops quickly from 7% to 3% when λ increases from 0.01 to 10 but remains around 3% regardless of further increases in λ. The same pattern exists for a long-short strategy, though the range of risk is wider, as expected.

Adding returns and risk of long-only portfolios to Fig. 7.3 yields Fig. 7.5. We see that the frontier lies outside the set with the highest return for each risk level (value of λ). This shows that if we know expected return and risk, the mean–variance approach does provide the optimal solution. The out-of-sample returns are volatile, indicating that the accuracy of α is critical to the performance of an MV portfolio. The long-short portfolio exhibits a similar pattern.

Table 7.4 Return (%) and risk (%) for long-only and market neutral MV portfolios

Lambda	Port std	In-sample ret	Out-sample ret	Port std	In-sample ret	Out-sample ret
	Long-only	Long-only	Long-only	Long-short	Long-short	Long-short
0.01	7.27	1.81	0.20	10.35	1.63	0.20
1	6.65	1.78	1.38	8.08	1.50	1.38
5	4.27	1.47	1.70	3.44	0.88	1.52
10	3.64	1.30	1.17	2.54	0.64	0.57
100	3.14	0.97	1.24	2.34	0.40	0.57
1000	3.14	0.95	1.20	2.40	0.38	0.45

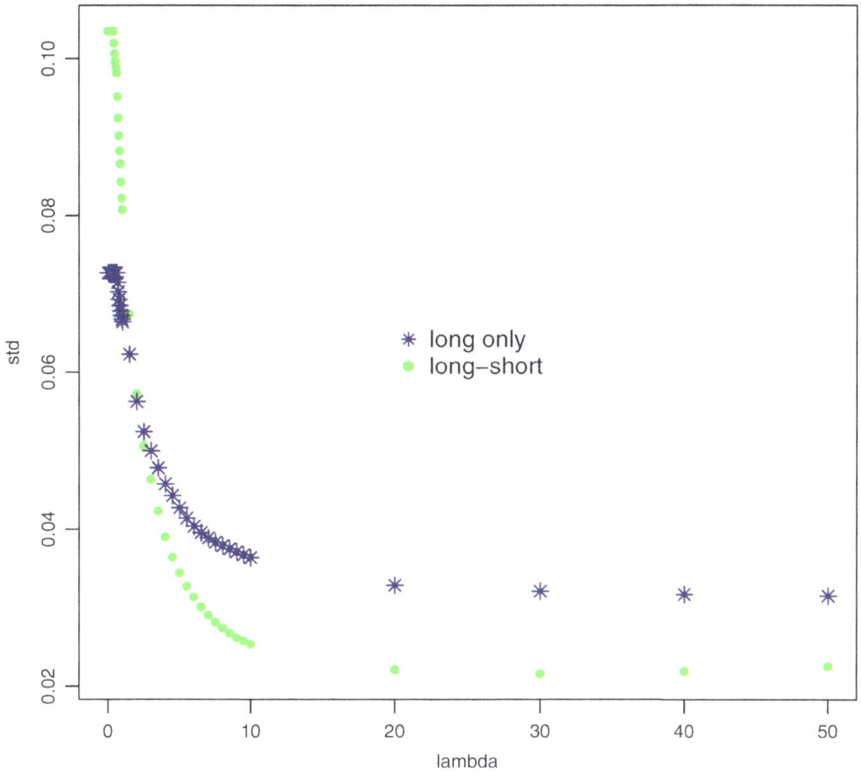

Fig. 7.4 The values of λ and corresponding risk levels for both a long-only (solid circle) and a long-short portfolio (star)

7.4 Variations of the Mean–Variance Portfolio

We discussed the mean–variance approach in the previous section. We have observed that the accuracy of expected return or alpha is very important for portfolio performance, though this is always a challenging task. One possible solution is that, if we do not have confidence in alpha, we can just ignore it in the mean–variance approach and focus on minimizing the volatility. This is called the min-volatility optimization. Another solution is to add constraints to limit the role of alpha. This is called lasso. We introduce these two variations of MVP in the following.

7.4.1 Min-Vol Portfolio and Smart Beta

Since the 2008 financial crisis, there has been a surge in one type of quantitative product: min-volatility, or minvol for short. It has also been extended and called

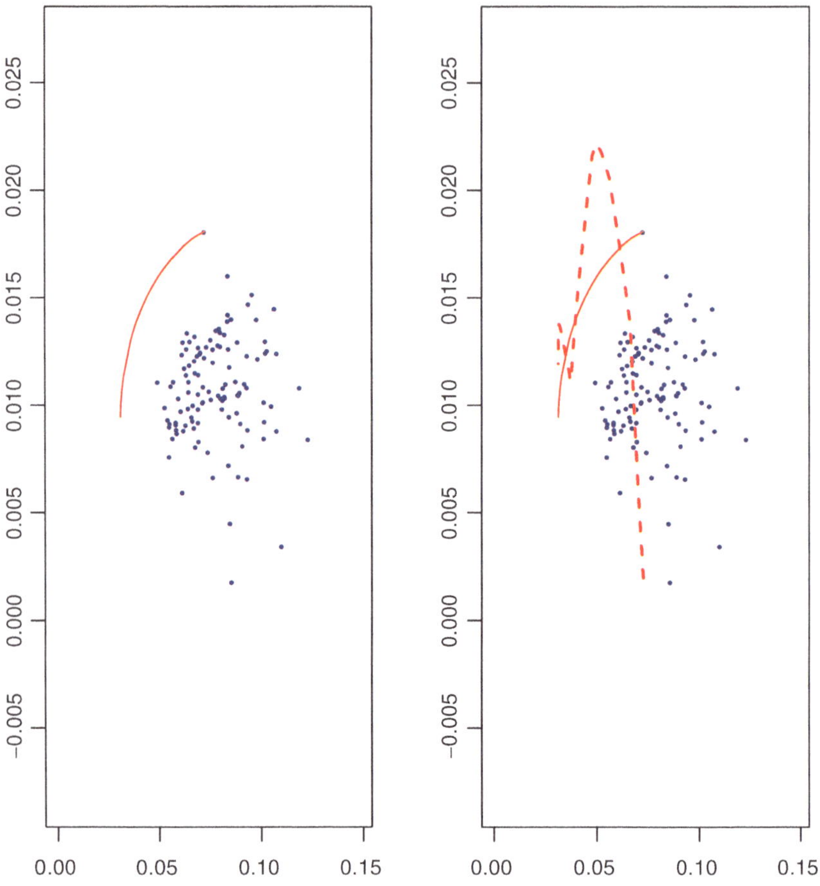

Fig. 7.5 The efficient frontier of a long-only MV portfolio. The left plot represents in-sample data, the right plot represents the out-of-sample results (dashed line)

smart beta. These products are constructed by minimizing the volatility matrix, and alpha is deliberately left out of the portfolio.

The plots in Fig. 7.6 show the growth of assets under management for minvol products in the USA between 2000 and 2017. The left plot presents the AUM of US smart beta products from 2000 to 2017, where we see that AUM increased from zero in 2005 to USD ten billion during the financial crisis and to about USD 600 billion in 2017.[3] Note that US low-vol products represent about 80% of all minvol products globally. The right plot shows the average annual AUM growth rate for minvol ETFs, 2009–2015, together with other quantitative products. We see that during this period,

[3]Data source: Morningstar Research. Data as of June 30, 2017.

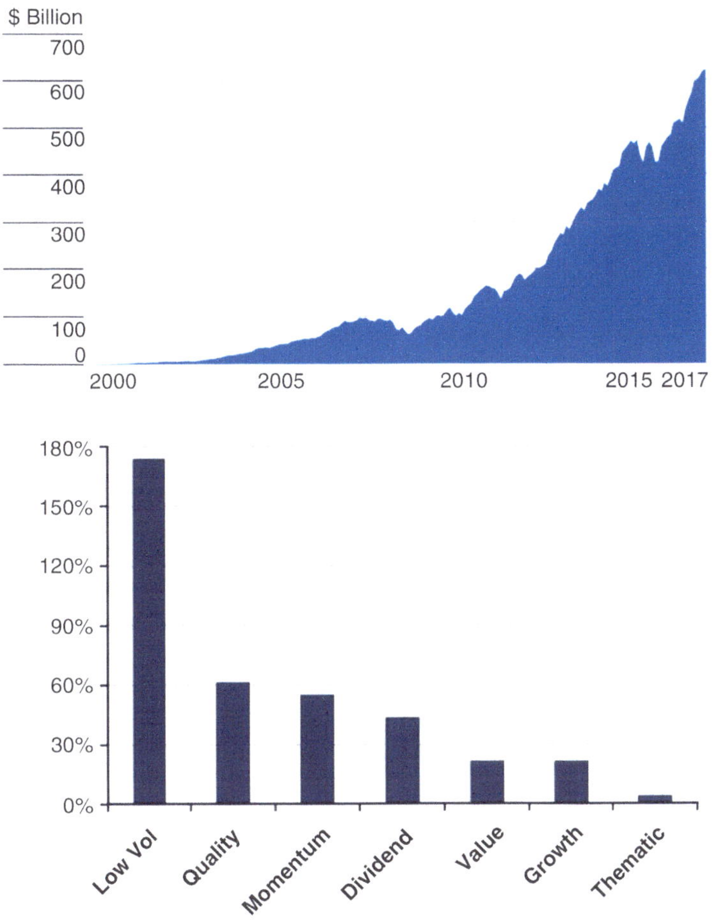

Fig. 7.6 The AUM of US smart beta products from 2000–2017 (left plot) and the average annual AUM growth of smart beta products relative to other quant products from 2009–2015 (right plot)

the annual growth rate of AUM was about 180% for minvol products but only about 30–60% for other quant products.[4]

The mean–variance approach had been working well before the financial crisis, so what changed suddenly? Min-volatility products address investors' suspicion about MPT's relying on only the first two moments, i.e., of the mean–variance framework. MPT is a great theory when things are "normal" but has a disconnect from the real world. The assumptions of the mean–variance approach are based on many factors. Three of the most significant factors are (1) the forecasting power of alpha for future returns; (2) the validity of characterizing returns by the first two moments of a return distribution, mean and variance; and (3) the measurement of risk by standard deviation.

[4]Data source: Bloomberg, BofA Merrill Lynch US Equity and US Quant Strategy.

Consider the first assumption. Quant became popular in the 1990s and gained momentum in the early 2000s when replication of similar strategies started to create crowding and similarities between quant strategies: Alpha became beta due to similar alpha factors, risk models, and portfolio holdings. During the financial crisis, the so-called alpha factors failed and the quant funds' performance collapsed.

Regarding the second assumption, we know that a normal distribution can satisfy this. However, if returns of assets display fat tails or skewness, MPT and the mean–variance approach will not work well. In other words, the validity of MPT in reality depends on how close the asset returns are to a normal distribution.

For the third assumption, according to MPT, risk can be measured by volatility, standard deviation. However, variance is directional blind and high volatility does not necessarily result in high returns (see Sect. 5.1). Taking extra risk at the downside may well get lower returns.

During normal periods when there is no significant turmoil in the financial markets, the mean–variance approach has delivered products that do not exhibit the drawbacks mentioned above. During the financial crisis, there were days with large downturns that caused skewness and fat tails, making the return distribution deviate from normal. In such circumstances, a portfolio generated using MPT would deliver an inefficient combination of securities. More importantly, high-betas deliver higher returns only when the market is up and can deliver much lower returns during a financial crisis. Conversely, given a long enough period with downside financial markets, low-beta securities may deliver better returns than high-beta securities.

During and after the 2008 financial crisis, investors focus on how to best use beta. A low-beta portfolio, given a long enough period, will yield a similar level of returns to the index but with much lower risk and therefore a much higher Sharpe ratio.

In terms of portfolio construction, low-volatility products can be considered a special case of a mean–variance portfolio that allows the expected return to be the same or zero. Using a long-only and fully equitized portfolio as an example,

$$\min_W W^\top \Omega W, \quad s.t. \sum_i w_i = 1 \text{ and } w_i > 0. \tag{7.12}$$

Using the data based on the variance of historical monthly returns for stocks in the S&P 500 from January 1965 to May 2014, we construct a minvol portfolio based on (7.12). The minvol portfolio has a standard deviation of 3.13%, the in-sample return is 0.94%, and the out-of-sample return is 1.19%.

Minvol portfolio

```
## long only constraints and full equitization
A1=rep(1,nstock)
b1=1
A2=matrix(0,nstock,nstock)
diag(A2)=1
b2=rep(0,nstock)
```

```
Amat=t(rbind(A1,A2))
bvec=c(b1,b2)
## the alpha or expected return is set to zero
dmat=var(xx.in)
dvec=rep(0,nstock)
## derive the optimal weights
opt=solve.QP(dmat,dvec,Amat,bvec,meq=1)
```

We present the minvol portfolio returns in the efficient frontier figure (Fig. 7.7). The triangle is the minvol portfolio based on the in-sample average return, and the square is the minvol portfolio based on the out-of-sample data for June 2014. As expected, the minvol portfolio (in-sample) is at the left end of the long-only portfolio efficient frontier.

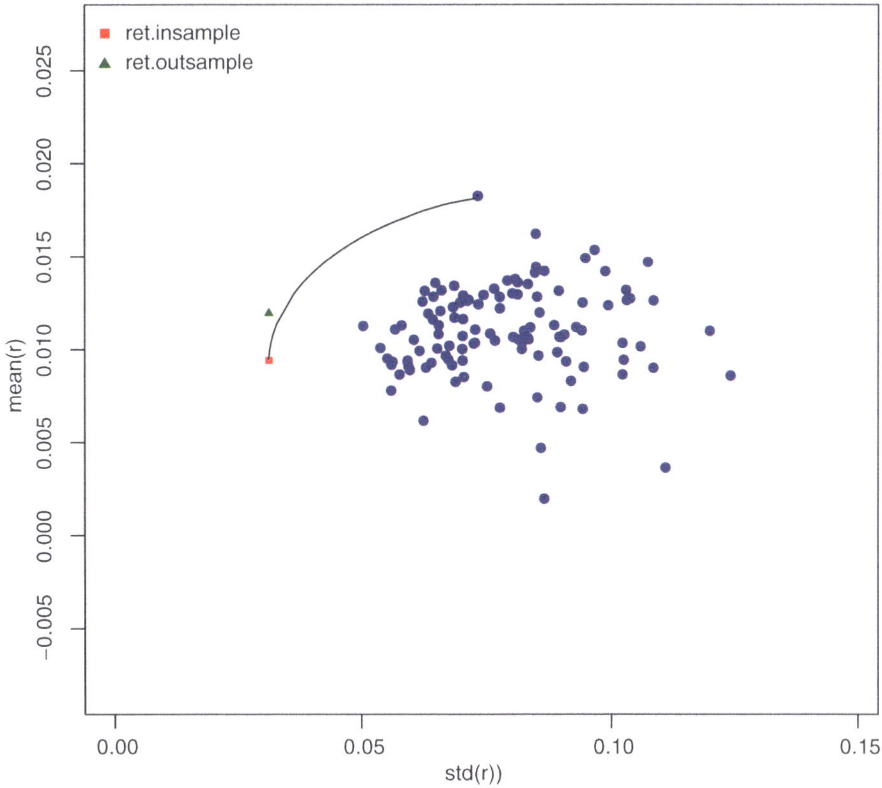

Fig. 7.7 The triangle is the minvol portfolio using in-sample average return, and the square is the minvol portfolio using the out-of-sample data for June 2014

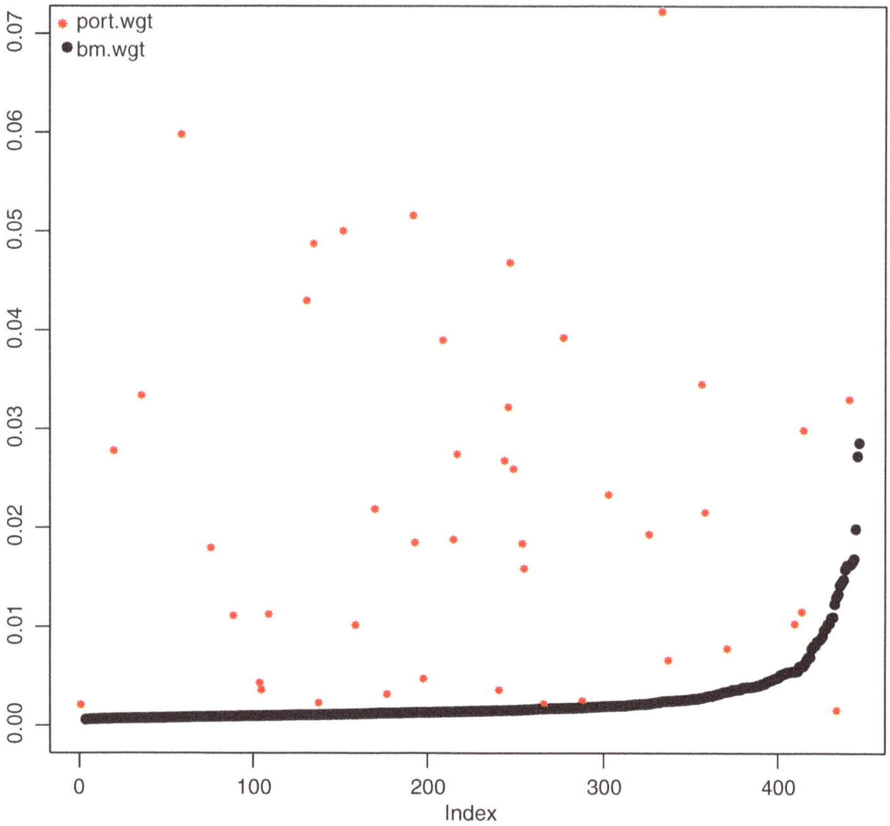

Fig. 7.8 The minvol portfolio: portfolio weights and benchmark weights

What are the characteristics of these low-volatility stocks? To answer this, we explore market capitalization and sector distribution. In Fig. 7.8, we present the distribution of portfolio weights across the index weights. We see that the weights of the minvol portfolio are distributed evenly across the index weights of selected S&P 500 constituents (investment universe), indicating no market cap bias. Furthermore, there are only 44 stocks in the minvol portfolio, with the largest weight being 7.23%, implying that the minvol portfolio can be highly concentrated.

What kinds of stocks are in a minvol portfolio? We employ GICS sector classifications and add portfolio weights for each sector. GICS sectors and their corresponding numbers are listed in Table 7.5. Note that because our data is from 1964 to 2014, the GICS had ten sectors back then. As of 2019, the GICS has eleven sectors. It turns out that most stocks are in the consumer staples and utilities sectors. The bar plot in Fig. 7.9 displays the portfolio and index weights for each of the ten GICS sectors. We see that consumer staples and utilities are the two sectors with the largest weights for the minvol portfolio. This makes sense because these two sectors

Table 7.5 GICS sector numbers and names as of December 31, 2014

Sector number	Sector name	Sector number	Sector name
10	Energy	15	Materials
20	Industrials	25	Consumer discretionary
30	Consumer staples	35	Health care
40	Financials	45	Information technology
50	Communication services	55	Utilities

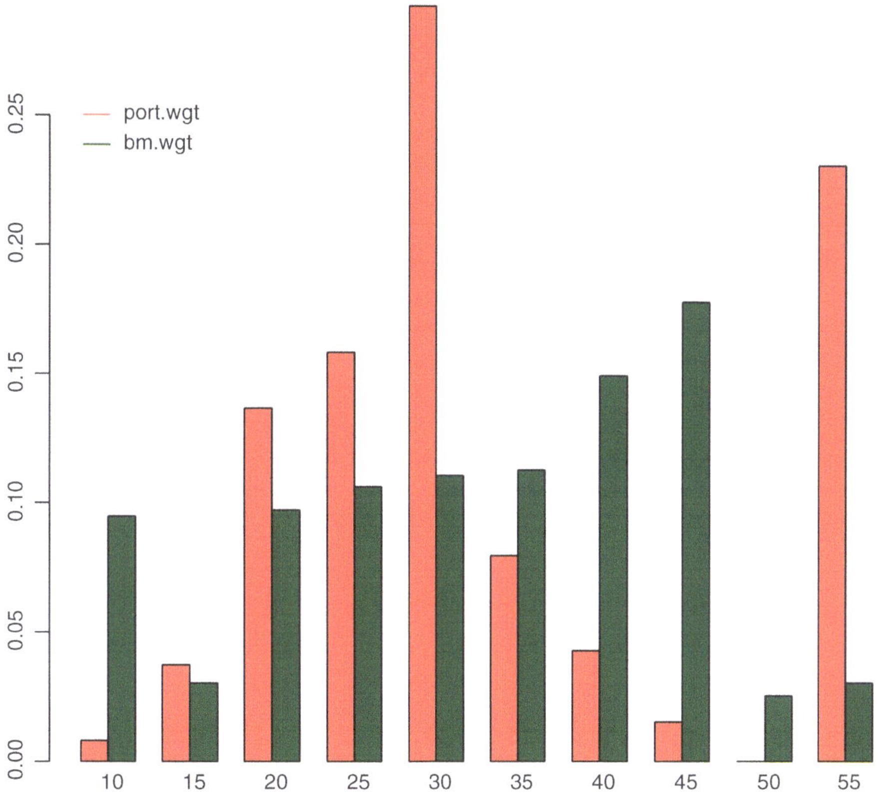

Fig. 7.9 The minvol portfolio weights and benchmark weights by sectors

have companies offering necessary goods and services, such as retail drugs, heating services, etc. Those goods and services do not change much over different business cycles and cross-periods of ups and downs in stock markets. In other words, when the market is up, those stocks will not be hot, while on the other hand, those stocks will not lose value as much as other stocks when the market declines significantly. Therefore, those stocks tend to have lower volatility and much lower correlation with overall market performance. They can provide a safer option (compared with other sectors) when the equity market is in turmoil, such as the financial crisis in 2008. We show next that how the minvol portfolio can stand out to deliver better returns using only beta.

Low Beta Portfolio For the minvol type of portfolio, an alternative approach is to use beta (as defined in the CAPM) to construct a portfolio.

To construct a low beta portfolio, we first calculate the beta value for each stock from 1964 to 2008 using monthly returns data.

$$\text{stock return}_i = b_{0,i} + \beta_i \times \text{SP500RET} + \epsilon_i$$

Because beta is a fundamental trait of a public company, its computation relies on long-term historical data, and it is not expected to change much over a short period of time, except in the case of structural changes in the company. For example, the company changes its core business from electrical engineering to credit cards.[5]

We plot the values of beta and standard deviation for each stock in the S&P 500 in Fig. 7.10. We see that the values of beta stay in the range of 0 to 2.5, with most ranging from 0.5 to 1.5. For the values of volatility, measured by σ, the estimates are in the range of 5–25%, with most stocks around 5–15%. It is clear that values of beta and standard deviation are closely related, with a correlation of about 75%. Note that some investors refer to low volatility as low beta, and they are interchangeable in most cases. However, even though they overlap, there are significant differences between the two definitions: low beta really means that the stock price does not closely follow the market or the co-movement is rather weak; while low volatility (measured by standard deviation), from the MVP perspective, means the stock price does not change much and/or has a low or negative correlation with the price movements of other stocks. In mathematical terms,

$$\hat{\beta} = (F^\top F)^{-1} F^\top r$$
$$Var = \sigma^2 + \sum_{i \neq j} \rho_{ij}\sigma_i\sigma_j.$$

To see how the financial crisis triggered the popularity of low-beta products, we show two portfolios constructed based on beta: one is at the year end of 2007 and the other is at the year end of 2008. We employ a simple nonparametric approach, decile analysis.[6] We group stocks into ten groups based on the values of beta from lowest to highest. The stocks with the lowest 50 betas are in group 1, while the stocks with the highest 50 betas are in group 10. For each group, we calculate equally weighted monthly average returns and the standard deviation of those returns. This allows us to see how each group performs at the same time. We present Sharpe ratios in Fig. 7.11. We see that for the portfolio at the end of 2007 (left plot), the high-

[5]One such example is GE.

[6]For details about nonparametric approaches, please refer to Chap. 5.

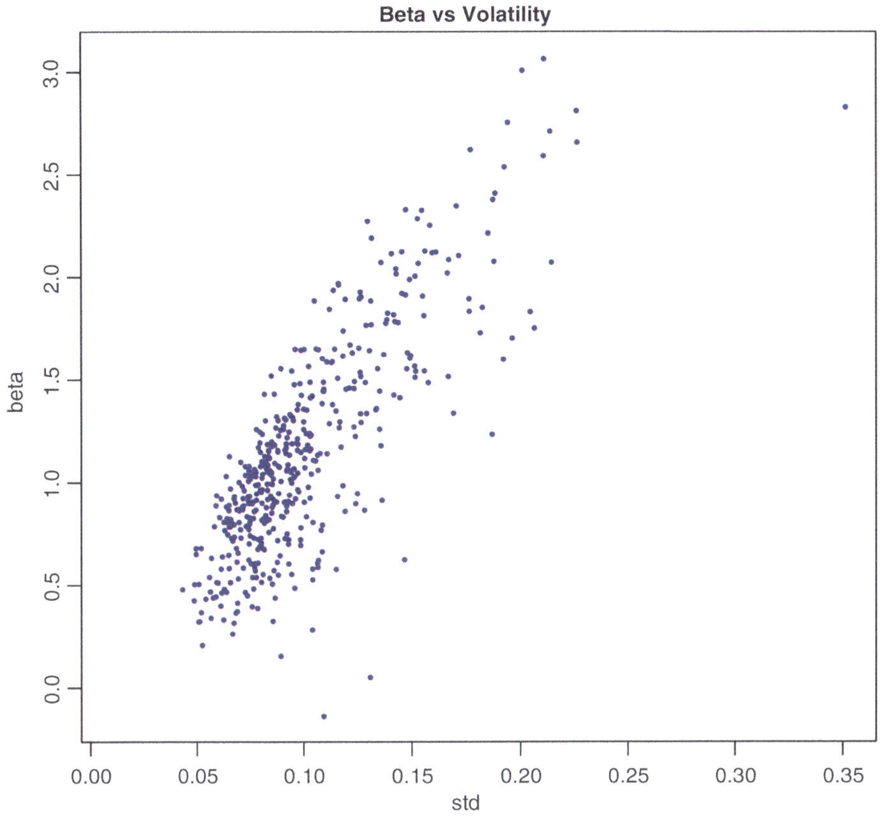

Fig. 7.10 The scatter plot of beta and standard deviation of monthly returns for S&P 500 constituents from 1964 to 2008

beta stocks underperform all other deciles and hence the overall market.[7] For the 2008 portfolio (right plot), we see that low beta deciles indeed outperform high-beta portfolios. The monotonicity of Sharpe ratios indicates that, contrary to conventional thinking, it is the low or below average betas that deliver superior performance when the market is down.

We also present in Table 7.6 the return and standard deviation for each decile portfolio based on the 2008 beta values. When moving from decile 1 to decile 10—from lower beta to higher beta—returns decrease, while standard deviation increases. It is the second lowest decile that has both the highest return and lowest volatility.

The financial crisis occurred in 2008. This was the moment when low volatility or low-beta products gained recognition from institutional investors. Because of

[7]This phenomenon triggered much research in 2007 and 2008 regarding the validity of beta as a predictor of higher returns.

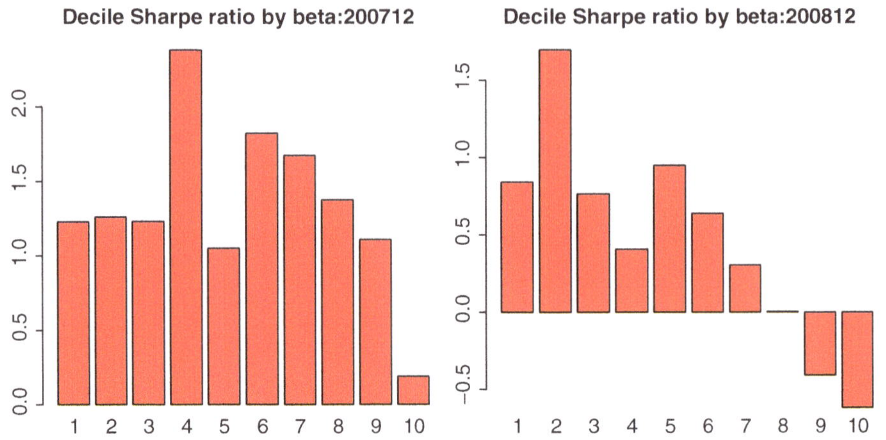

Fig. 7.11 The Sharpe ratio of decile portfolios at the year of 2007 (left plot) and at the year end of 2008 (right plot). Bar 1 corresponds to the lowest beta and bar 10 corresponds to the highest beta

Table 7.6 Decile returns and volatility for beta portfolios in 2008

Decile	Equal-weight return (%)	Volatility (std, %)
1	0.57	0.68
2	0.97	0.57
3	0.84	1.09
4	0.57	1.39
5	0.71	0.74
6	0.78	1.22
7	0.35	1.14
8	0.01	2.05
9	−1.27	3.11
10	−4.78	7.72

their perceived safety for investments, there was a wave of launching these kinds of products back then. The products started with low vol, then expanded to smart beta. A smart beta strategy is simply a portfolio with stocks of minimum volatility combined with high-quality characteristics, such as sustainable earnings growth, stable cash flow, and financial strength. As mentioned in the beginning of this section, smart beta strategies experienced the most growth in quantitative investing during the last 10 years: they grew from under \$1 billion in 2007 to about \$600 billion in 2017. They were first developed in the USA and then expanded into other regions and markets, such as Europe and emerging markets.

7.4.2 Lasso and Shrinkage

The mean–variance optimization based on MPT relies completely on the first two moments of the return distribution. The optimal weights derived from this approach rely on the accuracy of estimates of the expected return and covariance matrix. Usually, the risk based on the historical returns is stable, but the expected return can change dramatically over time. This can be confirmed by applying simple computations as follows.

1. The stability of the covariance matrix over time, especially for a large number of stocks over time t, can be measured by:

$$\text{distance} = abs(B - A)/sum(abs(A))$$

2. Optimal weights are very sensitive to the expected return, where the covariance matrix acts as the enlarger of alpha for the weights:

$$\frac{\partial w}{\partial \alpha} = \frac{1}{2\lambda}\Omega^{-1}$$

   which implies that any small change in α will result in large changes in weights.

   This sensitivity is illustrated by the change in the efficient frontier. We pick 3 year-end dates, 2007–2009. Note that the monthly data has only twelve data points between the two adjacent years. For these data sets, we use the same values of λ and obtain long-only and long-short portfolio results. For adjacent years, the difference will be reflected in in-sample average returns and portfolio weights. We plot the portfolio return and risk in Fig. 7.12 for long-only and long-short portfolios separately. For the three periods, we can see that the portfolio weights are very different. Of course, these are driven by the differences in average returns, as the frontier will stay right outside of the area of return and volatility pairs.

   The non-robustness of mean–variance optimization is inherent in the framework. For this reason, the MV approach is called the error maximization solution by some scholars, such as Michaud (1989). However, there have been counterarguments, such as Kritzman (2006).

3. Solutions. An immediate technical solution is to add constraints to the MV approach, such as a lasso or ridge methods. These shrinkage methods function as an extra penalty to ensure portfolio weights bound to a pre-specified set. For example, the lasso (least absolute solution) is to add an $\mathcal{L} - 1$ condition for the weights:

$$\max W^{\top}\alpha - \lambda W^{\top}\Omega W, \quad s.t. \sum_i |w_i| = h. \tag{7.13}$$

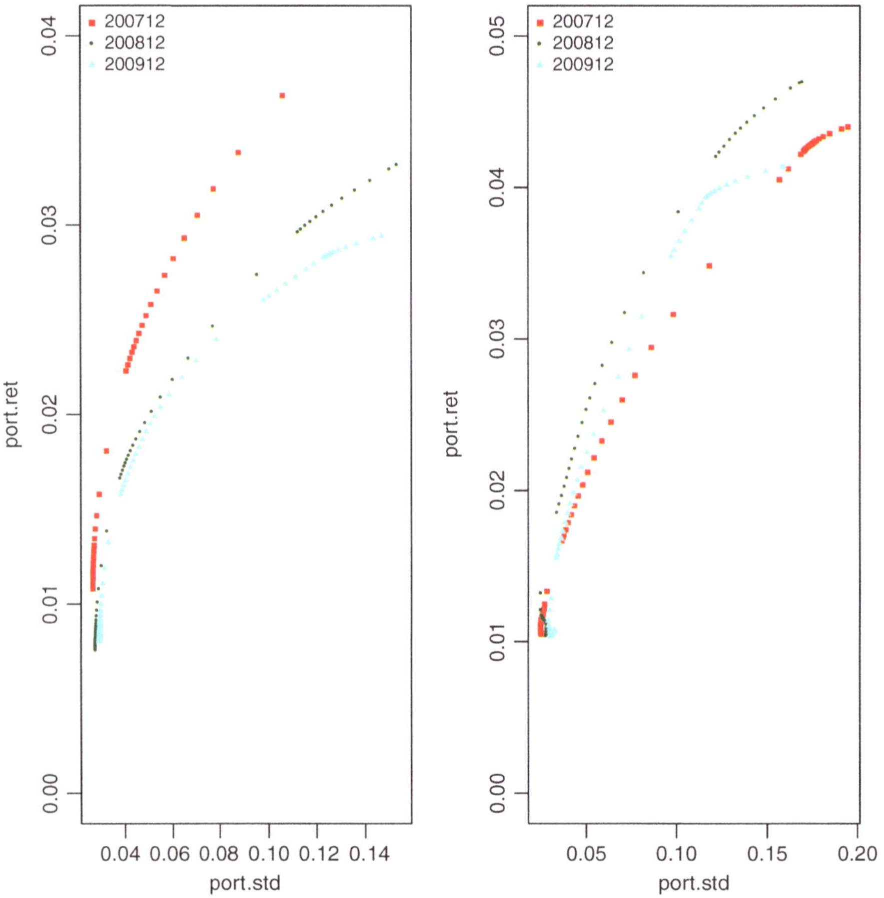

Fig. 7.12 The efficient frontiers for data ending in 2007, 2008, and 2009, respectively. The left plot is for long-only and the right plot is for long-short portfolios

From the portfolio construction perspective, lasso can be interpreted as a gross exposure constraint, such as a way to account for transaction costs (Brodie et al. 2009).

In reality, investors rarely have an unconstrained mean–variance solution in quantitative investing. Instead, there are usually constraints on transaction costs, tracking error, bets on individual stocks and sectors, etc. Actually, these constraints we put on the portfolio all act as a robustness shield during the optimization process: one small change in a security will not cause large changes in overall portfolio weights. We discuss these further in the section on industry insights, where we use industry approaches for portfolio construction.

Of course, a more fundamental solution is to seek (1) a more value-added alpha and/or (2) a more robust optimization methodology. We discuss the second issue in Chap. 8 and the alpha issue in Chap. 9.

7.5 Portfolio Backtesting

In quantitative investing, to validate an idea or investigate how a strategy might work, we need to generate a series of portfolios for multiple periods. This process is called backtesting. Backtesting is used widely in quantitative investing to, for example, test a new factor, change a portfolio feature such as constraints, test a new strategy, etc. We should keep in mind that backtesting is just one of many stages in a quantitative strategy. In general, the following items need to be considered when we design a backtest:

1. Major focus of the backtest: what to test for
2. Backtest set up: optimizer, alpha, risk, and constraints
3. Running the backtest with R codes: dealing with special cases, outliers, etc.
4. Analyzing backtest results: return, risk, distribution, sector, etc.
5. Conclusion and business implications: What is so special about the testing period? To what degree is it repeatable?
6. Backtesting and resampling methods

We briefly review each item in the following.

7.5.1 Review of Backtest Procedure

In this section, we briefly describe the full spectrum of a backtest for a quantitative strategy. We discuss various important aspects of backtesting, including design, procedure, evaluation, etc. We also provide suggestions for when disparity occurs between the result and planned target.

Main Focus of a Backtest While the main function of a backtest is to test how well a quantitative model works with historical data, the target and focus can vary a lot based on the main purpose. We classify backtest purposes into several categories and list major items to consider for each category.

- New strategy
  - efficacy of alpha and matching risk
  - overall performance
- New model
  - efficacy, areas of strength, and weakness
  - performance across business cycles

- New factor
  - IPRAE, interaction with constraints
- New constraint
  - effects on performance, where and how it kicks in
  - unwanted consequences
  - sensitivity
- New process
  - initial evaluation
  - implications for live product

We see from the above list that a backtest will have very different focuses for different purposes. Many times, the effects are not linear, and one target may conflict with another. For example, suppose we have a new constraint on certain industries to limit portfolio exposure to those industries. After we impose the new constraint, we may find that the portfolio has a much lower return or higher risk.[8] However, regardless of variations of backtests, the procedure for a general framework or setup will be similar.

Backtest Setup In general, a backtest requires optimization for multiple periods. Therefore, to set up a backtest, we need the following basic parts: values for alpha and risk, quantifiable constraints, and an optimizer. Often nowadays, the data for a backtest can be huge, and a database may be needed for convenient interaction between the optimizer and the database: importing inputs from and exporting outputs to the database.

Backtest Procedure We illustrate the backtest procedure through an example. Suppose we backtest an alpha model for a Chinese stock selection strategy with an investment universe of the CSI 300. We have month-end data from 2000 to 2018. We also have a risk model and a turnover constraint of no greater than 25% (monthly). The annual target return is 3–5%. We decide to use the 2000–2010 data as the first in-sample period, then proceed with a moving window for each new month.

The flow chart below shows a backtesting procedure. The sequence flows from the main focus of the backtest, to backtest design, then preparing data, setting up the optimizer, and running optimization for each period in a multi-period framework. For each period, the optimizer will generate the optimal portfolio weights based on the "in-sample" data, then an "out-of-sample" performance is calculated, saved, and exported. After all periods are done, we evaluate the overall performance of the backtest and infer business implications.

[8]For example, certain funds require a portfolio to have no exposure to alcohol and tobacco products.

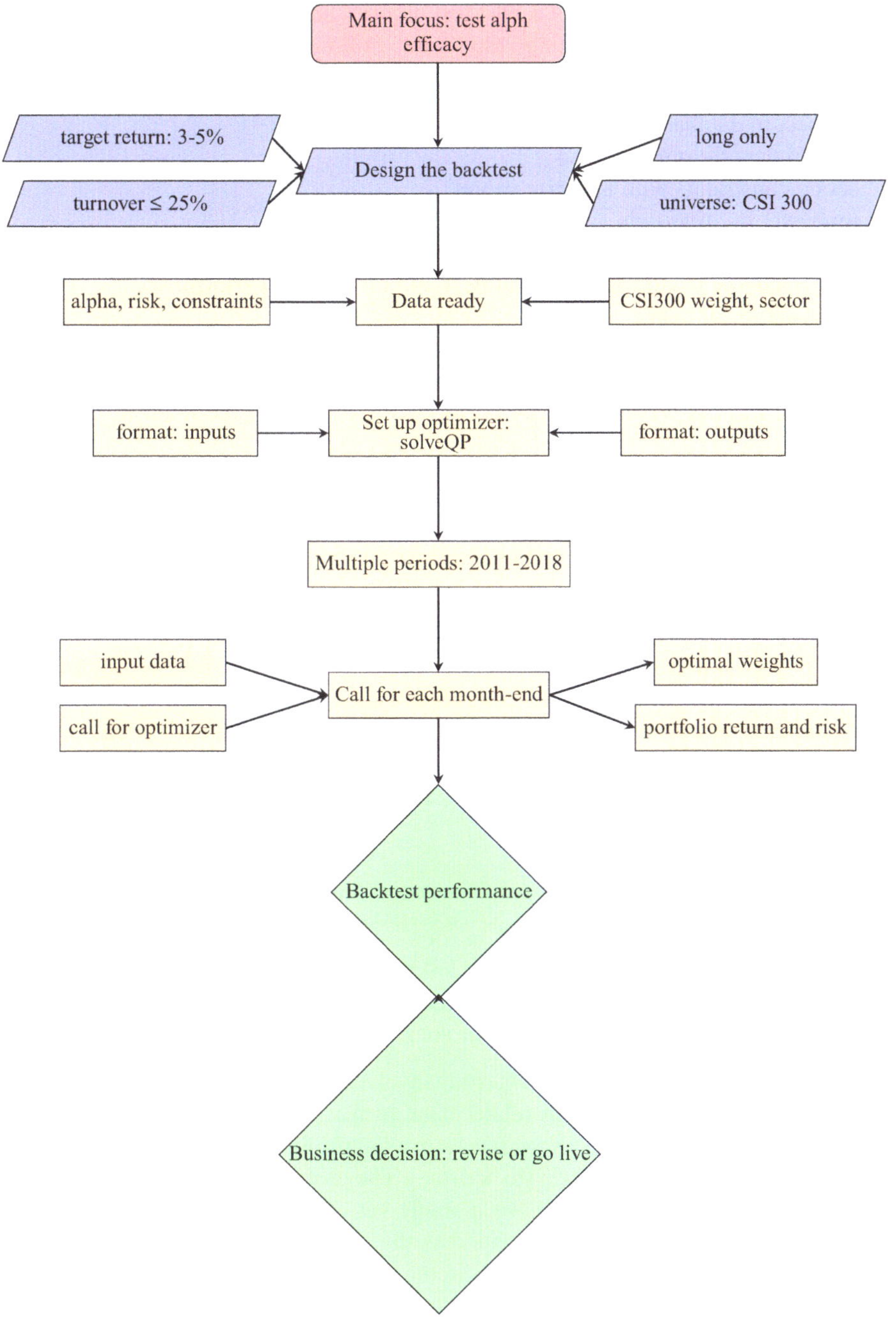
Main focus: test alph efficacy
target return: 3-5%
turnover ≤ 25%
Design the backtest
long only
universe: CSI 300
alpha, risk, constraints
Data ready
CSI300 weight, sector
format: inputs
Set up optimizer: solveQP
format: outputs
Multiple periods: 2011-2018
input data
call for optimizer
Call for each month-end
optimal weights
portfolio return and risk
Backtest performance
Business decision: revise or go live

Coding The implementation of a backtest requires statistical language for computation, reading the data, exporting results, etc. Following the flow chart above, we list the corresponding coding functions.

- connect to database and optimizer
- read in data for optimization
- set up constraints with numeric values
- loop over multiperiods
- for each period: call data, gather inputs for optimizer, get optimal solution
- save results to a database: constraints, optimal weights, performance
- evaluate performance

We discuss how to organize R functions in Sect. 7.8.

Evaluating Results This should be performed according to the main focus of the backtest.

Disparity Between Planned Targets and Backtest Results What if the backtest results do not agree with the original design or planned target? Unfortunately, this is usually the case. We present some suggestions here for further diagnostics.

- The issue requires deeper thought than the original design. For example, the planned target based on conventional thinking may not work.
- Check the data. Sometimes, a disparity may be due to the data. Many missing values or outliers, for example, distort the results.
- Constraints have conflicts. The constraints may conflict with each other, causing unexpected results.
- The process may have significant flaws. This includes both the data treatment and the portfolio construction process.
- The codes may contain errors.
- The optimizer may not deliver results as expected.

Of course, the investigation should be done case-by-case because each case is different. One common mistake that is hard to resist for many investors is backtest fine tuning: when the results of a backtest do not meet with the original targets, they twist the design to fit the returns data for areas where the backtest did not do well.

Resampling Methods Portfolio backtesting involves using samples of historical data. In finance, we observe investment-related data, such as price movements and statistics about company performance. These data form a sample for us to estimate a population parameter (e.g., effects of B/P on stock price). The problem is that at a specific time or for a given period, we only have a single set of observations, and this is the only data we can use. One way to address this is by resampling, to estimate the population parameter multiple times from the sample data. In statistics, resampling is a quantitative method that consists of drawing samples from the original data.

In the previous section, we mentioned the term "in-sample" and "out-of-sample" many times. A general practice is to obtain estimates using the in-sample data and then explore how these estimates work for the out-of-sample data. This is because everything we have from the in-sample data will be tested by further data that we have not observed yet. Resampling methods help to explore the robustness of the in-sample estimates and mimic out-of-sample analysis. Thus, resampling can play a significant role in backtesting. Two commonly used resampling methods are bootstrapping and cross-validation.

- Bootstrapping
  Bootstrapping is a resampling method for population parameter estimation: a large number of small samples of the same size are repeatedly drawn, with replacement, from a single original sample. Then the summary statistics of the estimates for each small sample will be used as an estimate for the original sample to infer population parameters.
- Cross-validation
  Cross-validation is a resampling method for model validation: one set of data is chosen for validation, while the remaining is used for the model building. Backtesting is one type of cross-validation where the out-of-sample period does the validation for the in-sample portfolio. While a typical backtest is in time sequence order, this does not need to be the case. In fact, for robustness check, cross-validation is more effective without the time order. For example, when performing a backtest, we can partition a data set into k groups, where each group has a chance of being used as a held-out test set, leaving the remaining groups as the training set.

7.5.2 From Simulation to Live Product

The purpose of backtesting is to mimic a real portfolio as closely as possible with historical data, such as by using the same setups and constraints on securities and industries. However, there are always differences between a simulated portfolio with known information and a live product with uncertainties and unknown consequences. Below we discuss challenges when transforming a backtest into a live product (Fig. 7.13).[9]

History Repeats Itself, But Not Exactly the Same This principle provides the fundamental rationale for backtesting. However, across historical cycles, there are differences between events. The sensitivity of a backtest setup to those differences can be critical for the performance of a live portfolio. The robustness of backtest results is very important.

One such example is the recent recession of 2009–2011 following the 2008 financial crisis and the Great Depression of 1929–1933. These two extreme periods exhibited similar patterns of stock market performance: the stock market was in a bubble before the crisis and then declined dramatically during the crisis, and it

[9]Photo source: https://www.northinfo.com/documents/389.pdf.

Fig. 7.13 The results of a backtest can be very rosy (and misleading!)

took many years for the economy to recover. However, the triggers and post-crisis performance were different.

Strength and Weakness We should address here that although we can build a successful quantitative investing strategy, the information a quant model can capture is very limited, regardless of the virtues of quantitative models and their associated investment process. A correlation of 5–10% between alpha and realized returns is considered very decent for an alpha model. The other 90–95% of information in returns is not known and should be the purview of risk management.

One useful function of backtesting is to identify areas where alpha is strong. For example, if a model performs well for IT companies or small companies, we should have the portfolio weights rely more on alpha in these areas and limit the use of alpha in other areas.

Implementation Shortfall There will be a significant difference between the calculated performance of a paper portfolio and the realized performance of an identical real portfolio based on actual market transactions. One cause of the difference is transaction costs, including not only the direct monetary cost of buying and selling securities but also the opportunity cost of delaying the sale of a security intended for disposition or the purchase of a security targeted for acquisition.

Live Events A backtest assumes we know all the events and the impacts of these events reflected in price changes. However, for a live portfolio, we only know the events after they occur, and we take action based on our decisions. Usually, these decisions are not model-driven. Rather, they are handled case-by-case with quick on-the-spot judgments.[10] For a live portfolio, proper handling of events may have dramatic impacts on portfolio performance because events usually have extreme impacts on stock prices. We discuss this further in Sect. 7.7.

[10]Except for quantitative event-driven strategies.

7.6 Portfolio Performance Attribution

We have discussed how to construct a portfolio with information about alpha, risk, and constraints. One critical component of quantitative investing is to understand the performance of a portfolio. Performance attribution interprets how investors achieve their performance by measuring the sources of value added to a portfolio. For example, what are the sources of the portfolio return, and how much does each source contribute? How much value do constraints subtract from or add to the portfolio? In this section, we introduce major approaches used in quantitative investing for portfolio performance attribution.

Now, imagine you are a portfolio manager managing a quantitative fund with 100 million dollars. At the month end, you would like to understand (or are required to report to investors or the investment committee) the portfolio's performance for that particular month.[11] In the following, we present different angles for a PM to consider for portfolio performance attribution.

Factors and Alpha Performance First of all, for quantitative model-driven strategies, we need to understand how alpha and each factor perform during the performance reporting period. This can be done by either parametric methods such as regression, or nonparametric methods such as decile analysis. The model weights for each factor and impacts of alpha on the portfolio are then accounted for the performance of portfolio securities.

From Alpha to Portfolio Weights One approach is reverse engineering: starting from alpha without any constraints and then analyzing weight change and portfolio performance by adding constraints one by one in order of significance.

1. Add alpha to the optimizer without any constraints.
2. If the live portfolio is long-only, do not allow short sales.
3. Add constraints at the security level.
4. Add constraints at the sector and industry levels.
5. Add constraints on turnover and transaction costs.
6. Add other constraints that were used in the live portfolio.

Corresponding to each step, we obtain optimal portfolio weights and portfolio performance. This allows us to see portfolio changes due to each constraint and the full path from alpha to the final performance of the live portfolio. Moreover, this sequential analysis may help us to refine constraints and improve the portfolio construction process.

In quantitative investing, the transition from alpha to portfolio weights can be measured by the transfer coefficient, the correlation between alpha values and portfolio weights:

$$\text{Transfer Coefficient} = \text{correlation (alpha, portfolio weights)}.$$

[11] For many quant shops, it is routine to have team meetings about fund performance on a weekly or monthly basis.

For long-only stock selection strategies, a rule of thumb in the industry is that the correlation between the two should be about 40–70%.

Active Performance Attribution There are two quantitative approaches to performance attribution. One is nonparametric through decomposition, and the other is parametric with a regression framework, decomposition of total returns into market and active or alpha. The regression approach can be carried out by applying OLS to a multi-factor model. We introduce the nonparametric approach below.

For nonparametric decomposition, a well-known technique is the Brinson method proposed in Brinson and Fachler (1985) and Brinson et al. (1986). The basic structure of Brinson analysis is straightforward, as expressed in the following:

$$\begin{aligned}\text{Total Return} &= W^p R^p \\ &= (W^b + \Delta W)(R^b + \Delta R) \\ &= W^b R^b + \Delta W R^b + W^b \Delta R + \Delta W \Delta R, \end{aligned} \tag{7.14}$$

where $\Delta R = R^p - R^b$ and $\Delta W = W^p - W^b$. Hence, the active return of a portfolio consists of three components:

$$\begin{aligned}\text{Active Return} &= W^p R^p - W^b R^b \\ &= \Delta W R^b + W^b \Delta R + \Delta W \Delta R, \end{aligned} \tag{7.15}$$

where these three components are defined as follows:

$$\begin{aligned}\Delta W R^b &: \text{ Allocation or Timing Effect} \\ W^b \Delta R &: \text{ Selection Effect} \\ \Delta W \Delta R &: \text{ Interaction Effect.}\end{aligned}$$

Selection effect is the variation in return due to asset class returns in excess of benchmark returns. Allocation effect is the variation in return due to assets held in weights different from policy. Interaction effect is a blended effect of the other two. Usually, interaction effects are minimal and can be ignored. The sum of these three terms is the active portfolio return.

Note that the Brinson decomposition method works well for a single period. If we chain together multiple periods, the polynomial expansion becomes increasingly complicated, and the blended returns overwhelm the analysis.

7.7 Industry Insights: A Backtest Portfolio, a Global Portfolio, and a Live Portfolio

In this section, we show how industry professionals construct portfolios with practical constraints for both backtesting and live portfolios. For illustration purposes, we use examples of a long-only stock selection strategy in the S&P 500 investment universe. We first present industry approaches for backtest portfolios and then discuss related aspects when transitioning from simulation to a live product.

7.7.1 A Long-Only Portfolio: Practical Constraints

For practical purposes, we specify five constraint scenarios: unconstrained, very loose, loose, tight, and very tight. The constraints are applied at both stock and sector levels. At the sector level, following investment industry convention, we use the GICS industry classification system. The detailed specifications of active constraints are listed in Table 7.7. These five cases cover most scenarios of long-only stock selection strategies in the industry.

In the industry, and particularly in quantitative investing, the constraints on assets and at the sector/industry levels are usually the most important constraints imposed on a portfolio after short sales and tracking error (if it is benchmark relative). Usually, the total risk and TE target are achieved through constraints at the asset and sector levels. At the asset level, constraints on maximum and minimum holdings for each security ensure the portfolio is not concentrated on a few companies, thus potentially bringing in the benefits of risk diversification. Once the portfolio is specified as either long-only or long-short, the security level constraints are often regarded as the most critical and effective for portfolio construction, while other constraints are considered second-order or third-order.

Why do we need sector constraints? We list some major reasons below:

(1) An equity market reflects the natural size of economic components in that market, such as the size and competitive strength of different industries. For example, banking is always a big sector given its significance in most public equity markets, such as the USA, Japan, and China. IT companies like Google and Apple in the USA have a lead or competitive advantage in the market.

Table 7.7 Active constraints for portfolio construction

Constraint case	Constraint name	Sector level (%)		Stock level (%)	
		min	max	min	max
Case 1	Unconstrained	−100	100	−100	100
Case 2	Very loose	−10	10	−5	5
Case 3	Loose	−5	5	−2	2
Case 4	Tight	−2	2	−1	1
Case 5	Very tight	−0.50	0.50	−0.20	0.20

(2) Sectors have different characteristics. Some are more stable, such as utilities; some are seasonal, such as retail; and some are random and very volatile, such as biomedicals. It is challenging to timing all these sectors.
(3) Quant strategies stress breadth. Factors are identified to select the best stocks across sectors (except for sector based strategies). The skill set focuses on stock selection rather than sector selection or allocation.

For each constraint scenario, to ensure portfolios are representative and not concentrated in a narrow range of risk levels, we specify a wide range of values for the risk aversion parameter, λ. A wide and reasonable range of risk levels also ensures comparability between different MV portfolios.

For this study, we use S&P 500 stock-level returns data. The data is monthly from December 1994 to December 2013. For each company in the S&P 500 index at a given in time, we extend historical stock returns for that company back at least 5 years. This provides us with enough data to calculate the return statistics needed for optimization.

The MV portfolio is constructed using the MPT optimization framework, specified in Eq. (7.16). Since our long-only stock selection strategy is active-return based—that is, compared with a benchmark—the constraints on assets and individual stocks are all on a relative basis.[12] Defining active weights as $w_i^a = w_i^p - w_i^b$, we have the following setup:

$$\max W^\top \alpha - \lambda W^\top \Omega W \tag{7.16}$$
$$\begin{aligned} s.t. \quad & \text{long-only: } W^\top 1_n = 1,\, w_i > 0 \\ & \text{sector } j : \sum_i |w_i^a \times I(s_j)| \le s \\ & \text{security: } |w_i^a| \le h, \end{aligned}$$

where $I()$ is an indicator function for sectors. Note that the absolute values are nonlinear constraints, which can be decomposed into linear constraints. For example, for the case of *loose* constraints in Table 7.7, $|w_i^a| \le 0.02$ is equivalent to $-0.02 \le w_i^a \le 0.02$, the lower and upper bounds of active weights. Because this is a long-only portfolio, the most underweighted values we can obtain in terms of active weights are the benchmark weights, that is, $w_i^p = 0$ and $w_i^a = -w_i^b$.

For a medium investment horizon, our portfolio is rebalanced monthly at each month end. We use a 60-month moving window as the in-sample period for an out-of-sample study. Using a 60-month moving window from December 31, 1994, our first portfolio construction occurs on December 31, 1999. The inputs include expected returns, λ, and risk metrics. Expected returns are measured by the average returns for each stock during the most recent 60 months. The risk component is the covariance matrix of stock returns for the most recent 60 months. The same process

[12]This can easily be converted to the absolute version.

is repeated until the last month, November 30, 2013. Thus, the out-of-sample period is from January 2000 to December 2013.

We use the R function *solve.QP* in the *quandprog* package as the MV optimizer. We provide below a sample function for the constraints setup.

Industry approach: long-only portfolio

```
### sample codes on sector and asset constraints
 if(constraint.relativeBM==1)
 {
  minwgt <- if (longOnly==1) pmax(0,cap+constraint.asset[1]) else cap + constraint.asset[1]
  maxwgt <- cap + constraint.asset[2]

  # constraints: sector level
  sector.upper <- secwgt + constraint.sector[2]
  if(longOnly==1) sector.lower <- pmax(0, secwgt + constraint.sector[1])
  else sector.lower <- secwgt + constraint.sector[1]
 }
 else ###absolute sense
 {
   cap2=rep(0,length(cap))
   secwgt2=rep(0,length(secwgt))

   minwgt<-if (longOnly==1) pmax(0,cap2+constraint.asset[1]) else cap2+constraint.asset[1]
   maxwgt <- cap2 + constraint.asset[2]

  # constraints: sector level
  sector.upper <- secwgt2 + constraint.sector[2]
  if(longOnly==1) sector.lower <- pmax(0, secwgt2 + constraint.sector[1])
  else sector.lower <- secwgt2 + constraint.sector[1]
 }
# rename the asset level weights
 asset.lower=minwgt
 asset.upper=maxwgt
```

7.7.2 A Long-Only Portfolio: Empirical Results

Our focus will be on the performance of portfolios with different constraints at various risk levels. Before carrying out portfolio performance analysis, we investigate the number of holdings and turnover. This is to ensure that portfolios are practical and could potentially be used as a live strategy.

We present the number of names in the box plot in Fig. 7.14. We see that from Case 1, in which there are no constraints, to Case 5, with very tight constraints, the number of names increases gradually from 10 to 200. We also list more detailed information with λ values n Table 7.8. The number of holdings varies more across cases than λ, which indicates that the constraints on assets and sectors are more effective than λ on portfolio holdings. In general, the risk aversion parameter plays a more significant role when constraints at the asset and sector levels are loose.

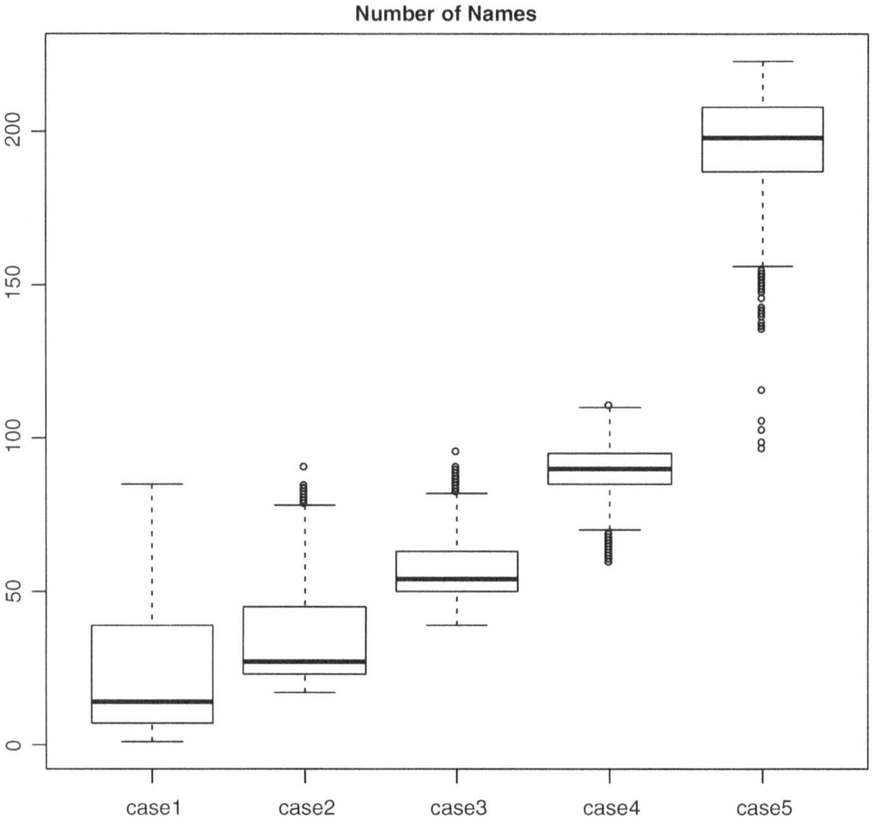

Fig. 7.14 Number of portfolio holdings across cases. The number of names in a portfolio increases from Case 1 to Case 5

Table 7.8 Number of portfolio holdings by λ for each case, λ is less effective when constraints become tight

Lambda	Case 1	Case 2	Case 3	Case 4	Case 5
0.25	6	22	49	86	193
0.5	7	23	50	86	193
1	9	24	50	86	193
2	13	26	51	86	192
10	29	37	59	89	192
100	60	62	72	98	199
1000	61	64	74	98	200

We know that Case 1 is more for theoretical exploration, while all other cases have practical numbers of holdings. For example, Case 2 has 22–64 portfolio holdings, and Case 3 has 49–74 holdings. Both cases are expected to produce portfolios with high tracking error, which we will see next.

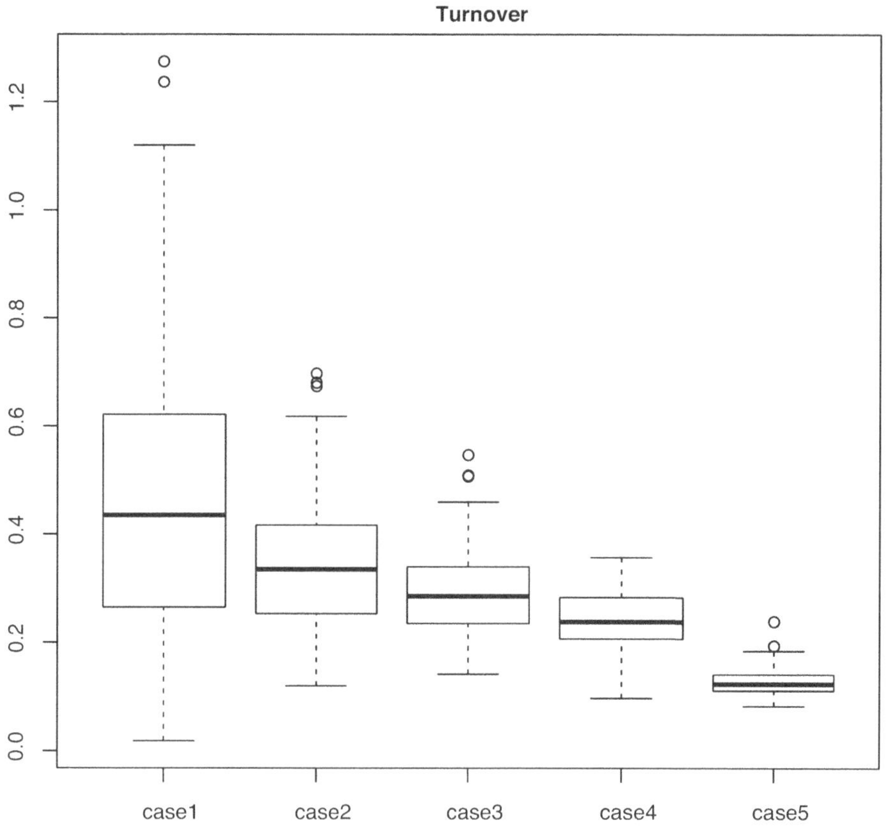

Fig. 7.15 Summary of portfolio turnover across cases. The turnover values are compatible with the holding numbers in Cases 2–5

Now consider turnover. We use the definition of two-way turnover: for a complete sell-off of all existing securities and purchase of new securities, the turnover is 200%. The portfolio turnover is summarized for each case in Fig. 7.15. We see that turnover decreases as constraints become tighter. For Cases 2, 3, 4, and 5, turnover is around from 50% to 20% on a monthly basis. For example, for Case 5, for each month, about 10% of the portfolio is sold off and replaced with new shares or additional shares of 10% portfolio weight. To see the turnover over time, we present the time series of monthly values in Fig. 7.16, which shows that turnover values are stable over time.

Based on the number of holdings and turnover, we see that for Cases 2–5, the MV portfolios are diversified and transaction costs are reasonable, which indicates that the portfolios are practical. Now we investigate the performance of portfolios in terms of return and risk.

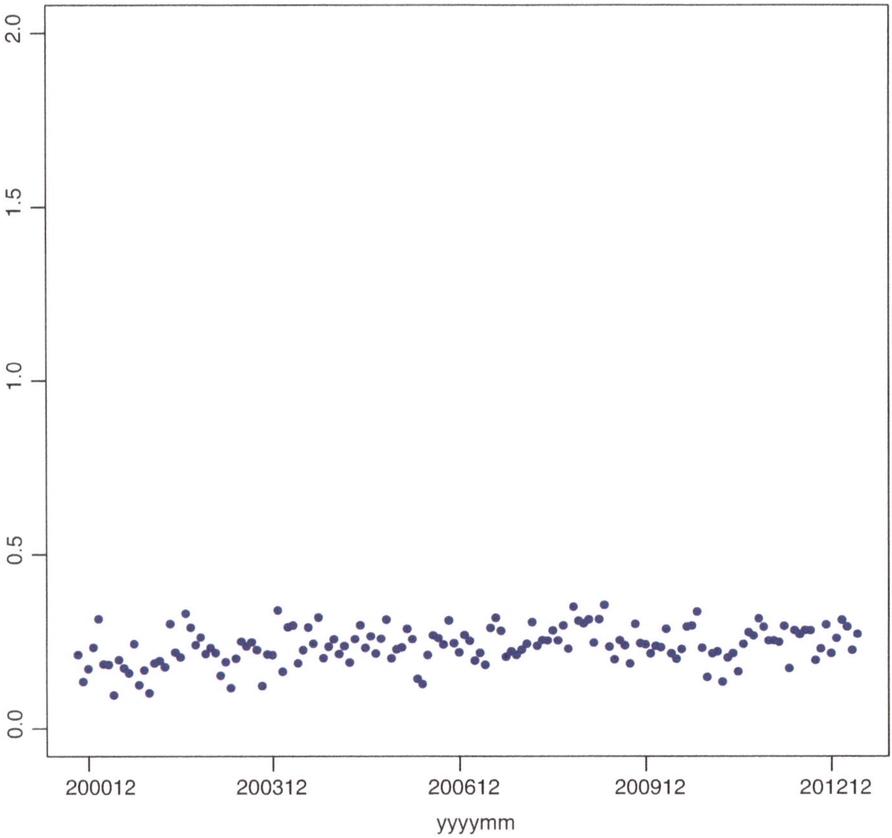

Fig. 7.16 Portfolio turnover over time for Case 4, 2000–2013

We investigate tracking error first. Recall that TE is measured by the standard deviation of active returns. TE is a very important measurement for long-only strategies because it defines how far the portfolio deviates from the benchmark and hence the associated management fee. As we pointed out early in this chapter in Table 7.3, high TE products usually entail high management or performance fees in the industry because those products are expected to deliver higher returns.

We want to determine the values of TE for portfolios across different cases and lambda values. In general, tighter constraints result in lower tracking error. We present annualized TE values across the four cases (Cases 2–5) in Fig. 7.17. We see that the TE values decrease from Case 2 to Case 5, reflecting the effects of constraints imposed on each case. Those TE values are in line with practical industry levels for quantitative long-only stock selection strategies in the developed markets:

$$\frac{\text{index plus}}{0-1\%} \quad \frac{\text{low}}{2-3\%} \quad \frac{\text{mid}}{3-5\%} \quad \frac{\text{high}}{5-8\%} \quad \frac{\text{very high}}{7-10\%}.$$

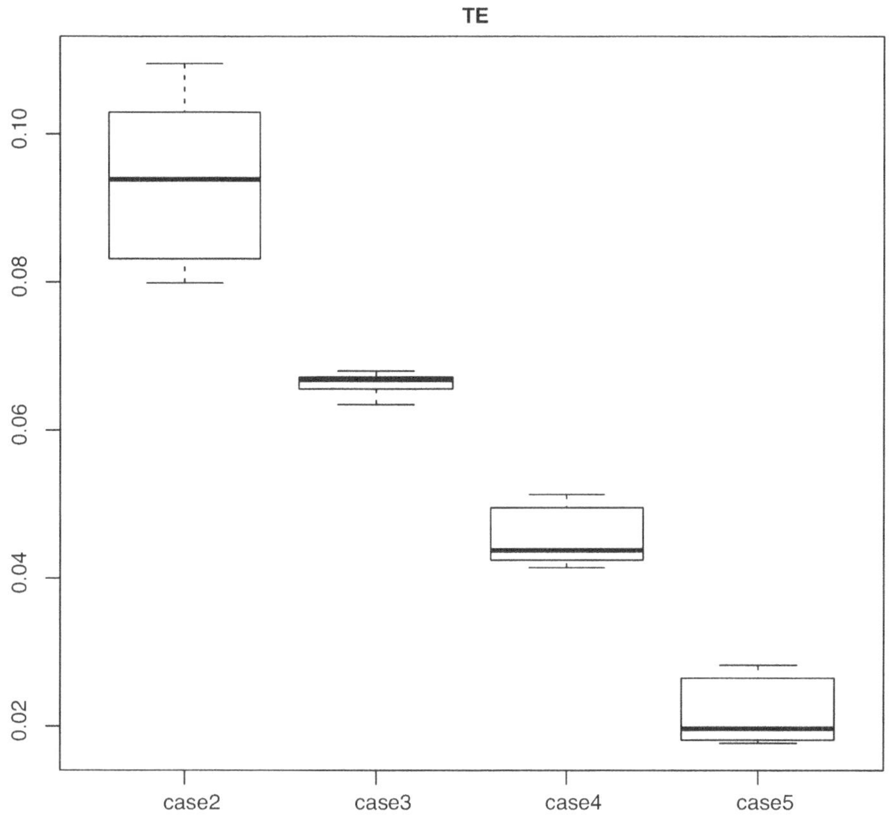

Fig. 7.17 Portfolio tracking error over time for Cases 2–5. The TE decreases from Case 2 to Case 5

Moreover, the narrow box indicates that values of TE do not change much over values of lambda once constraints on assets and sectors are fixed.

Now we present the performance of portfolios. In Table 7.9, we list annualized values of portfolio return, total risk, active return, and information ratio. Recall that the active return is the portfolio return minus the benchmark return, and the information ratio is active return divided by TE. We see that first, portfolios in Case 2 have similar returns and IR to those in Case 1, indicating that the constraints in Case 2 are very loose, and the tracking error is indeed very high. Moving from Case 2 to Case 3, while portfolio returns remain almost the same, IR increases about 20–30%, indicating a decrease in tracking error because of tighter constraints on assets and sectors. The IR increases from Case 3 to Case 4, indicating a further drop in TE. In Case 4 with $\lambda = 100$, the active return is 4.22%, and IR is 0.84. In Case 5, the portfolio returns drop because of the very tight constraints on assets and sectors, but the TE drops even further, resulting in a higher IR of around one. We see that the changes in return and tracking error values are not linear when the constraints and

Table 7.9 Annualized portfolio performance: total return (%), total risk (%), active return (%), and IR, 2000–2013

Scenario	Portfolio return	Portfolio risk	Active return	IR
Case 1				
$\lambda = 1$	0.61	29.03	−0.63	−0.03
$\lambda = 100$	5.12	10.79	3.87	0.42
$\lambda = 1000$	6.03	10.87	4.77	0.53
Case 2				
$\lambda = 1$	2.72	20.46	1.46	0.15
$\lambda = 100$	5.37	10.94	4.12	0.50
$\lambda = 1000$	6.31	11.02	5.05	0.63
Case 3				
$\lambda = 1$	2.56	18.02	1.31	0.20
$\lambda = 10$	5.44	11.43	4.19	0.62
$\lambda = 100$	5.99	11.51	4.74	0.71
Case 4				
$\lambda = 1$	3.18	16.24	1.93	0.47
$\lambda = 100$	5.48	12.28	4.22	0.84
$\lambda = 1000$	5.44	12.40	4.19	0.82
Case 5				
$\lambda = 1$	3.20	14.94	1.94	1.05
$\lambda = 100$	3.87	13.71	2.62	0.94
$\lambda = 1000$	3.86	13.79	2.61	0.92

The benchmark is the S&P 500 index

risk aversion parameters change. We observe from this backtest that when constraints relax (to a certain level), returns increase and tracking error increases, with the latter increasing faster. Overall, Cases 4 and 5 seem like profitable strategies if we use the average of historical returns as alpha and covariance of past returns as the risk matrix.

Now we want to investigate how portfolios perform when the market is up and down. For illustration purposes, we use the example of Case 4 with a lambda value of 100. We calculate annual portfolio performance across calendar years from 2000 to 2013 and plot both portfolio returns and benchmark returns in Fig. 7.18. We see that during the backtest period of 14 years, the portfolio outperforms the benchmark for 12 of those years. In terms of market performance, there are 4 years the market was down, 8 years the market was up, and 1 year with the market was flat. When the market is down in 2000–2013 particularly 2008, the portfolio outperforms the benchmark for each individual year. When the market is up, the portfolio outperforms the benchmark in most years. When the market is flat in 2011, the portfolio has a 3.67% annual return.

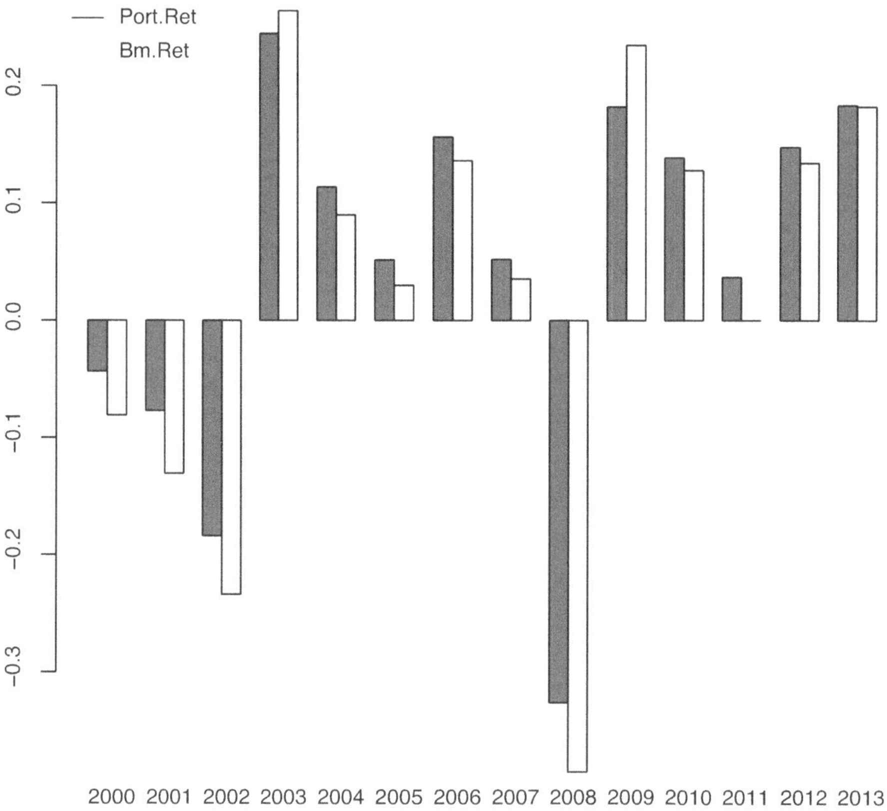

Fig. 7.18 Annual portfolio (Case 4 with $\lambda = 100$) returns from 2000 to 2013

7.7.3 A Global Portfolio

So far, we have considered only single-country portfolios. In the industry, there are large amounts of investments in international and global stock markets. For example, there are developed market funds, emerging market funds, global market all-country funds, etc.

For an investment across multiple countries, its investment process will be more complicated. The two additional aspects that need to be accounted for are country and currency.

(1) country: signal treatment, portfolio constraints at country level
(2) currency: hedging or not hedging, portfolio exposure

$$r_p = (1 + r_l) \times (1 + r_c) - 1.$$

Portfolio return in home currency is the compounding of local return (r_l) and currency return (r_c).

When investing in a foreign country, the following need to be considered:

- transparency of policy decisions by governments
- data reliability
- market efficiency
- financial reports: timeliness and accuracy
- cultural, economic, and legal features
- paper alpha versus obtainable alpha

For quantitative investing, the portfolio should be constructed with a good understanding of not only the market, but also the signals, alpha and risk models, and trading, in detail. For example, when constructing an alpha model, we need to consider the following items:

International portfolio

```
signals and alpha values
-- accounting rules
-- economic structure
-- industry competitive edge
-- corporate organization and governance
trading and portfolio management
 --rules on paper and in practice
 --holidays and trading hours
 --trade settlement
```

The items listed above will be discussed in detail in Chap. 9 in the context of a Japanese stock selection portfolio.

7.7.4 From a Paper Portfolio to a Live Portfolio

As we discussed earlier, before a live strategy is invested with money, it is usually backtested to mimic what the real portfolio would look like. From there, we can build a live portfolio. While the backtest portfolio serves as the basis and test version of a live portfolio, the two can be very different. Before getting into details, we list a few major differences between a paper portfolio and a live portfolio in Table 7.10.

Building a live portfolio requires a lot of effort. A quantitative strategy usually starts from a paper portfolio and is then expanded, tested, and revised many times. Below we list a basic workflow.

Table 7.10 Major differences between running a paper and live portfolio

Item	Paper portfolio	Live portfolio
Target	Return	(Active) risk
Cost	Ignore	Transaction cost model
Rebalance	Fixed backtest date	Schedule and market impacts
Constraint	Several scenarios	Soft, hard, and grandfather
Capacity	Ignore	Significant for small cap
Performance	Revisable	Record and accumulative

Cultivate a deep understanding of the market

$\Rightarrow$ identify drivers and indicators of stock price movements

$\Rightarrow$ acquire data for those drivers and indicators

$\Rightarrow$ construct signals $\Rightarrow$ signal treatment

$\Rightarrow$ build themes $\Rightarrow$ multi-factor alpha model

$\Rightarrow$ perform alpha diagnostics $\Rightarrow$ calibrate risk and constraints

$\Rightarrow$ specify portfolio parameters $\Rightarrow$ build portfolio

After making a business decision based on all the pieces of an investment strategy–alpha, risk, and portfolio parameters—it is usually time to conduct a dry run of the portfolio as a test for the live portfolio. After a reasonable period of dry runs, the strategy is ready to move forward to use real money.

The chart below describes some major common and different characteristics associated with a quantitative investment strategy at the research level, dry run stage, and production stage. For a live portfolio at the production stage, daily performance matters. A large drop in a single day could result in a critical blowup of the portfolio performance and the closure of the account and strategy. Whereas in research, this type of events is either ignored or taken as "known," and long-term performance is evaluated.

Portfolio characteristics: research, dry run and production

```
Research          Dry-run               Production
-- paper money     -- paper money         -- real money
-- research data   -- live data feed      -- live data feed
                   -- corporate events    -- corporate events
                   -- simulated trade     -- live trading
                   -- simulated account   -- client account
                                          -- performance record
```

7.8 Structure of R Functions

We observed that to run a backtest for a stock selection strategy, there are multiple stages and many functions involved in the process. It is therefore helpful to have an efficient and well organized structure for R functions.

In this section, we propose a structure of R functions with a focus first on components and flow then principles and organization.

7.8.1 Structure of R Functions: Components and Flow

In general, a package of R functions for a special purpose, such as alpha diagnostics or backtesting, consists of the following parts: (1) set up target and parameters; (2) prepare inputs, which may include data and scenarios; (3) run main functions; (4) export outputs; and (5) analyze results. Each part consists of one or more R functions. Given that there will be many R functions, an effective structure to organize these R functions in an efficient way is critical. Here we introduce a structure that many professional investors follow in quantitative investing:

$$\textit{Macro setup} \Rightarrow \textit{Prepare} \Rightarrow \textit{Run main functions} \Rightarrow \textit{Output} \Rightarrow \textit{Analysis of results}$$

We discuss each part with a focus on components and the connections between components.

1. *Macro set up* For this block, the purpose is to set up macro variables for the project, including the target clarification, system variables, configuration file, and log files.

   R environmental variable: This refers to the environmental variables that can affect an R session. Some variables pertain to the OS system functions that R uses, and they may affect add-on packages. The R command is *Sys.setenv* and *Sys.getenv*.

Set environmental variable

```
if(!grepl("ActiveQuantEquity",Sys.getenv("RESEARCH")))
  {Sys.setenv(RESEARCH="A:/QuantInvesting/Research/Chapter7")
}
```

   Configuration file: this deals with special designs or instructions that are expressed in a text or Excel file. The R codes will read the file and execute as instructed. Using a Japanese stock selection portfolio for illustration purposes, we present a configuration file in Fig. 7.19. The configuration file has weights

theme name	signal name	signal wgt(%)	theme wgt(%)	distribution permutation	note on treatment (single country)
VALUE			35		
	B2P	70			
	DIV2P	20			no industry neutral
	CFO2P	10			exclude Banks
PVM			20		
	PM1m	75			no industry neutral, truncation 7 winsorize 5
	IPM3m	25			no industry neutral, truncation 7 winsorize 5
PROF			15		
	FCF2EV	60			exclude Banks
	EBITDA2EV	20			
	FCF2NOA	20			exclude Banks
EQ			5		
	accrualsCF	60			exclude Banks
	CFg.left	20		left is handled inside the	exclude Banks, no industry neutral
	EPSg.left	20		left is handled inside the	exclude Banks, no industry neutral
MQ			10		
	Tag	30			
	xfinBS	50			
	foreignASSE	10		left is handled inside the	truncation 7 winsorize 5
	foreignSALE	10		left is handled inside the	truncation 7 winsorize 5
MS			15		
	EPSDiff	60			no trucation, winsorize 5, no industry neutral
	DPSDiff	25			no trucation, winsorize 5, no industry neutral
	SHI	15			no trucation, winsorize 5, no industry neutral

Fig. 7.19 The configuration file for a Japanese stock selection strategy

for signals to form themes, weights for themes to form alpha, and treatment instructions for signals. If later we need to change the weights for signals, we just need to change the number in the configuration file, and the R codes will remain the same. The configuration file adds flexibility and convenience to the process.

Log file: this is the file recording all the logs from running the R codes. This is very helpful for debugging when there is an error. It also records the production runs for live portfolios.

2. *Prepare* This block prepares data, parameters, and utility functions to get ready for executing the main functions.

   Utility functions: this includes functions that are used very often and called by other major functions, such as read in data, data treatment, general plots, dates, currency exchange, etc. Usually, all utility functions are included in one folder, which is called before the main functions. In general, each utility function usually targets only one thing with parameters as inputs for flexibility. For example, there could be a utility function to deal with missing values with options for filling in with the mean or median or removal by the user.
3. *Run main functions* This block uses the inputs from the previous step and runs the main functions with key algorithms. In this block, there should be one or several key functions for the most important computations to fulfill the target designed in block 1.
4. *Export outputs* This block exports outputs generated from the previous steps. The outputs can be text reports, tables, plots, or files in Excel, LaTex, or pdf formats. A good practice for R with quantitative investing is to output not only the results from main functions but also the key parameters from the inputs such that any change in the inputs will be recorded and associated with the outputs.
5. *Analyze results* This step analyzes the outputs generated from many different angles and generates a report for team review and business decisions by manager(s). For example, for a backtest, the analysis can focus on

   - Executability: portfolio holdings and transaction cost
   - Practicability: TE
   - Performance: return, IR, or SR; overall and over time, scenarios
   - Robustness: sensitivity to constraints
   - Risk exposure: exposure to standard risk factors

Last but not least, there will be a “run” function, which calls all other functions in sequence and executes the structure. Sometimes, there are two “run” functions: one for the major run, the other to analyze results. Besides the log file, it is helpful to have execution time in the “run” function. This can be done by using the R command *Sys.time*.

Record time

```
start.time <- Sys.time()
...
end.time <- Sys.time()
```

7.8.2 Structure of R Functions: Principles and Organization

We learned from the previous section that the five parts consist of many R functions, and most of them are functions of functions. We suggest here some basic principles and organization tips to structure R functions for a quantitative investing project.

A Few Principles of Structuring R Projects Every R project is different. For example, a project for a new factor will be different from a backtest for a risk model. However, there are a few common principles followed by professionals when organizing R functions.

1. The project determines the structure.
   A structure is designed to achieve the goals of a project. The characteristics of the project, for instance, the purpose, size, and scope, should decide the structure, not vice versa.
2. Structures should be simple and meaningful.
   Avoid complexity and instead try to make the structure simple and meaningful.
3. Structures should be self-explanatory for teamwork.
   Even if this is a solo project for now, try to make the structure self-explanatory for potential collaborators. In quantitative investing, the analyst, portfolio manager, and trader all work together at some point.
4. Structures should be capable of expansion.
   Your project may change objectives; it may evolve. For example, a small project on a new alpha factor may end up turning into a big team project for a quantitative strategy driven mainly by that factor. Therefore, the ability to adapt and expand should be considered from the beginning. For example, try to avoid fixed codes as much as possible and instead use a parameter passing on as an input through successive functions.

Organization of an R Project A standardized physical organization of folders for R projects usually proves to be very efficient, not only for individuals, but also for teamwork. Since most projects in quantitative investing involve data, R functions, output, and analysis, a minimal organization chart will be something like the following:

Project XYZ

– – source codes

  – – *utility functions*

  – – *main functions*

– – data

  – – *raw*

  – – *treated*

– – output

– – analysis

– – archive

– – development

– – read me

In this structure, we use "–" as the symbol for a folder, the indented space stands for a subfolder. The root folder is the project with subfolders of resource codes, data, outputs, read me file, etc. Each subfolder may contain subfolders, and so on.

The data folder is where your data is stored. It is crucial to separate raw data from treated data. The former should be kept as is (i.e., read only) with information on data sources and the latter should include an information file with treatment details.

A development folder is for anything that is still in the development stage and not "officially" proven and released yet. For example, any update or extension, such as $XYZ2.0$, is appropriate to put in this folder. The archive folder can be very useful. Usually, outdated or older versions are stored here, and they are often revisited later either for reference or to be used again.

The analysis folder stores codes and tools for analysis, and the output folder is for all the outputs. Of course, if needed, each folder or subfolder can have other layers of subfolders as long as they are clear and meaningful.

Keywords, Problems, and Group Project

Part I. Keywords

Mean–variance optimization, lasso, variance decomposition

Efficient frontier, diversification, min-vol portfolio, tracking error

Sector and asset constraints, backtesting, simulated/dry/live portfolio

R structure of functions

Part II. Problems

Problem 7.1 Using the data in Problem 4.5, calculate mean and standard deviation of returns for each security. Plot the efficient frontier.

Problem 7.2 Using the data in Problem 4.5, estimate covariance of returns.

(1) Make sure $T \gg N$, and calculate the covariance matrix of N securities.
(2) Using the alpha model in 4.3, recalculate covariance with the multi-factor model.

$$R_F = b_0 + b_1\, PROF + b_2\, EQ + b_3\, VALUE + b_4\, PM + b_5\, MQ + \epsilon$$

Problem 7.3 Using the alpha from Problem 5.3 and covariance data from Problem 7.2, build mean–variance portfolios.

(1) A long-only portfolio
(2) A long-short market neutral portfolio

Part III. Group Project

Problem 7.4 Continue from Problem 7.3. Using practical constraints as described in this chapter, conduct a backtest and analyze portfolio performance.

(1) Calculate the correlation between optimal weights and alpha.
(2) Compare the portfolio performance with those from Problem 7.3.
(3) If you decide to go live, what else do you need for production?

References

Brinson, G.P., and N. Fachler. 1985. "Measuring Non-U.S. Portfolio Performance." *Journal of Portfolio Management* 11(3): 73–76.

Brodie, J., I. Daubechies, C. De Mol, D. Giannone, and I. Loris. 2009. "Sparse and stable Markowitz portfolios." *Proceedings of the National Academy of Sciences of the USA* 106: 12267–12272.

Brinson, G.P., L.R. Hood, and G.L. Beebower. 1986. "Determinants of Portfolio Performance." *Financial Analysts Journal* 42(4): 39–44.

Goldfarb, D., and A. Idnani. 1982. "Dual and Primal-Dual Methods for Solving Strictly Convex Quadratic Programs." In *Numerical Analysis*, edited by J.P. Hennart, 226-239. Berlin: Springer.

Goldfarb, D., and A. Idnani. 1983. "A Numerically Stable Dual Method for Solving Strictly Convex Quadratic Programs." *Mathematical Programming* 27: 1–33.

Kritzman, M. 2006. "Are Optimizers Error Maximizers? Hype Versus Reality?" *Journal of Portfolio Management, Summer* 32: 66–69.

Markowitz, H.M. 1952. "Portfolio Selection." *The Journal of Finance* 7(1): 77–91.

Michaud, R. 1989. "The Markowitz Optimization Enigma: Is Optimization Optimal?" *Financial Analysts Journal* 45(1): 31–42.

Chapter 8
Quantitative Investing with Tail Behavior—A Distributional Approach

Abstract In previous chapters, we introduced classical mean–variance methodologies. Classical methodologies have been used widely in risk management, such as the use of standard deviation for risk; alpha models, such as the use of OLS for weighting schemes; and modern portfolio theory, such as the mean–variance optimization. However, these are all based on the assumption that the first two moments will capture most information about asset returns, which is usually not true of real-world finance data, where fat and long tails are often the case. In this chapter, we present a distributional approach to capture tail behaviors for quantitative investing. Quantile regression (QR), a frontier methodology that extends beyond the median into tail percentiles, provides a useful tool for incorporating tail information into portfolios. We explore how QR can be employed for risk management, alpha modeling, and portfolio construction. The last section introduces R codes and packages for QR.

8.1 Tails Matter: Distributions of Asset Returns

We mentioned numerous times in previous chapters that distributions of asset returns in financial markets are usually not normal and often have fat and long tails. This poses challenges for alpha forecasting and portfolio construction in quantitative investing because classical approaches (e.g., OLS and MPT) rely on mean and variance for their efficiency and desired properties. In this section, we present empirical distributions of representative asset classes: equity, commodity, and currency.

8.1.1 Non-normal Distributions of Asset Returns

Regarding public equity, we have learned that the daily returns of the S&P 500 (in the US stock market) and daily returns of the CSI 300 (in the Chinese stock market) are not normally distributed. To show that this is generally the case, we present the distributions of daily returns for the Japanese and Russian stock markets.

L. Ma, *Quantitative Investing*, https://doi.org/10.1007/978-3-030-47202-3_8

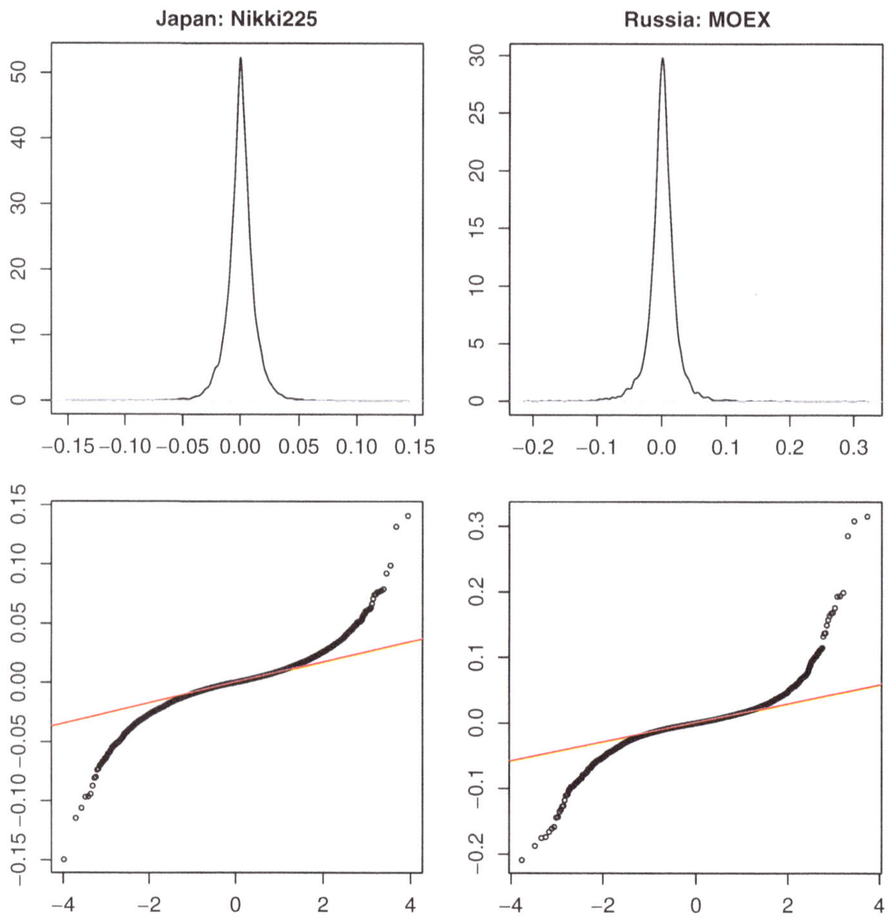

Fig. 8.1 Asset returns of equities. Daily data for the Japanese Nikkei 225 (January 5, 1965–September 20, 2019) and the Russian MOEX index (Sept 20, 1997–Sept 20, 2019). The top panel has density plots, and the bottom panel has the qqnorm plots

Return Distributions for the Nikkei 225 and MOEX We use the major stock index for each country: the Nikkei 225 for Japan and the MOEX for Russia. The Nikkei 225 consists of the 225 largest and most liquid public companies across different industries in Japan. The MOEX measures the performance of about the 45–50 most liquid large companies in Russia. The performance of both indices is measured in local currency. We present the density plots and normality plots of daily returns for each index in Fig. 8.1. The normality plots indicate that the distributions of both indices are not normal.

We also carry out the Kolmogorov–Smirnov test of the sample against a normal distribution. The test of daily returns indicates that neither the Japanese nor the Russian stock index is normally distributed. On the contrary, there are fat and long tails for both indices. Below are the R scripts with computation results.

Normality Test: Nikkei 225 and MOEX Daily Returns

```
> ks.japan=ks.test(x=japan$return,y='pnorm',alternative='two.sided')
> print(ks.japan)
One-sample Kolmogorov-Smirnov test
data:  japan$return
D = 0.47838, p-value < 2.2e-16
alternative hypothesis: two-sided

> ks.russia=ks.test(x=russia$return,y='pnorm',alternative='two.sided')
> print(ks.russia)
One-sample Kolmogorov-Smirnov test
data:  russia$return
D = 0.46294, p-value < 2.2e-16
alternative hypothesis: two-sided
```

Return Distribution for Commodities In the commodities space, we use WTI and gold (London PM fixed). We have monthly price data for WTI and gold from April 1964 to June 2019. In Fig. 8.2, we present densities and normality plots for both commodities. Again, we see that the returns of commodities do not follow a normal distribution.

Return Distribution for USD/EUR and USD/JPY For the currency market, we use two largest pairs in terms of trading volume: USD/EUR and USD/JPY. We have daily data for the returns of both pairs from January 5, 2000 to September 20, 2019. Figure 8.3 shows the density plots and normality plots for USD/EUR (left panel) and USD/JPY(right panel). Again, we see the same phenomenon: financial asset returns do not follow a normal distribution; rather, they have long and fat tails.

8.1.2 When Asset Returns Are Not Normally Distributed, First Two Moments Are Not Enough

We now use various metrics to measure tail behaviors of asset returns discussed above and compare them to a normal distribution. We show that for quantitative investing, the normal distribution-based measures underestimate the potential risk and gains across all asset classes studied in this chapter: equity, currency, and commodity.

First of all, both the mean and variance are sensitive to outliers whereas extreme returns have never been short in asset returns. Secondly, asset returns are very dynamic, return distribution varies over time creating exciting and challenging investing opportunities. The real world of financial markets is too colorful and dynamic to be characterized by a plain normal distribution.

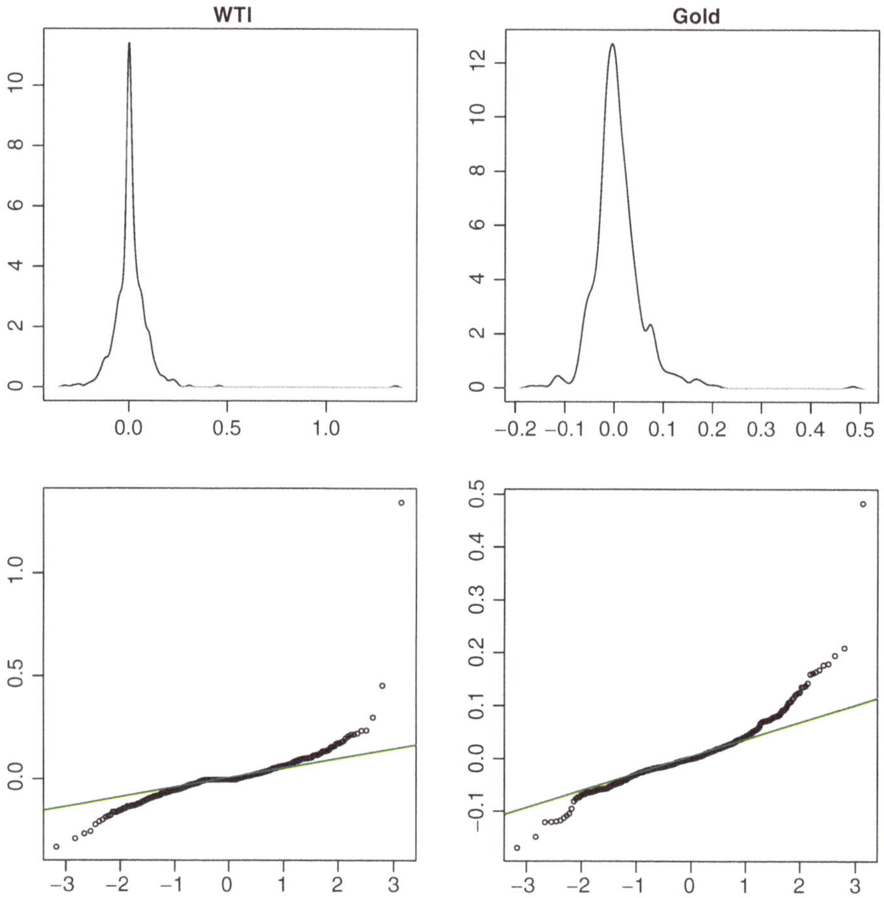

Fig. 8.2 Asset returns of commodities. Monthly data for WTI and gold (London PM fixed): April 1968 to June 2019

We employ the interquantile ratio to measure the deviation of a distribution from normal:

$$IQR = \frac{Q_x(\tau) - Q_x(1-\tau)}{Q_x(0.75) - Q_x(0.25)}, \tag{8.1}$$

where the denominator conventionally takes the 3rd quartile. Table 8.1 shows that for asset returns from all listed asset classes, their interquantile values are much larger than the values for a normal distribution, indicating fat and long tails of asset returns across the three asset classes.

To show how critical tail behavior matters for a portfolio, we form two hypothetical portfolios, one avoids the left 1% and the other misses the right 1% of returns for each year. The results are presented in Table 8.2 with benchmarks. We see that for

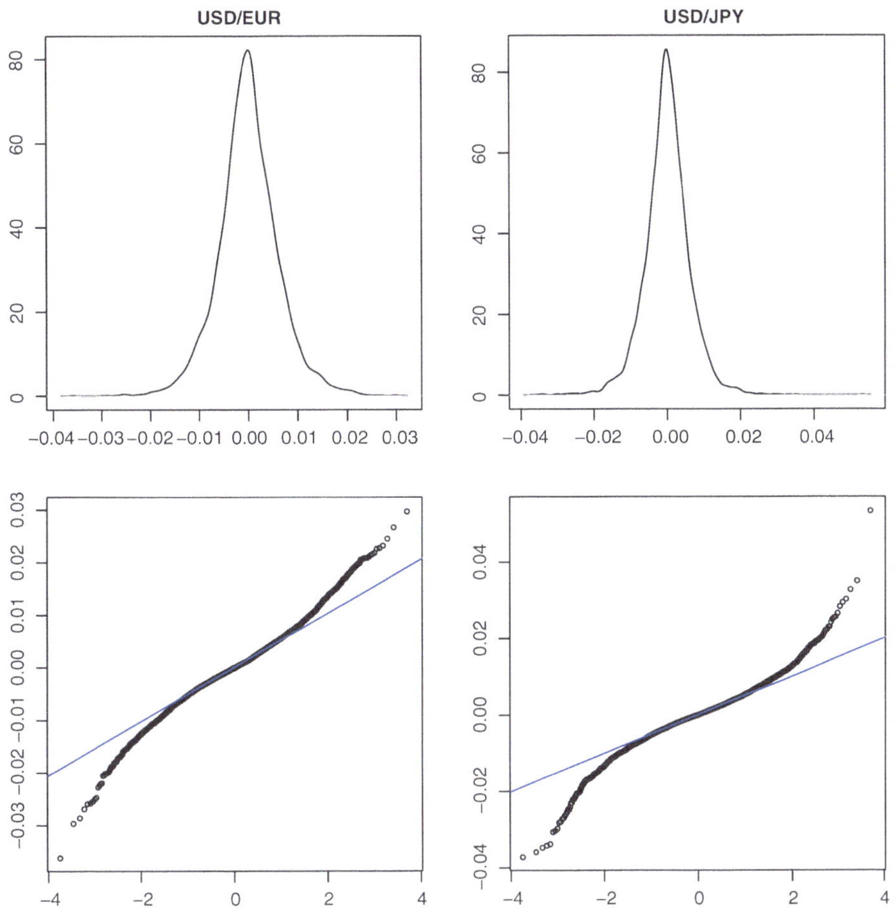

Fig. 8.3 Asset returns of currencies. Daily data for USD/EUR and USD/JPY, from January 5, 2000 to September 20, 2019

all asset classes, if a portfolio avoids the extreme left-tail (1%) returns each year, its annualized return will increase by 1.5 to 20 times! On the other hand, if a portfolio misses the right-tail (99%) returns each year, its annualized return will drop by 5 to 10%. These magnitudes of out- and underperformance caused by tails are far beyond what a normal distribution indicates. In his well-known book on black swan, Taleb (2007) addressed the importance of tail behaviors because of more often than less "our blindness with respect to randomness, particularly large deviations." This is well confirmed by the portfolio performance in Table 8.2. Clearly, tails matter![1]

[1]Tails matter not only in finance but also in many other fields, such as the dangers of the worst-case scenarios for climate change (Weitzman 2009).

Table 8.1 Statistics of interquantile values at 90%, 95%, and 99% across different asset classes. A normal distribution is used for comparison

Asset name	$\tau = 0.90$	$\tau = 0.95$	$\tau = 0.99$
Normal	1.86	2.38	3.28
Equity			
Japan	2.31	3.35	5.93
Russia	2.26	3.51	7.23
Currency			
USD/EUR	2.09	2.89	4.59
USD/JPY	2.13	2.90	4.92
Commodity			
WTI	2.70	3.76	6.53
Gold	2.44	3.22	6.33

Table 8.2 Annualized returns (%) excluding tails across different asset classes

Asset name	Benchmark	Excluding $\tau = 0.01$	Excluding $\tau = 0.99$
Equity			
Japan	4.79	17.56	−5.34
Russia	14.24	46.71	−10.50
Currency			
USD/EUR	−0.38	4.18	−4.62
USD/JPY	0.22	5.50	−4.51
Commodity			
WTI	5.75	10.12	1.07
Gold	7.25	9.40	4.39

8.2 Tails Matter a Lot: Conditional Value at Risk

The first two moments fail to capture tail behavior of non-normal distributions. In this section, we introduce measures of tail behavior: value at risk (VaR) and conditional value at risk (cVaR).

8.2.1 Rule 1: Don't Lose Money—VaR

Suppose we have one million dollars. We can either deposit the cash in a savings account or invest in a global stock market.

Not only for Warren Buffett but for all investors, the most important thing is not to lose money. If we hold cash in a savings account (assuming there is insurance for the bank's deposits), then the loss is zero, or the probability of losing money is zero. However, there is not much positive return either. If we choose instead to invest the money in stocks, then given the ups and downs of the stock market, there are chances of both big profits and big losses. Therefore, it would be very useful to know the probability of potential losses.

Value at risk (VaR) measures the amount an investment might lose with a given probability during a given period, for instance, a day or a month. In mathematical terms, VaR can be expressed as follows. Let X be a random variable with CDF $F(\cdot)$, the loss with probability of τ is defined as:

$$VaR_\tau(x) = F_X^{-1}(\tau),$$

which is a quantile function. We discuss this in detail in the next section.

There are two ways to calculate VaR: using historical data or simulations such as Monte Carlo. The historical method simply uses historical returns, ordering them from lowest to highest and calculating the empirical loss. For example, for the S&P 500 daily returns distribution from 2000 to 2019, we know that the 1% and 5% loss during a typical day are −2.56% and −1.43%, respectively. The simulation method, such as Monte Carlo, relies on a data generation process assuming some scenarios and then applies VaR to the hypothetical distribution.[2]

Note that since most distributions (except the uniform distribution) are not linear in x, the VaR is not additive and thus not coherent.

$$\begin{aligned} VaR_{\tau_1+\tau_2}(x) &= F_X^{-1}(\tau_1 + \tau_2) \\ &\neq F_X^{-1}(\tau_1) + F_X^{-1}(\tau_2). \end{aligned} \tag{8.2}$$

8.2.2 Rule 2: Don't Forget Rule 1—cVaR

For investors, risk means losing money. VaR calculates the potential loss of an investment with a given probability. VaR has critical shortcomings: it is a point estimate and it does not give any information about the severity of losses beyond the VaR level. Investors prefer to know the forest (the whole area of the left tail) over a tree (a single point value).

If a return distribution changes over different time periods, then a 5% loss during the last year may be very different from a 5% loss this year. Moreover, the loss beyond the left tail of 5% may also look very different even though a 5% loss is the same for two return distributions. Hence, simply using a point estimate can be misleading, particularly if a return distribution is not stable.

For quantitative investing, a single value for the probability of losing a specific amount of money does not seem realistic. Instead, the probability of losing a range

[2]See Chap. 5 for more on nonparametric methods.

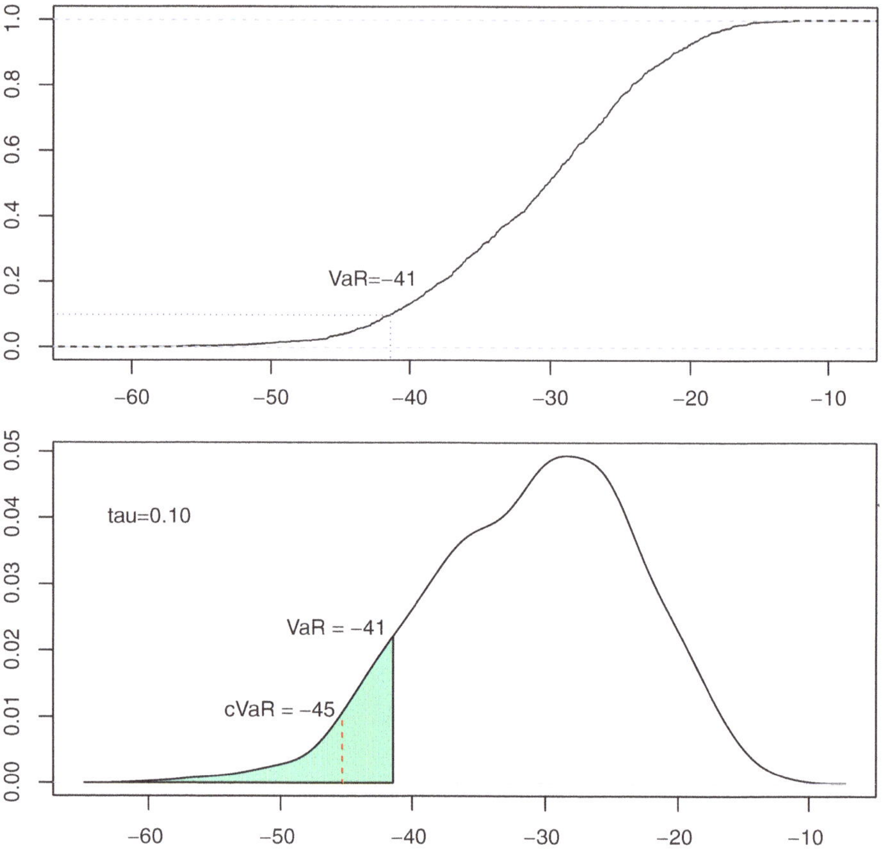

Fig. 8.4 VaR and cVaR. The top CDF plot shows the VaR and the bottom density plot shows the VaR and corresponding cVaR (the red dashed line)

of amounts of money is more realistic. The latter is called conditional value at risk (cVaR) or expected shortfall. In mathematical terms,

$$\begin{aligned} cVaR_\tau(x) &= -\frac{1}{\tau}\int_0^\tau VaR_t(x)dt \\ &= -\frac{1}{\tau}\int_0^\tau F_X^{-1}(t)dt \end{aligned} \tag{8.3}$$

Thus, cVaR does not consider only the single most catastrophic outcome, as VaR does; rather, it considers the whole range, from the worst to the VaR level (Rockafellar and Uryasev 2000). For a given portfolio, cVaR is greater than or equal to VaR at the same quantile (q) level. In Fig. 8.4, we present the values of both VaR and cVaR.

Note that cVaR is coherent. It is convex for all possible portfolios, which means that it always accounts for diversification effects.

It seems natural for investors to minimize the expected shortfall. The optimization problem can be transformed into a linear program and solved with the optimal solution. In fact, cVaR optimization is equivalent to finding the solution to a quantile regression, the subject of seminal work by Koenker and Bassett (1978). Recall that VaR is nothing but a sample quantile. We discuss quantile regression in detail in the next section.

8.3 History of QR and Roger Koenker

Before going into details of quantile regression, we present a brief history of quantile regression and introduce Roger Koenker.

8.3.1 Prelude

Recall that the least squares method was proposed by Gauss in 1795 (Gauss 1809). The mean and median are both common statistics, used in people's everyday life and scientific research long ago. It does not sound unreasonable to believe that our intelligent ancestors considered, perhaps, median regression as well during the same period. This is indeed the case, and actually, median regression was proposed by Roger Joseph Boscovich in 1760, about a half century earlier than the least squares method (first published by Legendre in 1805). Boscovich was interested to measure the earth's ellipticity suggested by Isaac Newton when he developed the least absolute criterion.

Following Boscovich, the research on median regression continued to evolve. For example, Pierre-Simon Laplace developed the so-called methode de situation, a blend of mean and median, in 1789. Built on the work of Boscovich and Laplace, Francis Edgeworth developed a "double median" method for linear regression in 1888—mostly close to the quantile regression (Koenker 2005). However, compared with the least squares method, computation (by hand) for median regression can be very burdensome, especially with multiple factors and a large data set. This was the major reason for historical unpopularity of median regression, even it has a clear robustness advantages over the conditional mean method.

The situation changed in the twentieth century when computers emerged and computation was no longer an issue. Meanwhile, ground breaking work by Koenker and Bassett (1978) generalized the median regression to the whole distribution of random variables.

8.3.2 Roger Koenker and Quantile Regression

Roger William Koenker (Fig. 8.5) is an American econometrician who is best known for his contributions to quantile regression. The quantile regression methodology and analysis tool he developed are widely used across many disciplines.

In the late 1970s, Koenker and his co-author, Gilbert Bassett, proposed quantile regression. Since the 1980s, Koenker has collaborated with others to develop QR methods. We present below a brief timeline on the development of this new frontier approach during the last forty years. For details, please refer to He (2017).

1970*s*—quantile regression proposed by Koenker and Bassett (1978)
1980*s*—inference and computation methods were developed
1990*s*—QR for single equation models become well developed
QR applied in labor economics and other fields
QR developed for time series models
2000*s*—QR evolves in identification, multiple equation models,
Ma and Koenker (2006) and longitudinal data
Koenker publishes his book, *Quantile Regression*, in 2005
application in quantitative investing, Ma and Pohlman (2004), etc.
2010*s*—applied widely in almost all fields
gradually gains recognition as a general method
becomes part of the mainstream econometric methodology
complementary to least squares method

Fig. 8.5 Roger Koenker (1947–)

In the field of quantitative investing, quantile regression was studied by a few pioneers in the early 2000s, such as return forecasting in Ma and Pohlman (2004, 2008, 2010), risk modeling in Engle and Manganelli (2004), and asset allocation in Cenesizoglu and Timmermann (2008). During and after the 2008 financial crisis, motivated by wealth preservation and loss avoidance, left-tail behavior became an important topic in both industry and academia, and both sides started to conduct more studies applying quantile regression to quantitative investments. For example, Adrian and Brunnermeier (2008) apply QR to study cVaR and tail behavior in hedge fund performance; Goldberg et al. (2010) apply quantile regression to study portfolio exposure to Barra risk models; and Ma (2015) employs QR for optimization across quantiles as an alternative approach to classical mean–variance portfolio construction.

Given the natural link between the distributional approach and tail behaviors of asset returns, it is reasonable to assume that there will be more and promising studies of QR in quantitative investing.

I close this brief historical overview with Koenker's conclusion when he reviewed the forty years of development of quantile regression.

> Gaussian models and methods have encouraged the misconception that all things empirical are revealed by conditional means, and perhaps one or two more moments. Quantile regression offers a set of complementary methods designed to explore data features invisible to the inveiglements of least squares. As data sources become richer and awareness of the importance of heterogeneity increases, quantile regression methods have become more relevant. The scope of quantile regression methods has broadened considerably in recent years, thanks to the efforts of numerous researchers. I hope that this constructive process will continue.
>
> —Roger Koenker, Quantile Regression: 40 Years On (2018)

8.4 A Distributional Approach: Introduction to Quantile Regression

Suppose we have a random variable, Y. What is the best way to describe the outcomes and associated probabilities of Y? Of course, the most complete way is to list every possible outcome with its probability. However, we know that for a continuous or even a discrete variable with millions of records, this is an impractical task. One approximation is to use a density or cumulative distribution function.

Recall from Chap. 2 that for a random variable Y, its CDF is defined as

$$F_Y(y) = P(Y \geq y).$$

Now, let τ be the value of $F_Y(y)$, we have

$$\tau \equiv F_Y(y) = P(Y \geq y),$$

and $\tau \in (0, 1)$. So, there is a one-on-one mapping between $Y = y$ and $F(\cdot)$: for every value of Y, there is a CDF value, F. We present the relationship in Fig. 8.6. We see that when $y = 1$, $F_Y(1) = 0.8431$. Similarly, for each value of F, there is a corresponding value of Y.

Suppose there exists an inverse function F_Y^{-1}. The τth quantile function is defined as

$$\begin{aligned} Q_\tau(y) &= F_Y^{-1}(\tau) \\ &= inf\{y, F_Y(y) \geq \tau\} \\ &= inf\{y, P(Y \geq y) = \tau\} \end{aligned} \tag{8.4}$$

Using the example in Fig. 8.6, we have $Q_{\tau=0.8431} = 1$ and $Q_{\tau=0.5} = 0$, where the latter is called the median, a special quantile value.

8.4.1 Sample Quantiles

We just defined a theoretical quantile function in (8.31). Now we show how to construct an empirical quantile function with data. Suppose we have a data sample for a random variable X with unknown population parameters. All we have are the observed values of X: $x_1, x_2, \ldots, x_{n-1}, x_n$. We can order the values of X from smallest to largest,

$$X_{(n)} = x_{(1)}, x_{(2)}, \ldots, x_{(n-1)}, x_{(n)},$$

then the kth percentile of $X_{(n)}$ divides $X_{(n)}$ in such a way that $k\%$ of the values lie below and $(100 - k)\%$ of the values lie above. For example, $k = 0.50$, the median, lies in the middle such that half of the values lie below and half of the values lie above.

In statistics, it is more common to refer the value of k to quantiles. These are the same as percentiles, but are indexed by sample fractions rather than by sample percentiles.

The values $x_1, \ldots, x_{(i)}, \ldots, x_{(n)}$ are the order statistics of the original sample. We now transform the order statistics to quantiles:

$$\tau_i \equiv F_n(x_{(i)}) = \frac{i-1}{n-1}, \quad i = 1, \ldots, n$$

where $\tau \equiv F_n$ is the sample CDF of X. Following the definition of quantile function, we derive the τ sample quantile as the adjacent average of two points: Defining $Q_\tau(x_i)$ as the sample quantile, we have

$$Q_\tau(x_i) = (1-s)Q_{\tau_i}(x_i) + sQ_{\tau_{i+1}}(x_i).$$

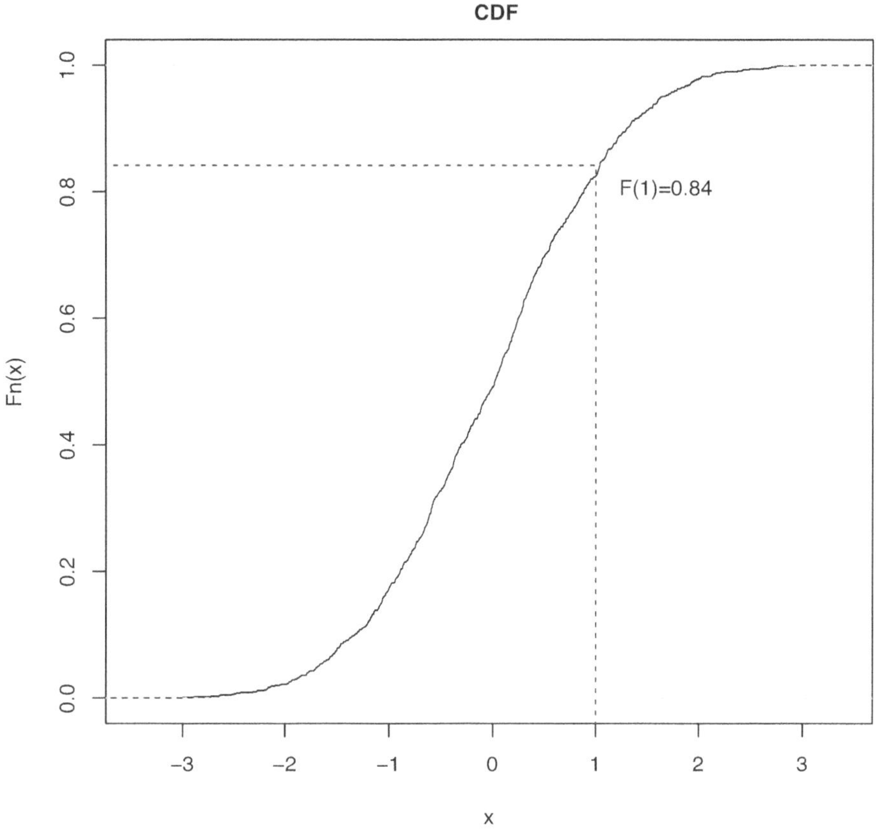

Fig. 8.6 Quantile and cumulative distribution function

In an optimality framework, we can get the sample quantile, τth, by solving the following minimization problem,

$$\hat{q}_\tau = \operatorname{argmin}\left[\tau \sum_{y_i \geq q}(y_i - q) + (\tau - 1)\sum_{y_i < q}(y_i - q)\right]$$

$$= \operatorname{argmin}\left[\tau \sum_{y_i \geq q}(y_i - q) + (1 - \tau)\sum_{y_i < q}(q - y_i)\right] \tag{8.5}$$

Equation (8.5) is piecewise linear with two parts, the positive part being $y_i \geq q$ and the negative part being $y_i < q$. Let $e_i = y_i - q$ as the error for the ith observation, we rewrite (8.5) as the follows:

$$
\begin{aligned}
\hat{q}_\tau &= \operatorname{argmin}\left[\tau \sum_{e_i \geq 0} e_i + (\tau - 1) \sum_{e_i < 0} e_i\right] \\
&= \operatorname{argmin}\left[\sum_{e_i \geq 0} e_i \tau + \sum_{e_i < 0} e_i (\tau - 1)\right] \\
&= \operatorname{argmin}\left[\sum_{e_i \geq 0} e_i (\tau - I(e_i < 0)) + \sum_{e_i < 0} e_i (\tau - I(e_i < 0))\right] \\
&= \operatorname{argmin} \sum e_i (\tau - I(e_i < 0)) \\
&= \operatorname{argmin} \sum \rho_\tau (e_i),
\end{aligned} \tag{8.6}
$$

where $I(.)$ is an indicator function and the function

$$\rho_\tau(e) = e(\tau - I(e < 0)) \tag{8.7}$$

is the loss function at the τth quantile. Thus, the sample quantile can be derived by finding the optimal solution for the loss function $\rho_\tau(e)$ (Fig. 8.7).

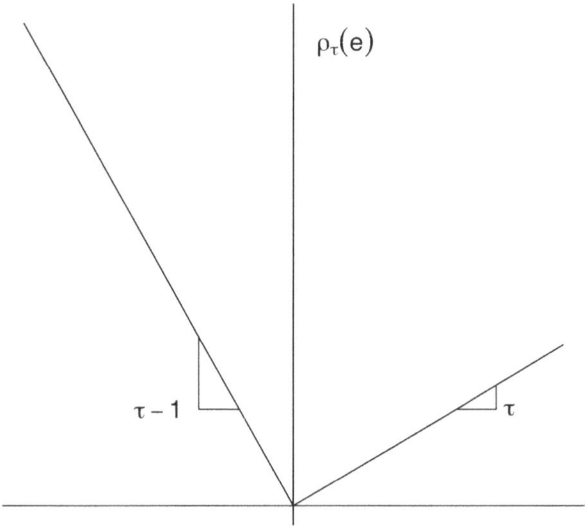

Fig. 8.7 Quantile regression ρ function

We now show briefly how to find the solution for (8.6). We know that the τth sample quantile can be obtained by solving the following minimizing problem:

$$\min_{q \in \mathbf{R}} \sum_{i=1}^{n} \rho_\tau(y_i - q) \tag{8.8}$$

Equation (8.8) is equivalent to a linear programming with "slack" variables $\{u_i, v_i : 1, \ldots, n\}$

$$\min_{(q,u,v) \in \mathcal{R} \times \mathcal{R}_+^{2n}} \{\tau 1_n^\top u + (1 - \tau) 1_n^\top v | 1_n q + u - v = y\}, \tag{8.9}$$

where 1_n is a vector of ones, u is for the positive residuals, and v is for the negative residuals. As elaborated in Koenker (2005, pages 7–8), in (8.9): "we are minimizing a linear function on a polyhedral constraint set, consisting of the intersection of the $(2n + 1)$-dimensional hyperplane determined by the linear equality constraints and the set $\mathcal{R} \times \mathcal{R}_+^{2n}$." Thus, we have thus transformed the solution for a sample quantile to a standard linear programming optimization. Interested readers can refer to linear programming textbooks on procedures and methods for detailed solutions.

We present sample quantiles and CDF of daily returns for the S&P 500 index and CSI 300 index in Fig. 8.8, where the left plot is for the S&P 500 and the right plot is for the CSI 300. At $\tau = 0.75$, we have the values of sample quantile $\hat{q} = 0.0067$ for S&P 500 and $\hat{q} = 0.0054$ for CSI 300. We also present other values of sample quantiles of $\tau = (0.01, 0.05, 0.25, 0.50, 0.75, 0.95, 0.99)$ delivered by the R command *quantile* below.

Sample Quantiles

```
> taus=c(0.01,0.05,0.25,0.50,0.75,0.95,0.99)

> ## S&P 500 daily returns in 2018
> round(quantile(xx2$return.sp5, taus),4)
     1%      5%     25%     50%     75%     95%     99%
-0.0327 -0.0206 -0.0042  0.0006  0.0057  0.0151  0.0225

> ## CSI 300 daily returns in 2018
> round(quantile(xx2$return.csi, taus),4)
     1%      5%     25%     50%     75%     95%     99%
-0.0402 -0.0233 -0.0088 -0.0009  0.0065  0.0213  0.0301
```

By specifying different quantiles, we obtain distributional information about returns, especially the tails with extreme quantile values, such as 1% and 99%.

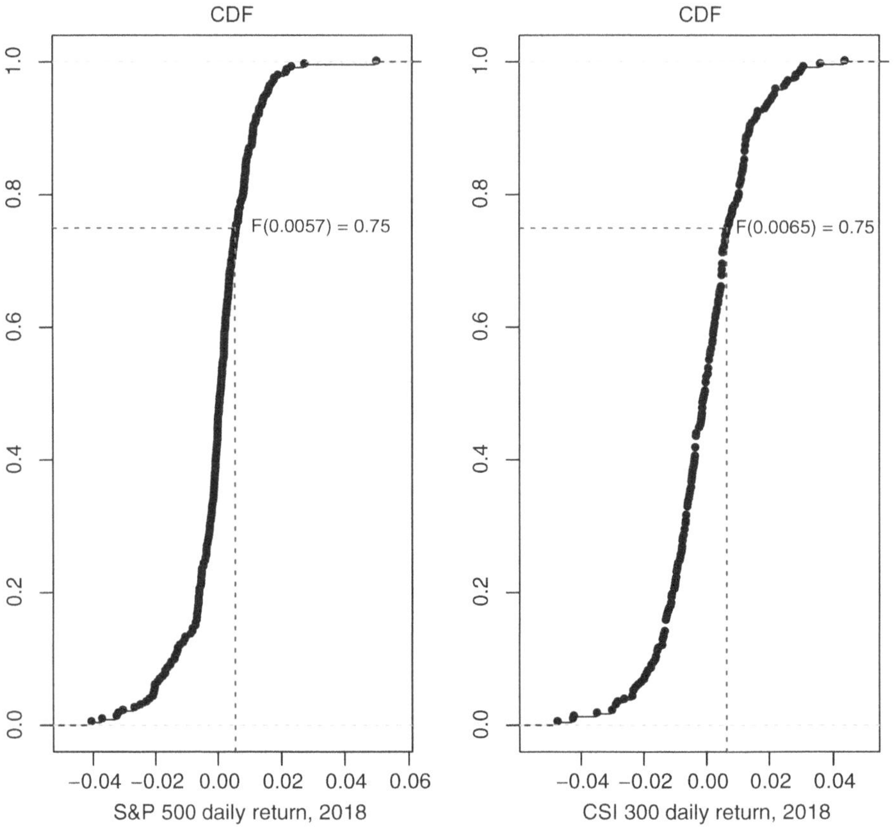

Fig. 8.8 Sample quantiles of S&P 500 and CSI 300 index daily returns for 2018

Recall that sample quantiles are also called value at risk. For example, with the probability of 1%, S&P 500 can lose 3.27% and CSI 300 can lose 4.02%, in a single day.

8.4.2 Conditional Quantile Regression

So far, we have focused on population and sample quantiles. Applying the same principle, we can extend the sample quantile to obtain conditional quantile functions. In quantitative investing, alpha or risk models are usually based on multi-factor models,

$$y = b_0 + b_1 x_1 + b_2 x_2 + \epsilon,$$

where y is stock return and $x = (1_n, x_1, x_2)$ are factors. Suppose the distribution of error term is F_ϵ, we have the conditional quantile regression model

$$Q_y(\tau|x) = b_0 + b_1x_1 + b_2x_2 + F_\epsilon^{-1}(\tau)$$

In terms of the impacts of x on y, there are two representative linear quantile regression models: location- and scale-shift. We discuss each in detail below.

1. *A pure location-shift QR model.* Consider a linear model where the values of x only impact the location of y,

$$y = b_0 + b_1x + \epsilon.$$

The corresponding quantile regression model can be written as

$$\begin{aligned} Q_y(\tau|x) &= b_0 + b_1x + F_\epsilon^{-1}(\tau) \\ &= (b_0 + F_\epsilon^{-1}(\tau)) + b_1x \\ &= b_0(\tau) + +b_1x \end{aligned} \tag{8.10}$$

where the intercept part changes with τ,

$$b_0(\tau) = b_0 + F_\epsilon^{-1}(\tau),$$

while x has constant impacts on y, hence the pure location-shift of x on y.

Following the model setup above, we specify a data generating process

$$y = 1 + 2x + e$$

the corresponding quantile regression model is

$$Q_y(\tau|x) = (1 + F_e^{-1}(\tau)) + 2x.$$

We specify $\tau = (0.10, 0.25, 0.50, 0.75, 0.90)$, and obtain quantile estimates at each τ (Fig. 8.9). Below are the R scripts for the model specification and QR estimates.

Pure Location-Shift Model

```
set.seed(10)
e=rnorm(100)*5
x=rt(100,df=5)
y=1+2*x+e
```

```
taus=c(0.1,0.25,0.50,0.75,0.90)
ntau=length(taus)
mm1=rq(y~x,tau=taus)$coef

out.dir="/.../quantInvesting/chapter8/"
pdf(paste(out.dir,"pureLocation.pdf",sep=""))
plot(x,y,cex.lab=0.7,cex.axis=0.7)
for(i in 1:ntau)
{
    abline(a=mm1[1,i],b=mm1[2,i])
}
```

The value of the intercept, $1+F_e^{-1}(\tau)$, increases with τ. The slope is 2. In Fig. 8.9, we see that the lines are parallel with each other, indicating the shift of the intercept and a constant slope.

2. *A scale-shift QR model.* The pure location-shift model has the restrictive assumption that the error term is only additive to the factors, and thus factors only impact the location of y. In other words, the values of x have the same impacts across the entire distribution of y. This usually does not hold in the real

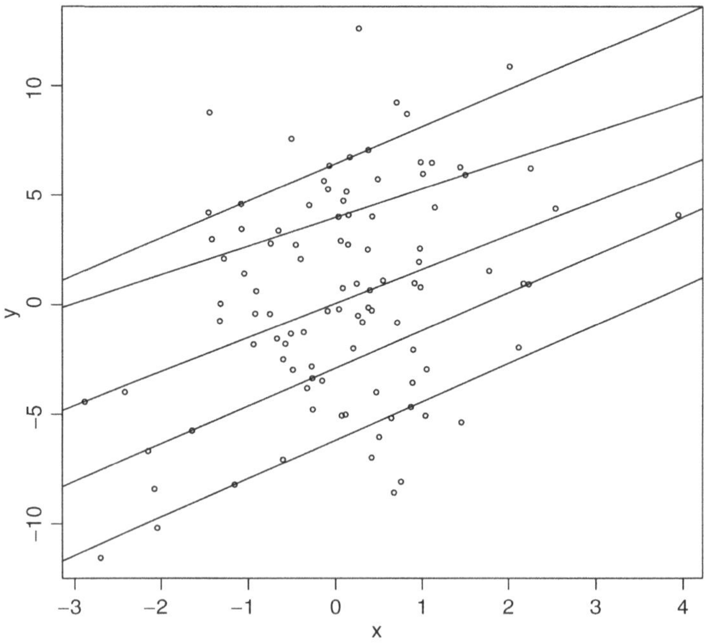

Fig. 8.9 Pure location-shift model

world. Consider a simple model of stock returns and a value factor, B/P. It would be reasonable to think that the effects of B/P would be different for stocks at the different parts of the return distribution. For example, we would expect B/P to be more effective for low-return stocks than for high-return stocks. Therefore, we would expect that

$$Q_r(\tau|x) = b_0 + b_1(\tau)\, B/P \tag{8.11}$$

and $b_1(\tau)$ varies with τ. This is a scale-shift model.

In real-world data, the error term usually contains much unknown information. The information may be related to x and may impact the whole shape of y. Consider a linear model,

$$y = b_0 + b_1 x + \gamma x \epsilon.$$

The corresponding quantile regression model can be written as

$$\begin{aligned} Q_y(\tau|x) &= b_0 + b_1 x + \gamma x F_\epsilon^{-1}(\tau) \\ &= b_0 + (b_1 + \gamma F_\epsilon^{-1}(\tau))x \\ &= b_0 + +b_1(\tau)x \end{aligned} \tag{8.12}$$

where the intercept is constant, while the slope changes with τ,

$$b_1(\tau) = b_1 + \gamma F_\epsilon^{-1}(\tau),$$

hence x impacts y with a scale-shift. Or in another word, the effects of x on y is heterogenous, depending on the value of y. For example, factor BP has different impacts for stocks at different parts of return distribution.

Pure Scale-Shift Model

```
x=rt(100,df=5)+5
y2=1+2*x+ e*x

mm2=rq(y2~x,tau=taus)$coef
pdf(paste(out.dir,"locationShift.pdf",sep=""))
plot(x,y2,ylab="y", cex.lab=0.7,cex.axis=0.7)
for(i in 1:ntau)
{
     abline(a=mm2[1,i],b=mm2[2,i])
}
```

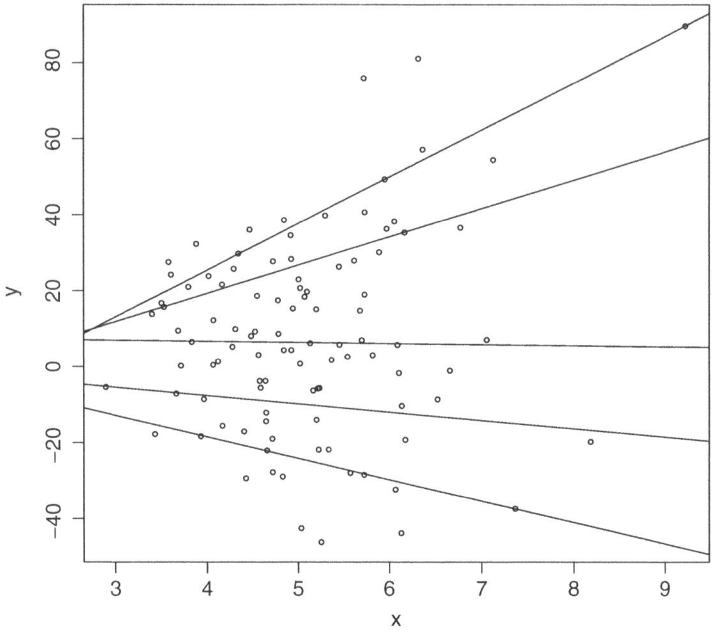

Fig. 8.10 Scale-shift model

In Fig. 8.10, we see that the lines have different slopes and a constant intercept. The value of the slope, $2 + F_e^{-1}(\tau)$, increases with τ. The slope is 1. Depending on the quantile of the conditional distribution of y, the impacts of x can be very different. One application of such models is treatment effects. Policy effects are usually not constant across all entities; rather, they are heterogenous. For example, consider a policy change, such as an interest rate change or class size reduction, where the former may have more effects on some stocks, such as the banking industry, than other stocks, and the latter may have more effects on math for lower-achieving students. Thus, the quantile treatment effect captures such heterogeneity via the value of $\beta(\tau)$ across $\tau' s$.

8.4.3 Interpretation of Quantile Effects

In the real world, the impacts of factors on a response variable can affect both location and scale. Consider a general linear model:

$$y = b_0 + b_1 x + \epsilon + \gamma x \epsilon \tag{8.13}$$

With the zero mean error, we have the impacts of x on y as

$$\frac{\partial E(y)}{\partial x} = b_1.$$

That is, all else being equal, the marginal effect of one unit change of x on the conditional average value of y is b_1.This is what delivered by the least squares method. Note that the least squares model of (8.13)

$$y = b_0 + b_1 x + \nu, \quad \text{where } \nu = \epsilon + \gamma x_1 \epsilon,$$

has the error term ν that is not homogenous.

The quantile regression model of (8.13) can be written as

$$\begin{aligned} Q_y(\tau|x) &= b_0 + F_\epsilon^{-1}(\tau) + b_1 x + \gamma x F_\epsilon^{-1}(\tau) \\ &= (b_0 + F_\epsilon^{-1}(\tau)) + (b_1 + \gamma F_\epsilon^{-1}(\tau))x \\ &= b_0(\tau) + b_1(\tau)x \end{aligned} \tag{8.14}$$

where the quantile effects are

$$\begin{aligned} b_0(\tau) &= b_0 + F_\epsilon^{-1}(\tau) \\ b_1(\tau) &= b_1 + \gamma F_\epsilon^{-1}(\tau). \end{aligned}$$

The impact of x on the τth quantile values of y is

$$\frac{\partial Q_y(\tau|x)}{\partial x} = b_1(\tau) \tag{8.15}$$

Note that for illustration purposes, we have only one variable in (8.14). If there are K factors, (8.15) becomes

$$\frac{\partial Q_y(\tau|x)}{\partial x_k} = b_k(\tau) \tag{8.16}$$

that is, all else being equal, the marginal effect of one unit change of x_k on the τ-th conditional value of y is $b_k(\tau)$. Compared to the conditional mean effects of classical least squares models, quantile regression models present conditional distributional effects. With $\tau \in (0,\ 1)$, quantile effects offer the full-picture effects of a factor on the response variable. Recall that in the model with B/P in (8.11), OLS would tell how the value signal impacts returns on average. Quantile regression would tell how this value signal impacts returns at different parts of a distribution: $b_1(\tau = 0.9)$ shows the effects for stocks with returns at the 90th percentile, the right tail of the return distribution, while $b_1(\tau = 0.1)$ shows the effects for stocks with returns at the 10th percentile, the left tail of the return distribution. Thus, we see that QR provides a general framework to explore tail behaviors.

Note that when $\tau = 0.50$, $b_1(\tau = 0.5)$, the median effects, may be close to the conditional mean effects if the conditional distribution is not skewed too much.

From the perspective of quantitative investing, distributional effects equip investors with more information to make correct decisions about investments. This frontier approach has the potential to add value to quantitative investing due to the following characteristics:

1. Robust
2. Distributional effects
3. Distribution free

The robustness derives from the order statistics that connect quantile regression with ranking. Outliers will not impact quantile regression results as much as they do for the classical least squares models. The benefit of being able to discern distributional effects is clear: As we change the values of τ, the quantile effects reveal rich information about how factors impact asset returns, instead of a one-size-fits-all approach. This is very exciting. Finally, there is the distribution-free trait. QR only assumes that the error term follows a distribution, but that does not need to be Gaussian. Actually, distributions with fat and long tails make quantile regression a natural tool to use. We discuss the distribution aspect in more detail in the next section.

It is worthwhile now to clarify some misunderstandings and misconceptions about QR. The following are two of the most common, particularly among those who have just started to learn quantile regression and those who may be resistant to new ideas.

Subsample Approach "Similar results can be obtained by applying traditional methods, such as OLS, to a subsample of the data." Yes, you may apply OLS to a subsample, but this is different from QR. QR uses the whole sample. The line is obtained such that τ points are below the line and $1 - \tau$ points are above the line. The OLS for a subsample is the conditional mean estimate within that subsample. Thus, it is a tree rather than a forecast and potentially creates a selection bias.

Robust Method for Heteroscedasticity "Heterogeneity, robustness, and heteroscedasticity can all be addressed by classical mean methods if needed." Yes, that is true. However, these methods are still about the conditional mean, which may make the conditional mean estimates more efficient or consistent, but they do not reveal distributional effects. Rather, they tell the robust version of conditional mean effects.

Thus, this frontier approach—quantile regression—does present new information and can be regarded as a complementary approach to classical mean methodologies, such as least squares, which are still very useful for exploring quantitative relationships of average effects in multi-factor models.

8.5 Quantile Regression: Estimation, Inference, and an Example

We introduced quantile regression in the previous section. Now, we focus on estimation and inference. We also present an empirical study using QR to illustrate the relationship between the price of gold and US unemployment.

8.5.1 *Estimation: Linear Programming*

Consider a linear quantile regression model,

$$Q_y(\tau|x) = x^\top b(\tau), \tag{8.17}$$

where x is a vector of factors and $b(\tau)$ is a vector of coefficients at the τth quantile of y given x. In the context of quantitative investing, y is the returns of stocks and x is a vector of themes, such as value, momentum, profitability, and management quality. For example, consider the value theme, b(0.90) and b(0.10) tell how much the value theme impacts the 90th percentile and 10th percentiles of stock returns. If b(0.90) is significantly different from b(0.10), we know that value theme has different effects for stocks with higher returns and lower returns. Treating them the same may be misleading when building an alpha model or a portfolio.

The question now is how to estimate $b(\tau)$? Since we are familiar with least squares methods, it is now worth comparing quantile regression with least squares regression. For the least squares method, its sample mean problem is:

$$\min_{\mu \in \mathcal{R}} \sum_{i=1}^{n} (y_i - \mu)^2 \tag{8.18}$$

Replacing μ with $\mu(x) = x^\top \beta$, we have the conditional mean problem:

$$\min_{\beta \in \mathcal{R}^p} \sum_{i=1}^{n} (y_i - x_i^\top b)^2. \tag{8.19}$$

Similarly, for quantile regression method, we have the sample quantile at τ:

$$\min_{q \in \mathcal{R}} \sum_{i=1}^{n} \rho_\tau (y_i - q), \tag{8.20}$$

Replacing q with the *conditional* quantile function $q(\tau|x) = x^\top b(\tau)$, we obtain the conditional quantile problem:

$$\min_{b \in \mathcal{R}^p} \sum_{i=1}^{n} \rho_\tau(y_i - x_i^\top b). \tag{8.21}$$

The estimation of $b(\tau)$ is analogous to the estimation of sample quantiles. Recall that to estimate the τth sample quantile, we have the error term

$$e_i = y_i - q$$

and then derive the estimate $\hat{q}_\tau$ by minimizing the loss function

$$\hat{q}_\tau = \text{argmin}_q \sum \rho_\tau(y_i - q)$$

For the conditional quantile regression model (8.17), we apply the same loss function $\rho_\tau(e)$ with

$$e_i = y_i - x_i^\top b(\tau)$$

and then obtain the estimate $\hat{b}(\tau)$:

$$\hat{b}(\tau) = \text{argmin}_{b(\tau)} \sum \rho_\tau(y_i - x_i^\top b(\tau)) \tag{8.22}$$

Following the same principle for sample quantiles as in (8.9), we reformulate conditional quantile regression as a linear programming problem but now with $\xi = Xb$. Everything else stays the same.

$$\min_{(b,u,v) \in \mathcal{R}^p \times \mathcal{R}_+^{2n}} \{\tau 1_n^\top u + (1-\tau) 1_n^\top v | Xb + u - v = y\}. \tag{8.23}$$

The solution to (8.23) is the quantile regression estimate $\hat{b}(\tau)$. For technical details on finding a solution for a classical LP problem, refer... We now use a numerical example to show what a QR estimate look like.

A Numerical Example: How a QR Line Is Obtained Following (8.22) and (8.23), we demonstrate how to obtain a QR line with a simple numerical example for the model below:

$$y_i = 1 + 2x_i + x_i\, e_i, \quad i = 1, \ldots, N$$
$$Q_y(\tau|x) = 1 + x\,(2 + F_e^{-1}(\tau)) \tag{8.24}$$

Let N^+ be the number of positive and zero residuals and N^- the number of negative residuals. The QR optimality indicates that, the τ-quantile QR line is obtained by minimizing the sum of the positive distances (on or above the line) times τ plus the sum of negative distances (below the line) times $(1-\tau)$. If $\tau = 0.75$, then the

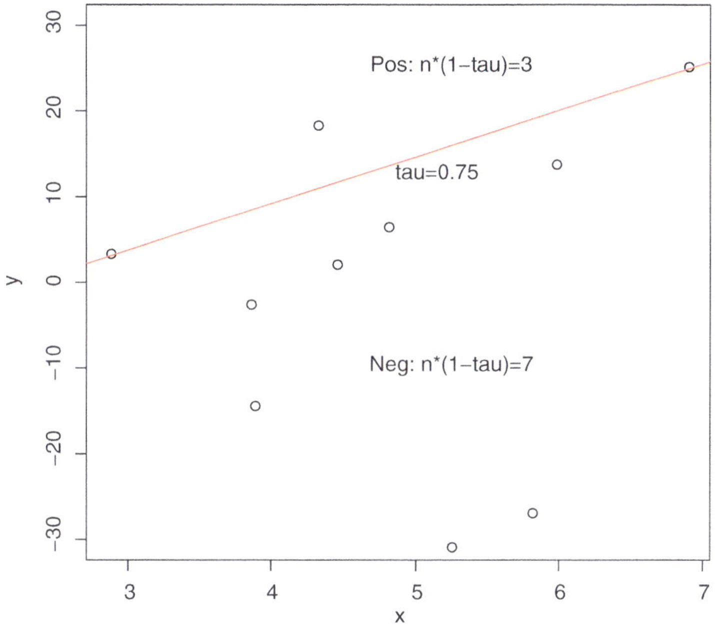

Fig. 8.11 An numerical example of QR at $\tau = 0.75$ for (8.24) with $N = 10$

positive errors are weighted by 0.75, while the negative errors are weighted by 0.25. Intuitively, to have the minimum total of the weighted negative and positive sums of errors, a large τ should be assigned with a small positive sum value, resulting in a relatively high position of the QR line (for a positive relationship between $Q_y(\tau|x)$ and x). Reflected in the optimal solution, there should be $N^+ = N \times (1 - \tau)$ points above or on the line and $N^- = N \times \tau$ points below the line. For a general case with formal proofs, see Theorem 2.2 and Corollary 2.1 in Koenker (2005, pp 36–37).

We illustrate how QR works with N=10 and $\tau = 0.75$ in (8.24). Below are the R scripts. The data and QR line at $\tau = 0.75$ are displayed in Fig. 8.11. Clearly, we see that there are 3 points above or on and 7 points below the QR line at $\tau = (0.75)$.

A Numerical Example of QR

```
qrnz <- function()
{
library(quantreg)
set.seed(10)
e=rnorm(10)*5
x=rt(10,5)+5
y=1+ 2*x + x*e
```

```
fit=rq(y~x,tau=0.75)
pdf("/Users/l.../qrnz.pdf")
plot(x,y,ylim=c(-30,30),cex.axis=0.8,cex.lab=0.8)
abline(fit,col="tomato")
text(5,12.5,"tau=0.75",cex=0.6)
text(5,25,"Pos: n*(1-tau)=3",cex=0.7)
text(5,-10,"Neg: n*(1-tau)=7",cex=0.7)
resid=round(fit$resid,2)
graphics.off()

reg.tab=data.frame(y,x,resid)
return(round(reg.tab,2))
}
```

We now make variations of data and investigate their impacts on the QR regression. The first variation is changing the sixth observation value of y from 18.02 to 50. We then rerun QR at $\tau = 0.75$ and find that the QR estimates remain the same (middle plot in Fig. 8.12). The second variation is changing all the value of y to $\log(y + 50)$. We then rerun QR at $\tau = 0.75$ and find that the QR line remains the same (right plot in Fig. 8.12). For comparison purposes, we also have the OLS estimates for each case, we find that the OLS estimates change for each variation. The first variation verifies the robustness of QR. The second variation indicates the monotonicity of QR which we discuss in detail in the following.

Properties of Quantile Estimator The QR estimates are equivariant and monotonic in τ. The equivariance properties are shared by least squares estimates, but monotonicity is not.

The QR estimator being monotonic in τ is rooted in the definition of quantile functions, the one-on-one mapping between τ and the cumulative distribution function. For example, in the equation (8.14),

$$b_1(\tau) = b_1 + \gamma F_\epsilon^{-1}(\tau)$$

$b_1(\tau)$ is monotonic in τ because $F_\epsilon^{-1}(\tau)$ is a nondecreasing function of τ. Moreover, for a more general case, suppose there is a monotonic function $m(\cdot)$, we have

$$P(Y \leq y) = P(m(Y) \leq m(y))$$

The following holds immediately,

$$Q_{m(Y)}(\tau) = m(Q_Y(\tau)). \tag{8.25}$$

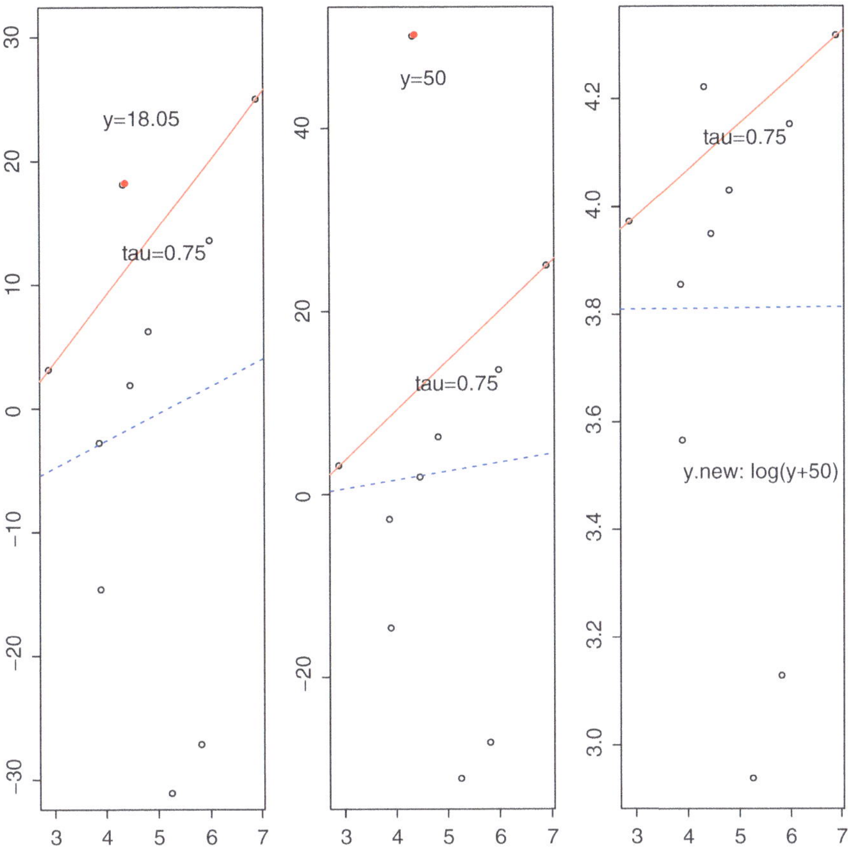

Fig. 8.12 Two variations of the numerical example of QR at $\tau = 0.75$ for (8.24) with $N = 10$. Left plot: original example. Middle plot: y[6] is changed from 18.02 to 50. Right plot: values of y change to $\log(y + 50)$. Dashed lines are the OLS estimates

The monotonicity property provides many benefits for empirical work. For example, when a nonlinear transformation makes the data less problematic or meaningful, we do not need to worry about the changes of estimates for QR estimation as we do for OLS $(E(m(Y)) \neq m(E(Y))$.

The monotonic property of quantile regression estimators has many important implications for investment. For example, asset pricing for derivative products or structured products can be reformulated as a monotonic transformation of the base financial product, which offers great convenience.

8.5.2 *Inference: Finite and Asymptotic Properties*

In this section, we discuss inference for quantile regression estimates. We focus on the asymptotics on consistency and convergence rate, from which we show how to proceed to the significance test of a factor at a specified value of τ and the heterogeneity test for the effects of a factor at two different values of τ.[3]

Now consider a linear quantile regression model. The linearity is defined as being linear in parameters. For asymptotic properties of QR estimates, we cite the elegant discussion of Theorem 4.1 in Koenker (2005, pp 120–121):

> Let $Y_1, Y_2, \ldots, Y_n$ be independent random variables with distribution functions $F_1, F_2, \ldots, F_n$, and suppose that the τth conditional quantile function
>
> $$Q_{Y_i}(\tau|x) = x^\top \beta(\tau)$$
>
> is linear in the covariate vector x. The conditional distribution functions of the Y_i's will be written as $P(Y_i < y|x_i) = F_{Y_i}(y|x_i) = F_i(y)$, so
>
> $$Q_{Y_i}(\tau|x_i) = F_{Y_i}^{-1}(\tau|x_i) \equiv q_i(\tau).$$
>
> We will employ the following regularity conditions to explore the asymptotic behavior of the estimator,
>
> $$\hat{\beta}_n(\tau) = \text{argmin}_{b \in \mathcal{R}^p} \sum \rho_\tau(y_i - x_i^\top b).$$
>
> A1 The distribution functions $\{F_i\}$ are absolutely continuous, with continuous densities, $f_i(q)$, uniformly bounded away from 0 and ∞ at the points $q_i(\tau)$, $i = 1, 2, \ldots$.
>
> A2 There exist positive definite matrices D_0 and $D_1(\tau)$ such that
>
> $$\text{(i)} \quad \lim_{n\to\infty} n^{-1} \sum x_i x_i^\top = D_0$$
>
> $$\text{(ii)} \quad \lim_{n\to\infty} n^{-1} \sum f_i(q_i(\tau)) x_i x_i^\top = D_1(\tau)$$
>
> $$\text{(iii)} \quad \max_{i=1,\ldots,n} \| x_i \| / \sqrt{n} \to 0.$$
>
> As has already been emphasized, the behavior of the conditional density of the response in the neighborhood of the conditional quantile model is crucial to the asymptotic behavior of $\hat{\beta}_n(\tau)$. Conditions A2(i) and A2(iii) are familiar throughout the literature on M-estimators for regression models; some variant of them is necessary to ensure that a Lindeberg condition is satisfied. Condition A2(ii) is really a matter of notational convenience and could be deduced from A2(i) and a slightly strengthened version of A1.
>
> **Theorem 8.1** *Under Conditions A1 and A2,*
>
> $$\sqrt{n}(\hat{\beta}_n(\tau) - \beta(\tau)) \rightsquigarrow \mathcal{N}(0, \tau(1-\tau) D_1^{-1} D_0 D_1^{-1}).$$

---

[3]For discussions about finite properties, see Koenker (2005).

For the sample quantiles, $x_i = 1$ and $D_1(\tau) = f_i(q_i(\tau))$. Thus, the asymptotic behavior becomes

$$\sqrt{n}(\hat{\beta}_n(\tau) - \beta(\tau)) \rightsquigarrow N(0, \omega^2),$$

where $\omega^2 = \tau(1-\tau)/f_i^2(q_i(\tau))$.

For the iid error case, we have $D_1(\tau) = f(q_i(\tau))D_0$. Thus, the covariance is

$$\sqrt{n}(\hat{\beta}_n(\tau) - \beta(\tau)) \rightsquigarrow N(0, \omega^2 D_0^{-1})$$

Note that $f_i(.) = f(.)$ for the special iid case.

For nonlinear models, the asymptotic properties follow the same format as for the linear models with the difference that x_i is replaced by the partial differentials of the nonlinear function with respect to the coefficient.[4]

The estimation of a local density function is critical for the properties of quantile regression estimates. Siddiqui (1960) proposes a method using a simple quotient of neighboring empirical quantile functions, whereas Hendricks and Koenker (1991) and Powell (1991) each propose estimation methods for non-iid scenarios.

On inference, we present below three cases of hypothesis testing that are used often in empirical studies: the significance test of a factor at a local quantile, the heterogeneity test across quantiles, and the overall model fit at a local quantile.

1. *Factor significance at certain quantiles* To test the local significance of a factor, we formulate the following hypothesis:

$$H_0 : b(\tau) = 0, \quad H_a : b(\tau) \neq 0$$

The test can be constructed using the t-value as introduced in Chap. 2,

$$\text{T} - value = \frac{\hat{b}(\tau)}{s.e.(\hat{b(\tau)})} \sim \text{T-distribution.}$$

We present an example of regressing the daily returns of the CSI 300 on the factor of S&P 500 daily returns at $\tau = 0.50$. The test is to determine if b(0.5) is significant, that is, whether the US stock market has significant impacts on the Chinese stock market at the median level. We see from the results produced from R codes that the t-value of $\hat{b}(0.50)$ is 12.31, indicating the median effect is significant.

[4]For a detailed discussion, please refer to Koenker (2005), pp. 123–124.

Factor Significance Test

```
> rq(return.csi~return.sp5,tau=0.5,data=sp5csi)
Call:
rq(formula = return.csi ~ return.sp5, tau = 0.5, data = sp5csi)
Coefficients:
(Intercept)  return.sp5
0.000505314 0.193236715

> summary(rq(return.csi~return.sp5,tau=0.5,data=sp5csi))
Call: rq(formula = return.csi ~ return.sp5, tau = 0.5, data = sp5csi)
tau: [1] 0.5
Coefficients:
            Value    Std. Error t value  Pr(>|t|)
(Intercept)  0.00051  0.00023    2.16326  0.03059
return.sp5   0.19324  0.01569   12.31472  0.00000
```

2. *Heterogeneity test* To test whether a factor has heterogenous effects across values of τ, we formulate the following hypothesis:

$$H_0 : b(\tau_1) = b(\tau_2), \quad H_a : b(\tau_1) \neq b(\tau_2).$$

The heterogeneity test requires the joint distribution of quantile regression estimates at different τs (Koenker and Bassett 1982a). This can be stated in a more general linear constraint setting as specified in Koenker and Bassett (1982b),

$$b(\tau) = (b(\tau_1), b(\tau_2), \ldots, b(\tau_k))^\top$$

$$H_0 : Rb(\tau) = r,$$

where R is the constraint matrix and r is a vector of constraint targets . For a special heterogeneity test of two values of τ, we have

$$R = (1, \ -1), \quad b(\tau) = (b(\tau_1), \ b(\tau_2))^\top, \quad r = 0.$$

Using the same data as above, we carry out a test of

$$H_0 : b(0.10) = b(0.90)$$

that is, whether the impacts of the US stock market are the same when the Chinese stock market is low and high. The test results from R scripts below show that the joint F-distribution has a value of 5.91 and p-value of 0.015, indicating that the two slopes are significantly different: the US market has significantly greater impacts when the Chinese stock market is bearish. Based on this simple bivariate model, a 1% increase in S&P 500 index return will cause a 16bps increase in the CSI 300 when the Chinese market is bullish (90%), while a 1% increase in S&P 500 index return will cause a 31bps increase in the CSI 300 when the Chinese market is bearish (10%).

Heterogeneity Test

```
> x1=rq(return.csi~return.sp5,tau=0.1,data=sp5csi)
> x2=rq(return.csi~return.sp5,tau=0.9,data=sp5csi)

> summary(x1)
Coefficients:
             Value     Std. Error t value   Pr(>|t|)
(Intercept)  -0.01780   0.00062   -28.65454   0.00000
return.sp5    0.31064   0.05258     5.90794   0.00000

> summary(x2)
Coefficients:
             Value    Std. Error t value  Pr(>|t|)
(Intercept)  0.01974  0.00059   33.18147  0.00000
return.sp5   0.15697  0.04137    3.79409  0.00015

> anova(x1,x2)
Quantile Regression Analysis of Deviance Table

Model: return.csi ~ return.sp5
Joint Test of Equality of Slopes: tau in {  0.1 0.9  }

  Df Resid Df F value  Pr(>F)
1  1      6709  5.9064 0.01511 *
---
```

3. *Model local fit at* τ. The goodness of fit test can be performed with something similar to R^2. We discuss this further in this chapter's section on R.

8.5.3 An Example: The Price of Gold and the US Unemployment Rate

Gold is the heaven during crisis. We apply QR to understand the relationship between the price of gold and the US unemployment rate. The traditional OLS method identifies *average* effects on gold price, that is, how independent variables impact the *average* gold price. When the current gold price is at the 90% of price distribution, for practical investment purposes, we are more interested in how the US unemployment rate (and other factors) impact the current price than the "average" gold price.

Consider a simple univariate model,

$$y = x\beta + \epsilon, \tag{8.26}$$

where y represents the price of gold, x represents the unemployment rate, and ϵ is an error term. The top plot in Fig. 8.13 shows the observed relationship between these variables from April 1968 to March 2012. We see that both the unemployment rate and gold price variables have enough variation to support the relationship identification.

How would one unit change in the unemployment rate affect gold price when the gold price is already at a high level? QR provides a tool to tackle such a problem. Corresponding to (8.26), the quantile regression model can be written as

$$\begin{aligned} Q_y(\tau|x) &= x(\beta + \lambda F_{\epsilon_i}^{-1}) \\ &= x\beta(\tau), \end{aligned} \tag{8.27}$$

where F_{ϵ_i} is the distribution function of ϵ_i and τ is the percentile (quantile) of the gold price distribution conditioning on the unemployment rate. Therefore, as long as $\lambda \neq 0$, $\lambda F_{\epsilon_i}^{-1}$ captures heterogenous effects of x on y.[5]By varying τ over the range of (0, 1), we would obtain a distributional view of unemployment effects on gold price. We use the expression $\beta(\tau)$ to indicate that β will depend upon the choice of τ, which is the percentile (quantile) of the conditional distribution of gold prices. To obtain the quantile estimates, we select a value for τ and then minimize the following expression:

$$\hat{\beta}(\tau) = \text{argmin}_{\beta(\tau)}[\tau \sum_{y \geq \beta x} (y_i - \beta x_i) + (\tau - 1) \sum_{y < \beta x} (y_i - \beta x_i)]. \tag{8.28}$$

In words, $\beta(\tau)$ is derived as a line ($y = \beta(\tau)x$) such that τ percent of observations lie on or below the line (the first sum), while $(1-\tau)$ percent of observations lie above the line (the second sum). Note that we can choose τ freely in the range of (0, 1).

In our example of the relationship between the unemployment rate and the price of gold, we provide QR estimates at $\tau = \{0.1, 0.5, 0.9\}$ on the top plot of Fig. 8.13,

[5]Note that λ is a scalar for the distribution F.

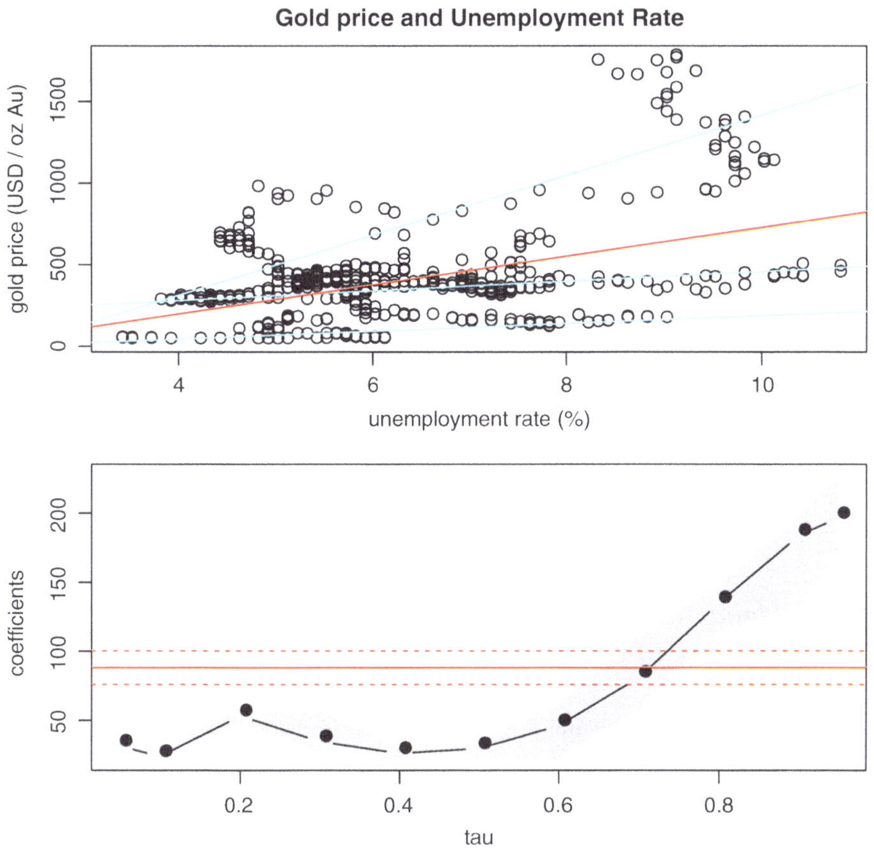

Fig. 8.13 An example of a univariate quantile regression model of gold price and the US unemployment rate. The top plot is the scatter plot of gold price and unemployment rate. The bottom plot describes the slope estimates from OLS and QR at different quantiles, where the shaded area is for the QR estimates with a 90% confidence band and the straight solid line is for the OLS estimate with a 90% confidence band

which shows how the unemployment rate affects the price of gold at low, median, and high levels. We see that the slopes are different and increase with values of τ, indicating a very different impact when the gold price is high or low. For comparison purposes, we also show the OLS estimate (red line), which is very close to the special median estimate. The bottom plot in Fig. 8.13 shows an alternative way to present the effects across quantiles, where quantiles are shown along the x-axis and the slope coefficient corresponding to each quantile is along the y-axis. The quantiles are set at $\tau = \{0.05, 0.1, 0.2, 0.3, 0.4, 0.5, 0.6, 0.7, 0.8, 0.9, 0.95\}$. The gray area indicates the 90% confidence interval. The OLS estimate and its confidence interval are also shown on the graph, represented by solid and dashed straight lines, respectively. The bottom plot provides a more colorful view of the unemployment rate's impact on gold prices. We see that in this univariate model, unemployment effects are significantly larger when gold prices are high. When gold prices are below the median level, for a one percent increase in the unemployment rate, the price of gold will increase by about \$20–60. However, when the gold price is high, the impact increases to \$100–200! Clearly, the impact of unemployment rate on gold price is very different when the price of gold is low or high.

8.6 Industry Approach: Capturing Tail Behavior with Quantile Regression

In this section, we show how to employ quantile regression to incorporate tail behaviors into quantitative investing. We first present the use of QR in alpha modeling, forecasting the price of gold. We then carry out a portfolio construction study using QR to construct portfolios across a return distribution.

8.6.1 Alpha Modeling: Employing QR to Forecast the Price of Gold

The price of gold provides unique insights about investor's preferences due to its unique position in history and current usage profile. Gold was historically used as a store of value, and today there remain few industrial applications for this material. We present the price of gold in Fig. 8.14 for the period from April 1968 to March 2012.[6] Clearly, gold price movements are closely related to the US economic outlook. We employ the quantile regression (QR) technique to estimate factor effects over the whole distribution of gold prices. Depending on where the current price sits in the gold price distribution, the distributional effects may have enormous practical value for real-world quantitative investing.[7]

8.6.1.1 Factors Impacting Gold Price Movements

As for all goods in modern markets, the price of gold is driven by supply and demand as well as speculation. Jewelry has consistently been the primary demand for gold. The World Gold Council reports that jewelry, investment, and industrial uses account for 57%, 31%, and 12%, respectively, of total gold demand over the

[6] In the US, the price of gold began to move freely in May of 1968.

[7] This subsection is based on an earlier paper by Ma and Patterson, "Is gold overpriced?," *Journal of Investing*, (2012).

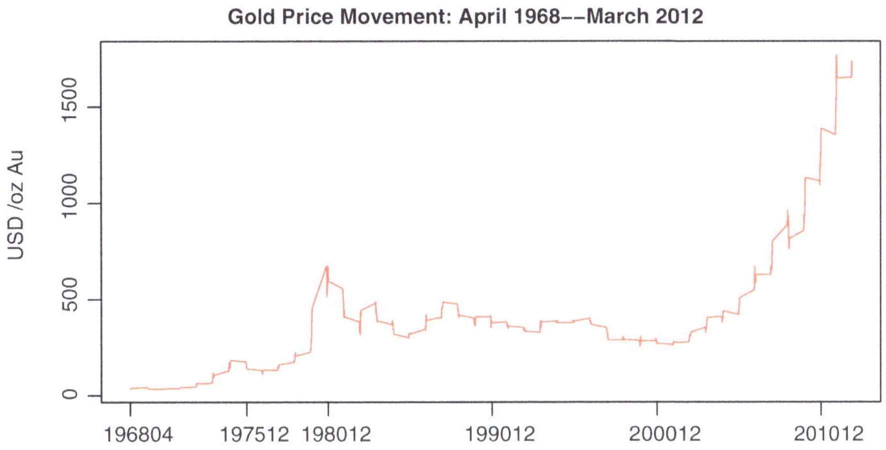

Fig. 8.14 Monthly gold price (USD/OZ Au) movements, April 1968–March 2012

last five years.[8]India is the largest consumer of gold for ornamental uses in terms of volume. Among the key drivers behind investor demand, there is one common thread: *gold's ability to hold value during periods of crisis.* For example, with the U.S. dollar losing its value, both individuals and governments' central banks have started to increase their holdings of gold. This increases the demand for gold. With regard to industrial applications, half of all the industrial use arises from its role in electrical components. Clearly, industrial demand is closely related to economic growth and development.

On the supply side, unlike most other commodities, hoarding (saving) and disposal play a very large role in the supply of gold. Most of the gold ever mined still exists today in the form of jewelry and bullion. It remains in an accessible form and thus is potentially able to re-enter the gold market at the right price. During the last five years, hoarding and disposal account for 35% of world gold supply, while mine production and government sales account for 59% and 6%, respectively, of the total gold supply. Given the scarcity of gold and unique supply features due to hoarding and disposal, the price of gold is clearly driven by the demand side. The demand for gold is influenced by a wide range of factors. In order to make our analysis as parsimonious as possible, we categorize the variables under consideration along the following dimensions:

- Macroeconomy It is generally believed that gold, particularly in physical form, will retain its value in the event of a major political, economic, currency, or social crisis that might otherwise destroy the real economy. We utilize the nominal GDP growth rate and the unemployment rate as potential factors that proxy for the overall strength of the US economy.

[8]See the supply and demand statistics at the World Gold Council web site: *www.gold.org*.

- Monetary System Gold has always been considered a good hedge against inflation, largely because the world supply of gold is relatively constant over time. Currencies, on the other hand, can be expanded or contracted over time by sovereign nations and are thus exposed to the phenomenon of inflation. The exchange rate of the U.S. dollar against other currencies reflects currency strength and investment opportunities. A weaker U.S. dollar exchange rate usually encourages an increase in world gold prices. This is because investors choose to sell their dollars and buy gold in the hope that gold can protect the value of their assets. To study effects of monetary policy on the price of gold, we employ the US inflation rate and US dollar index.
- Financial Market Gold is an unusual asset because of its scarcity and its unusual supply–demand characteristics. Gold is also an asset like stocks, bonds, and real estate, and we must consider its pricing relative to other asset classes. Historically, investors have favored gold when other investments are least attractive. We use the monthly returns of the Dow Jones Industrial Average and the yield on the U.S. treasury bill as our proxies for financial market performance.
- Oil Price The price of gold is also strongly correlated with the price of oil throughout modern history. The prices of these two commodities are so intertwined that they seem almost incapable of heading in separate directions over long periods of time. This relationship arises from both economic and structural linkages. Higher oil prices tend to slow the world economy as a whole and reduce disposable income for most people, which will eventually affect financial markets and the economies of the USA and other countries. In addition, the price of oil is a significant cost factor for gold production. Higher oil prices increase the cost of extracting gold and negatively impact the long-term gold supply.

It should be noted that the economy, monetary indicators, financial markets, and the price of oil are intertwined. Furthermore, there are structural changes throughout history, and gold is tied to historical events, economic health, and monetary policy more than any other commodity. Regime changes in economic policy, monetary instruments, and financial regulations often induce new relationships between the price of gold and the factors discussed. The identification of such structural changes and the effects of such changes on gold prices are critical. We specifically investigate the relationship between the inflation rate and the price of gold over different sub-periods due to changes in monetary policy.

8.6.1.2 A Gold Price Model and Empirical Results

Based on the discussion in the previous subsection, we build a QR model for the price of gold. The QR approach enables us to investigate the distributional effects of variables on the price of gold.

Data Description and Summary We use the London P.M. gold fix as our gold price. For the macroeconomic indicators, we use quarterly nominal GDP growth (GDP), monthly unemployment rate (UEM), and monthly expected inflation rate (INFL). We proxy for the strength of the US dollar with the US dollar index (USDX). The USDX measures the performance of the US dollar against a basket of major currencies: the Euro, Canadian dollar, Japanese yen, British pound, Swiss franc, Australian dollar, and Swedish krona. These seven foreign currencies trade widely in currency markets outside their respective home areas. We proxy the U.S. bond and equity market performance with the 3-month U.S. treasury bill (TBILL) yield and monthly returns of the Dow Jones Industrial Average (DJIA), respectively. We use monthly values of West Texas intermediate as our measure of oil prices (OIL).[9]

We construct a monthly data set from April 1968 to March 2012. Unemployment and GDP growth rate data are lagged by three months because the official numbers are released approximately two and a half months after each quarter end. The statistical characteristics of the data described above are summarized in Table 8.3.

Table 8.4 shows the Pearson correlation between all variables and the price of gold using monthly data from April 1968 to March 2012. The strongest correlation is between the price of gold and the price of oil. Material correlations are also observed between the price of gold and the US dollar index, the US unemployment rate, the US GDP growth rate, the expected inflation rate, and the 3-month treasury bill yield. There is virtually no correlation between equity market performance and the price of gold. This finding is robust to alternative broad measurements of equity performance. The high correlation between oil and gold prices (87.6%) can be attributed to many structural linkages between these commodities, as described above. At the other end of the spectrum, the gold price has almost no correlation with the monthly return of the DJIA, our proxy for stock market performance. In addition, the gold price is negatively correlated with treasury bill yield, implying a positive correlation with bond prices.

Table 8.3 Summary of variables from April 1968 to March 2012

Variable name	Min	Mean	Median	Max	Std
Gold price (USD)	34.94	398.37	356.38	1771.88	314.87
UEM (%)	3.40	6.24	5.90	10.80	1.64
GDP (%)	−5.37	6.91	6.35	25.51	4.17
INFL (%)	−2.10	4.47	3.64	14.76	2.94
USDX	69.07	95.78	95.01	143.91	14.29
DJIA (%)	−23.22	0.62	0.76	14.41	4.45
TBILL (%)	0.01	5.40	5.20	16.30	3.13
OIL (USD)	3.07	29.12	20.83	133.93	24.81

[9]We collected data from different sources. For the gold commodity price, we collected data from *www.measuringworth.com*. For the macroeconomic factors, we used data from both *www.federalreserve.gov* and *www.bea.gov*. The data for expected inflation rate is from the federal reserve at Cleveland.

Table 8.4 Pooled correlation of variables using monthly data from April 1968 to March 2012

Variable name	UEM	GDP	INFL	USDX	DJIA	TBILL	OIL
Gold price	0.462	−0.364	−0.235	−0.587	−0.021	−0.196	0.876
UEM		−0.063	0.057	0.142	0.091	−0.019	0.301
GDP			0.419	0.321	−0.026	0.437	−0.330
INFL				0.343	−0.037	0.690	−0.196
USDX					−0.005	0.551	−0.544
DJIA						−0.036	−0.045
TBILL							−0.387

The results in Table 8.4 indicate that the price of gold is negatively correlated with the expected inflation rate, which contradicts conventional wisdom. Note that this result is very sensitive to the time period under consideration.

Empirical Results Before proceeding to the multi-factor model, we first investigate the distributional effects for each individual factor. The factor effects in the univariate models are presented in Fig. 8.15 for the seven factors that we consider. The estimation is obtained using monthly data from April 1968 to March 2012 with quantiles set at $\{0.1, 0.2, 0.3, 0.4, 0.5, 0.6, 0.7, 0.8, 0.9\}$. Univariate OLS estimates are also provided for comparison purposes. Figure 8.15 reveals that the QR estimates are very different from the OLS estimates for most factors, indicating that distributional effects provide a more comprehensive picture than the "average" OLS effects. For every variable except the DJIA, QR estimates vary with percentiles of gold price distribution, implying that the impact of each variable is different depending on the level of the price of gold itself.

The unemployment rate has the largest impact on gold price measured by both QR and OLS estimates. The OLS estimate indicates that a 1% increase in unemployment rate translates into a \$90 increase in the gold price *on average*. The QR approach indicates that the impact of unemployment is dramatically different depending upon the gold price. At low or median gold price levels, a 1% increase in the unemployment rate produces a gold price increase of \$20. However, when gold prices are already high ($\tau = 90\%$), a 1% increase in the unemployment rate will result in a \$180 increase in the price of gold. The average effect yielded by OLS is indeed the "average" of quantile effects and fails to capture this detail. The GDP growth rate, another important indicator of the strength of the economy, also has a significant negative impact on gold prices. The QR estimates show that a 1% increase in the GDP growth rate will cause the gold price to decrease by \$10 when the gold price is low, while at high gold price levels the effect is −\$30.

We explore the impact of inflation on the price of gold before and after 1981, corresponding to the regime change in monetary policy that occurred in the USA. The rate of inflation has a significant impact on the price of gold prior to 1981, and the effects increase with the gold price. A 1% increase in expected inflation rate corresponds to a \$20 increase when the gold price is low and \$55 when the gold price is high. Low inflation became the main target of US monetary policy after 1981, and

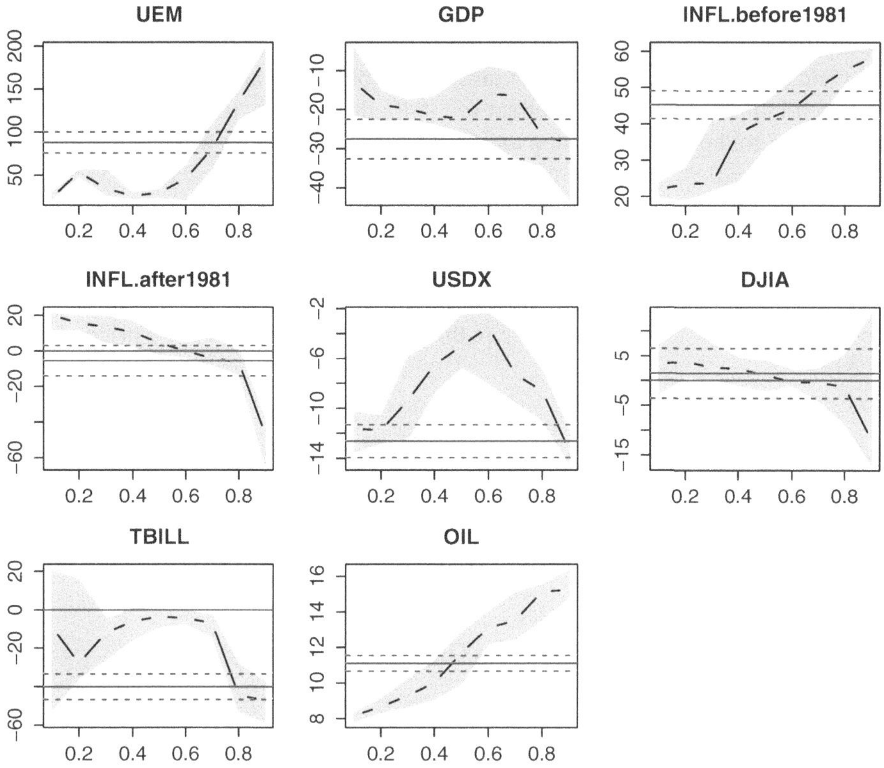

Fig. 8.15 Univariate QR and OLS results of variables on gold price from April 1968 to March 2012. The dot-connected line is for QR estimates with a 90% confidence band indicated by the gray area. The red line is for OLS estimates with a 90% confidence band

the observed inflation rate rarely exceeds 5% following this change in policy. For the post-1981 period, both OLS and QR show that the inflation effects are close to zero and insignificant when gold price is around the median level. Inflation appears to have an impact on the price of gold only when it is a dominant threat to economic stability and people's welfare. When inflation is low, an increase in inflation does not precipitate a level of fear that can induce a change in gold prices. In contrast to conventional wisdom, our results show that gold is not a hedge against inflation during this period, particularly for high gold prices.

The US Dollar Index (USDX) has the expected effect on the price of gold: gold prices increase when the US dollar is weak and decline when the US dollar is strong. It is interesting that USDX has a nonlinear impact on the price of gold and is most significant at either very high or very low levels of the gold prices. At these levels, one unit decrease in the USDX index results in a $12 increase in gold price.

Stock market performance, proxied by the monthly return of DJIA, has a very weak and statistically insignificant effect on the price of gold.[10]The 3-month treasury bill yield exhibits a significant effect only when gold prices are high. While the OLS approach indicates a significant effect of −\$40, the QR effects are insignificant for most of the gold price distribution and become significant only when the gold price is high. At $\tau = 0.90$, a 1% increase in TBILL yield causes a decrease of \$45 in the gold price.

As expected, the price of oil has a significant impact on the price of gold. The effect remains in the range of \$8–\$11 for $\tau < 0.5$, and then becomes more sensitive to the oil price when the price of gold is above the median level. In the present environment ($\tau = 0.9$), a \$1 increase in the price of oil will "push" the price of gold by \$15.

8.6.1.3 Forecasting the Price of Gold Using Quantile Regression

Built on the results of univariate models, we now conduct a multi-factor analysis for the price of gold. We employ the following linear model to study the factor effects on gold price:

$$\begin{aligned} G = {} & \beta_0 + \beta_1 * \text{UEM} + \beta_2 * \text{GDP} + \beta_3 * \text{INFL} + \beta_4 * \text{USDX} \\ & + \beta_5 * \text{DJIA} + \beta_6 * \text{TBILL} + \beta_7 * \text{OIL} + \epsilon, \end{aligned} \tag{8.29}$$

where G is the gold price and ϵ is the error term. Note that ϵ needs to be neither i.i.d. nor Gaussian because we are using the QR approach. The corresponding QR model is specified as

$$\begin{aligned} G(\tau) = {} & \beta_0(\tau) + \beta_1(\tau) * \text{UEM} + \beta_2(\tau) * \text{GDP} + \beta_3(\tau) * \text{INFL} \\ & + \beta_4(\tau) * \text{USDX} + \beta_5(\tau) * \text{DJIA} + \beta_6(\tau) * \text{TBILL} + \beta_7(\tau) * \text{OIL}. \end{aligned} \tag{8.30}$$

Using the coefficients estimated from the QR model for the period of 1968–2008, we generate out-of-sample forecasts of gold price from January 2009 to March 2012 on a monthly basis. The results are presented in Fig. 8.16.

We compare our forecasts with the actual gold price, which tests the efficacy of the QR model. We use QR estimates at the 90th percentile to extract the QR forecast of gold prices. We also provide forecasts based on the OLS estimates for comparison purposes. The actual gold price, QR forecasts, and OLS forecasts are shown in Fig. 8.16 on a monthly basis from January 2009 to March 2012. The gold prices forecasted by quantile regression follow actual gold prices closely, with an average deviation of 8% from actual gold prices. In addition, the actual gold price

[10]We also tried level values of the DJIA index and found the same results.

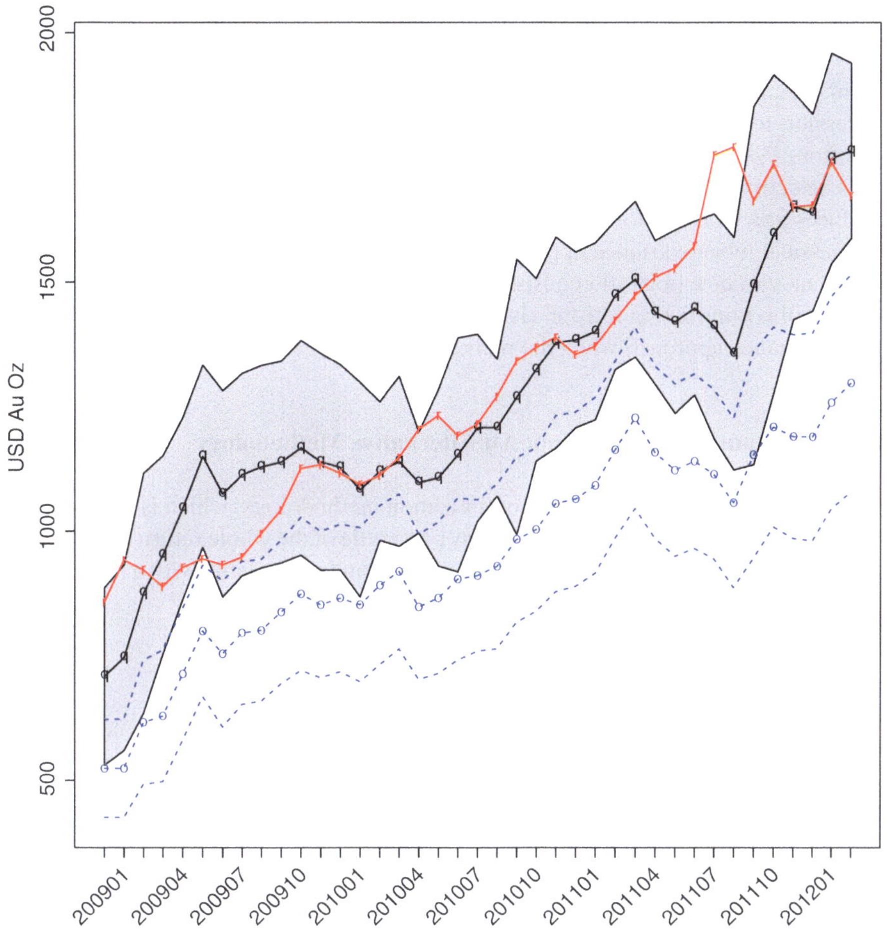

Fig. 8.16 The forecasted and real monthly gold price, Jan 2009–Mar 2012. The solid line with the letter "q" is for the QR forecasts with a 90% confidence band, while the dashed line with the letter "o" is for the OLS forecasts with a 90% confidence band. The solid line with the letter "r" is for the actual price of the gold

remains in the 90% confidence interval of QR forecasts for the entire period except for the months of August and September of 2011.

The capability of quantile regression to forecast tail behaviors with robustness, needless to say, has significant implications for an investing strategy involving gold.

8.6.2 *Portfolio Construction by Quantiles*

In this section, we demonstrate a frontier industry approach employing quantile regression to incorporate tail risk into portfolio optimization. Using the S&P500 data from 1985 to 2014, an empirical study was performed by constructing realistic stock selection investment strategies. The results indicate that quantile optimization produces practical portfolios with risk levels, diversity, and turnover comparable to the classical mean–variance approach. Moreover, the median portfolio outperforms the mean–variance portfolio consistently over the entire period of the study. Lastly, the portfolio minimizing left-tail risk outperforms those at other percentiles and the mean–variance approach when the market is trending down.

8.6.2.1 Quantile Optimization: An Alternative Methodology

In this section, we discuss quantile optimization methodology, which is based on the same principle as cVaR but expands to any percentile of the whole return distribution.

Recall from the previous section that if X is a random variable with a distribution function F_X, then the τth value at risk is defined as

$$VaR_\tau(X) = F_X^{-1}(\tau). \tag{8.31}$$

Clearly, the equation above is the sample quantile of variable X. The conditional value at risk is simply the expected value of VaR over the quantile range of $(0, \tau)$:

$$cVaR_\tau(X) = \int_0^\tau F_X^{-1}(t)dt. \tag{8.32}$$

Now, let $x = F^{-1}(t)$, then we have $F(x) = t$. Suppose there exists a left-continuous density function $f(.)$, and substituting $dt = f(x)dx$ into (8.32), we immediately get

$$cVaR_\tau(X) = \int_{-\infty}^{x_\tau} xf(x)dx, \tag{8.33}$$

where $x_\tau = F^{-1}(\tau)$. Because VaR is simply a quantile value, a natural question would now be, what is the relationship between cVaR and quantile regression? Bassett et al. (2004) offer an excellent exploration of this subject, and they show that cVaR is nothing but the objective function of the τ-th quantile regression. While this is not very surprising given that the MV optimization is an analogue to the least squares approach, the elegant illustration using algebra and thoughtful numerical studies in Bassett et al. (2004) provide a generalized perspective on cVaR as a special case of QR for portfolio construction. Having the generalization, many analyses from the literature of quantile regression can be applied directly to this subject.

Now, let us consider portfolio construction with QR approach. Assume that there are N stocks with period-end returns of $R = \{r_{it}\}$ for $t = 1, 2, \ldots, T$ periods. We would like to determine a set of optimal portfolio weights in order to minimize the cVaR at a specific quantile satisfying some constraints. Translating this into formal optimality algebra, we now have

$$\max_{W} W\alpha - \lambda Q_{\tau}(W, R), \qquad s.t. \quad constraints, \tag{8.34}$$

where $Q = cVaR$ for notational convenience, W is the $1 \times N$ matrix with optimal weights, $\alpha = \{\alpha_i\}$ is a set of quantitative views of future stock price movements, and λ is the risk aversion parameter. Bassett et al. (2004) show that (8.34) as

$$\min_{W} Q_{\tau}(W, R) = \min_{W} \sum_{i} \rho_{\tau}(WR - \xi), \tag{8.35}$$

where ξ is the shadow value of the τth sample quantile. We could leverage the quantile regression method to solve for the optimal quantile portfolio defined by (8.35).

At this point, it is worth comparing the quantile portfolio optimization to the classical MV optimization,

$$\max_{W} W\alpha - \lambda V(W, R), \qquad s.t. \quad constraints, \tag{8.36}$$

where $V(W, R) = W\Omega W^{\top}$, a quadratic form, and Ω is the covariance matrix of R measuring classical risk. Because covariance is not directional, the penalty is equally the same for the tails at both sides of the symmetric return distribution. However, in real-world investment, investors would rather have a portfolio that penalizes the left tail and rewards the right tail in order to get better portfolio performance.

The advantage of a quantile portfolio is that it provides investors a systematic tool for generating a wide selection of portfolios with a focus on different parts of the return distribution, thus resulting in a more customized portfolio to fulfill the specific needs of investors. For example, under extreme conditions, investors can be either defensive or go another way. Moreover, a portfolio based on quantiles may significantly outperform an MV portfolio.

8.6.2.2 Performance of Quantile Portfolios

We employ QR to construct portfolios for a long-only stock selection strategy. The investment universe is S&P 500. We have the same data and portfolios set up as in the industry insights section in Chap. 7. We constructed MV portfolios there. Now with the same data and constraints, we construct quantile portfolios.

$$\max_{W} W\alpha \;-\; \lambda Q_\tau(W, R),$$
$$s.t.\ \text{constraints at security level, and sector level, case 1–5}$$
$$\tau = \{0.10, 0.25, 0.50, 0.75, 0.90\} \tag{8.37}$$

Expected returns are measured by the average returns of each stock during the most recent 60 months, which are the same inputs for quantile and MV portfolio optimizations. The risk part of the quantile portfolio is constructed inherently corresponding to a percentile, the τth Q risk, of the return distribution. The risk part of the MV approach is the covariance matrix of stock returns for the most recent 60 months. For the quantile portfolios, we specify five percentiles, $\tau = \{0.10, 0.25, 0.50, 0.75, 0.90\}$. As a benchmark for quantile portfolios, a MV portfolio is derived by using the same constraints and data for each portfolio rebalance. The same process is repeated until the last month, November 30, 2014.

For each constraint scenario, in order to ensure the portfolios are representative and not concentrated in a narrow range of risk levels, we specify a wide range of values for the risk aversion parameter λ.[11] A wide and reasonable range of risk levels also ensures comparability between quantile portfolios and MV portfolios. Our primary focus will be on the performance of quantile portfolios across different parts of the return distribution, such as the left tail, median, and right tail, and comparison with classical MV portfolios. Before carrying out the portfolio performance analysis, we need to investigate the number of holdings and turnovers in quantile portfolios. This step is to ensure that the quantile portfolios are practical and can potentially be used as an alternative to the classical approach for stock selection strategies.

When comparing the performance of different portfolios, it is very straightforward for quantile portfolios because they are driven by the same methodology. However, it is not straightforward to compare quantile portfolios with MV portfolios, simply because quantile optimization and MV optimization are two very different methodologies. Even with the same set of constraints, expected returns, and risk aversion parameter, the portfolios derived from quantile and variance optimizations can be different in many ways (e.g., risk levels measured by standard deviation, portfolio returns, etc.). Thus, a logical question is, how can we compare two portfolios given that there is no algebraic mapping between two risk measurements in practice? Evidently, the most important characteristics of a portfolio are risk and return. Therefore, if we have two portfolios with similar levels of risk, then we could compare their returns or vice versa. Given our practical motivation, and to stay focused on what is practiced in the industry, we target risk levels in this section on portfolio construction. In practice, we use constraint scenarios and risk aversion parameters to generate a wide range of risk levels for quantile portfolios and variance portfolios. To measure the risk/return tradeoff, we employ the Sharpe ratio, defined as the total return divided by the total risk, which we find very helpful in this study when the risk levels are not exactly the same but fairly close.

---

[11]Without further specification, risk in this section refers to the classical measurement of standard deviation.

Table 8.5 Total risk levels measured by annualized standard deviation (%) for different constraint scenarios with different values of λ (risk aversion parameter) over the entire period

Constraint	λ	MV	Q(0.10)	Q(0.25)	Q(0.50)	Q(0.75)	Q(0.90)
Unconstrained	0.05	16.93	23.86	13.99	13.59	14.47	22.62
	0.1	13.66	14.39	12.10	11.82	12.67	16.22
	0.2	11.92	12.18	11.82	11.75	12.20	13.14
	0.5	11.02	12.49	11.94	11.51	12.06	12.30
	1	10.87	12.48	12.10	11.46	12.24	12.43
Very loose	0.05	15.91	15.85	13.25	12.80	13.76	17.44
	0.1	13.74	13.35	11.85	12.04	12.77	14.77
	0.2	12.12	12.09	11.66	11.75	12.45	13.14
	0.5	11.18	11.67	11.86	11.59	12.07	12.55
	1	11.04	11.75	11.95	11.66	11.89	12.41
Loose	0.05	14.94	14.28	13.20	12.70	13.36	15.39
	0.1	13.42	13.12	12.31	12.19	12.66	13.81
	0.2	12.30	12.30	12.18	12.13	12.41	13.00
	0.5	11.73	11.98	12.16	12.11	12.45	12.85
	1	11.61	12.08	12.21	12.11	12.44	12.73
Tight	0.05	14.28	13.75	13.17	12.86	13.27	14.58
	0.1	13.23	12.94	12.75	12.45	12.93	13.58
	0.2	12.56	12.56	12.69	12.51	12.82	13.21
	0.5	12.28	12.49	12.70	12.56	12.79	13.23
	1	12.24	12.47	12.78	12.53	12.76	13.22
Very tight	0.05	13.94	13.71	13.63	13.48	13.65	14.12
	0.1	13.59	13.47	13.54	13.39	13.53	13.96
	0.2	13.41	13.44	13.50	13.34	13.52	13.88
	0.5	13.26	13.40	13.49	13.32	13.53	13.84
	1	13.24	13.40	13.48	13.33	13.53	13.84

In the following discussion of empirical results, we first investigate the risk levels to ensure comparability of the two portfolios. Afterwards, we investigate the number of holdings in the portfolio over time for practicability and trading cost for executability. Then, we focus on portfolio performance such as returns and risk–return tradeoff.

Risk Levels of Quantile Portfolios By specifying a series of values for each constraint scenario, we obtain a reasonable range of target active risk, from 1.5% to 10%, which covers the majority of the active risk levels of long-only equity products offered by quantitative stock selection strategies in the US equity market.[12] The results of total risk, measured by annualized standard deviation, are listed in Table 8.5,

[12] In the quantitative investment industry, long-only stock selection strategies can be classified as index plus, active, and concentrated, corresponding to risk levels of about 1–2%, 3–6%, and 7–10%, respectively.

corresponding to each constraint scenario and λ value for both quantile portfolios and MV portfolios. The table immediately reveals that across quantile portfolios, the median has the lowest risk levels, and then the risk levels increase slightly as the portfolio moves away from the median and into the tails. Second, the risk levels of quantile portfolios are very close to those of MV portfolios. This is true for all constraint scenarios and values except for the extreme unconstrained case.[13]For example, for the loose constraints case, in which stock level and sector level weights are allowed to be ±5% and ±2% away from the benchmark, respectively, quantile portfolios have risk levels from 11.98% to 15.39%, very close to the levels of 11.61% to 14.94% for the MV portfolios. Another interesting fact is that the total risk levels of the median portfolios are much less sensitive to changes in λ values than those from MV and other quantile portfolios, indicating that the median portfolio is quite robust. It is worth mentioning that the unconstrained case serves the purpose of theoretical investigation only because it is barely used in practical investments.

The plot in Fig. 8.17 shows risk distribution for MV and quantile portfolios across different percentiles. The top plot represents the total risk, and the bottom plot represents the active risk relative to the benchmark of the S&P 500 index. It is clear that by total risk measurements, most portfolios range from 10–20%, while by active risk levels, most portfolios range from 2–10%. This shows that in terms of total and active risk levels, quantile portfolios can substantially achieve practitioners' targets. Thus, similar risk levels will make the comparison between different portfolios meaningful using the return/risk tradeoff.

Practicability and Executability Because the MV optimization method has been used for many years in investments, their portfolio characteristics have come to be considered "standard" by investors. We use MV portfolio features as a benchmark and focus on investigating the practicability of quantile portfolios. For example, are portfolio holdings diversified enough across different assets? How much is the trading cost resulting from a portfolio rebalance?

In regard to the portfolio holding concentration, we check the number of names in the portfolios for each period after the portfolio rebalance. Table 8.6 lists the average number of stocks for the portfolios from both quantile and MV optimizations over the study period from 1990 to 2004. Overall, we can see that quantile portfolios have a reasonable number of holdings that is comparable to the MV portfolios, indicating the practicability of quantile optimization. An exception is the unconstrained case, where quantile portfolios have fewer names than MV portfolios. However, we are aware that the purpose of the unconstrained case is to serve theoretical investigation rather than practical investment, so there is no practical consequence. When a constraint set gets to the point where it is practical, moving away from none, the number of holdings in quantile portfolios increases at a higher rate and is comparable to the MV portfolios. For example, in the tight constraints case, MV portfolios have approximately 100 names, while quantile portfolios have about 90. Interestingly, within each constraint

[13]See the first row in Table 8.5.

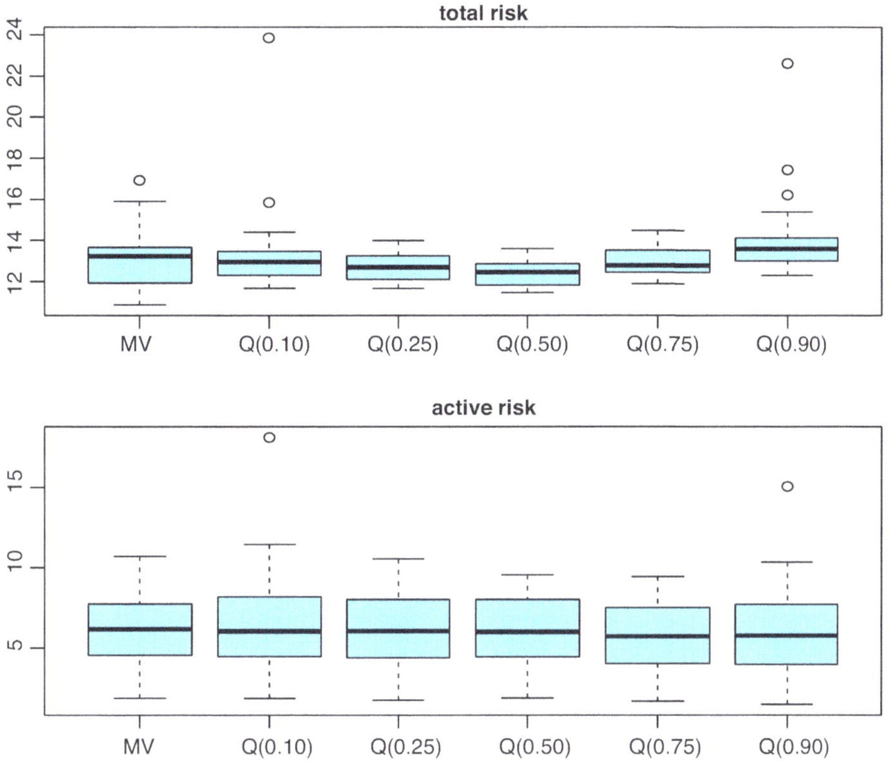

Fig. 8.17 Annualized total and active risk for quantile portfolios and a mean–variance portfolio. The top plot is for total risk (annualized standard deviation), while the bottom plot is for active risk against the benchmark of S&P 500 index returns

set, quantile portfolios have less variation than the MV portfolios even though both have more names when the values of lambda increase, as expected. In order to see portfolio holdings over time, we randomly choose the loose constraints case with lambda=0.05 and plot the number of names for each portfolio from 1990 to 2014 (Fig. 8.18). We can see that the number of names in each portfolio is consistent over time.[14]Overall, our results indicate that the number of holdings in quantile portfolios is diversified enough for practical investments for stock selection strategies.

Regarding executability, we investigate portfolio turnover and associated trading costs. Although it will not be an issue for S&P 500 constituents, as this is the most efficient trading area in the equity market, trading cost could potentially be a serious issue for other segments of equity markets. For example, there is the case where small market capitalization and emerging markets have liquidity playing a significant role in any investment strategies.

[14]A similar pattern holds for other constraint cases and λ values.

Table 8.6 Average number of portfolio holdings from different strategies with various risk levels over the entire backtest period from 1990 to 2014

Constraint	λ	MV	Q(0.10)	Q(0.25)	Q(0.50)	Q(0.75)	Q(0.90)
Unconstrained	0.05	27	10	18	23	21	13
	0.1	41	16	23	27	27	20
	0.2	54	20	23	27	28	25
	0.5	65	21	24	28	29	27
	1	67	21	23	28	29	28
Very loose	0.05	36	25	29	32	32	27
	0.1	47	28	31	35	35	32
	0.2	58	30	32	36	36	35
	0.5	68	31	32	36	37	36
	1	69	31	32	36	37	37
Loose	0.05	59	52	54	57	58	55
	0.1	66	54	55	58	60	58
	0.2	73	55	56	59	60	60
	0.5	78	55	56	59	60	60
	1	79	55	56	59	60	60
Tight	0.05	94	90	90	92	94	92
	0.1	99	91	91	93	95	94
	0.2	103	91	91	94	95	95
	0.5	106	92	92	94	95	96
	1	106	92	92	94	95	96
Very tight	0.05	218	218	219	220	222	220
	0.1	221	221	222	223	226	224
	0.2	224	224	224	225	227	227
	0.5	228	226	225	226	229	229
	1	230	227	226	226	229	229

Portfolio Return Performance We now compare return performance of portfolios from quantile and MV methodologies. To measure the risk/return tradeoff, we employ the Sharpe ratio. We also use quantile risk as an alternative risk measurement to calibrate performance of quantile portfolios against MV portfolios. To make a more thorough comparison, we also carry out t-tests for the returns from the backtest.

The values for the annualized portfolio returns and the Sharpe ratio are presented in Table 8.7 for the five constraint cases with a series of λ values. First, we can see that the left-tail portfolios, which focus on minimizing the risk from the far left tail of the return distribution, actually do not perform well over the entire backtest period from 1990 to 2014. More interestingly, the portfolios at the median and 75th percentile outperform both the left-tail and MV portfolios. This is a bit surprising at first, but a further analysis shows that this actually makes sense, which we discuss in detail in the next section.

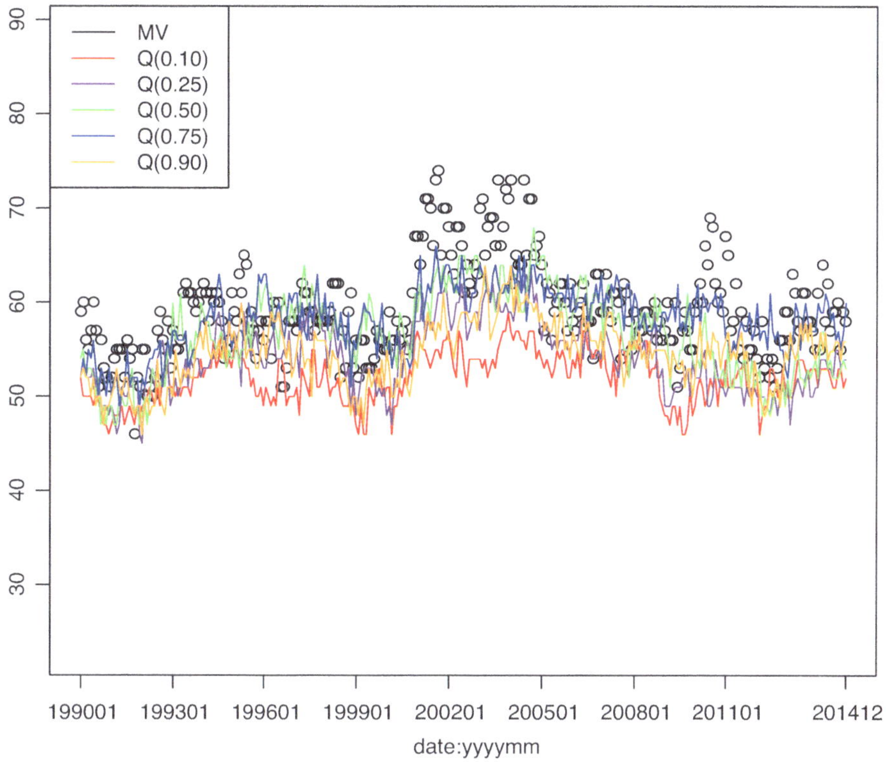

Fig. 8.18 Number of holdings for quantile portfolios and a mean–variance portfolio from 1990 to 2014 for the loose constraints case with $\lambda = 0.05$

Pooling all constraints and values together, the plots in Fig. 8.19 illustrate the performance of quantile and MV portfolios by return densities. The top plot represents the left tail and median, while the bottom plot represents the right-tail portfolios. The MV portfolio appears in both plots for comparison purposes. We can clearly see that the median portfolio is a strong competitor for the classical portfolio, outperforming the MV in both the lower and upper tails.

To see the comparison in detail, Fig. 8.20 displays the plots for the classical risk–return frontier using both the total and active risk levels. We can see that by using the same values of constraints and risk aversion parameters, the median portfolio delivers higher returns with lower risk for the majority of the cases than do the classical MV portfolios. Moreover, median portfolios outperform the minimum-volatility portfolio, the latter has been shown by both scholars and practitioners to outperform MV in the case that the investment period is long enough to include downside markets.

To see the performance of median portfolios across different constraint scenarios and various values of the risk aversion parameter, we plot returns of median portfolios

Table 8.7 Annualized performance of portfolios (return in %) from 1990 to 2014

Constraint	Portfolio	$\lambda = 0.05$		$\lambda = 0.10$		$\lambda = 0.20$		$\lambda = 0.50$		$\lambda = 1$	
		Return	Sharpe	Return	Sharpe	Return	Sharpe	Return	Sharpe	Return	Sharpe
Unconstrained	MV	10.01	0.59	9.02	0.66	8.58	0.72	8.89	0.81	9.31	0.86
	Q(0.10)	5.27	0.22	7.03	0.49	8.11	0.67	9.41	0.75	9.55	0.77
	Q(0.25)	4.71	0.34	8.32	0.69	9.64	0.82	9.16	0.77	9.06	0.75
	Q(0.50)	6.59	0.48	9.42	0.80	9.97	0.85	10.16	0.88	10.55	0.92
	Q(0.75)	8.21	0.57	8.46	0.67	9.35	0.77	10.14	0.84	10.54	0.86
	Q(0.90)	13.66	0.60	9.45	0.58	8.00	0.61	8.97	0.73	8.59	0.69
Very loose	MV	10.27	0.65	9.36	0.68	9.10	0.75	9.38	0.84	9.83	0.89
	Q(0.10)	8.43	0.53	9.01	0.67	8.68	0.72	9.08	0.78	8.89	0.76
	Q (0.25)	8.42	0.64	8.61	0.74	8.84	0.76	9.39	0.79	9.41	0.79
	Q (0.50)	8.13	0.64	9.91	0.82	10.79	0.92	10.69	0.92	10.75	0.92
	Q (0.75)	10.04	0.73	9.45	0.74	10.69	0.86	11.22	0.93	11.02	0.93
	Q (0.90)	10.86	0.62	9.91	0.67	10.23	0.78	9.96	0.79	10.71	0.86
Loose	MV	10.37	0.69	10.08	0.75	9.52	0.77	9.76	0.83	9.92	0.85
	Q (0.10)	9.37	0.66	9.40	0.72	9.48	0.77	9.55	0.80	9.77	0.81
	Q (0.25)	9.52	0.72	9.57	0.78	9.46	0.78	9.49	0.78	9.65	0.79
	Q (0.50)	10.51	0.83	10.50	0.86	10.68	0.88	10.85	0.90	10.74	0.89
	Q (0.75)	10.76	0.81	10.43	0.82	10.83	0.87	10.83	0.87	10.69	0.86
	Q (0.90)	10.40	0.68	9.87	0.71	9.95	0.77	9.71	0.76	10.02	0.79
Tight	MV	10.31	0.72	9.96	0.75	9.96	0.79	10.04	0.82	10.14	0.83
	Q (0.10)	9.33	0.68	9.39	0.73	9.56	0.76	9.71	0.78	9.82	0.79
	Q (0.25)	9.83	0.75	9.93	0.78	9.58	0.75	9.53	0.75	9.72	0.76
	Q (0.50)	10.01	0.78	10.08	0.81	10.06	0.80	9.98	0.79	9.86	0.79
	Q (0.75)	10.45	0.79	10.43	0.81	10.39	0.81	10.27	0.80	10.34	0.81
	Q (0.90)	10.26	0.70	10.53	0.78	10.62	0.80	10.45	0.79	10.46	0.79
Very tight	MV	9.72	0.70	9.75	0.72	9.77	0.73	9.78	0.74	9.78	0.74
	Q (0.10)	9.64	0.70	9.74	0.72	9.82	0.73	9.84	0.73	9.89	0.74
	Q (0.25)	9.58	0.70	9.62	0.71	9.70	0.72	9.63	0.71	9.63	0.71
	Q (0.50)	9.72	0.72	9.61	0.72	9.72	0.73	9.80	0.74	9.72	0.73
	Q (0.75)	9.91	0.73	9.94	0.73	9.99	0.74	9.97	0.74	9.98	0.74
	Q (0.90)	9.80	0.69	9.88	0.71	10.10	0.73	10.13	0.73	10.20	0.74

across both measures in Fig. 8.21. The left plot in gray represents the MV portfolio. In the right plot, we add the median portfolio, which is above the MV portfolio for most λ values and constraint scenarios.

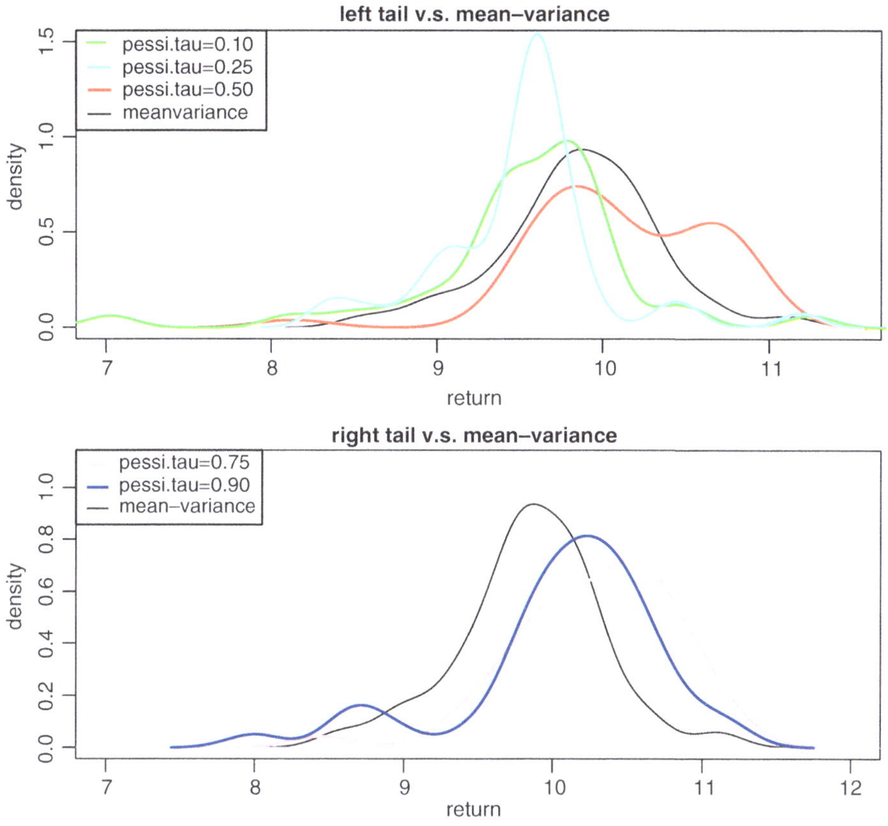

Fig. 8.19 Density of realized portfolio returns for quantile and MV portfolios. Portfolios are constructed with the same constraints across different levels of risk. Note that the median portfolio outperforms the portfolio in both the lower and upper tails

8.6.2.3 Quantile Portfolio Performance Across Different Periods

Given that equity markets are very volatile, it is worth investigating quantile portfolio performance in different market situations. We would expect that left-tail quantile portfolios will have a higher likelihood of outperforming both MV portfolios and right-tail quantile portfolios when stock markets are very bearish to the point where stock returns are significantly skewed with a very long and/or fat left tail. Conversely, right-tail portfolios will have a greater chance of performing better than low-percentile quantile and MV portfolios when stock markets are bullish, where the return distribution has a long/fat right tail. We demonstrate in this study that this is indeed the case. During the whole investigation period from 1990 to 2014, the S&P 500 index return was trending upward, which supports the idea that the median and high percentile portfolios would perform relatively better than the lower-percentile portfolios for the entire period. However, there were extreme ups and downs in the

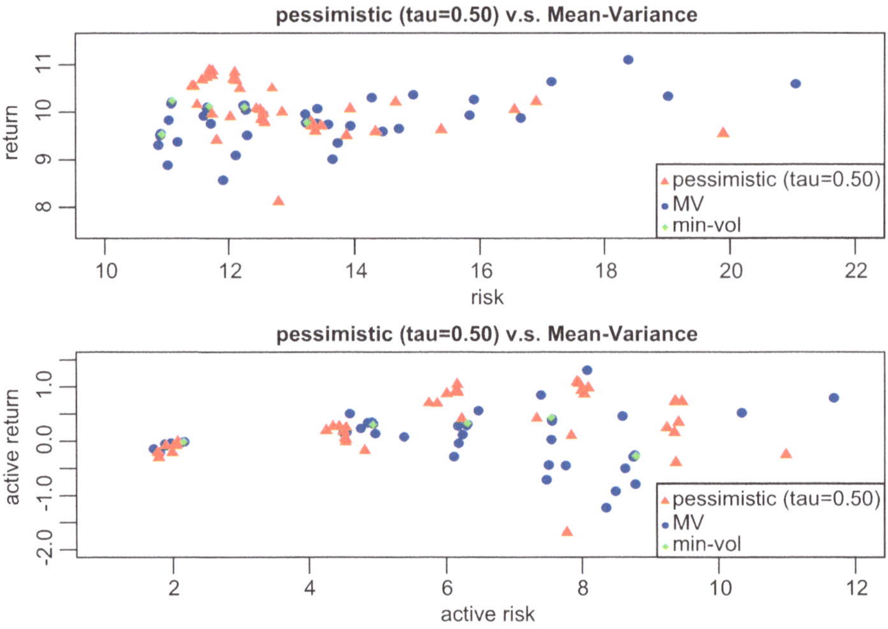

Fig. 8.20 Classical efficient frontier for realized portfolio returns and risks levels using standard deviation. The median portfolio exhibits better performance than the MV portfolio by both total and active risk measures

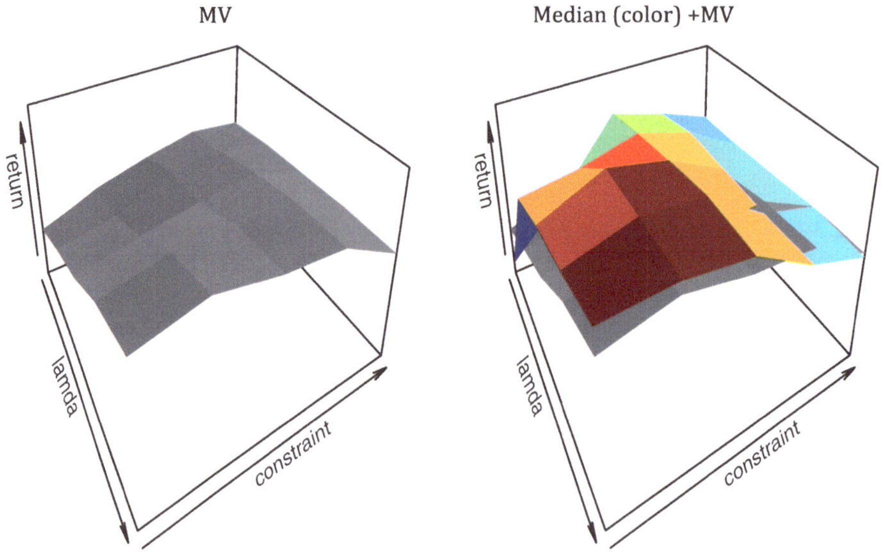

Fig. 8.21 Portfolio returns across the constraint levels and values associated with different risk levels. The median portfolio exhibits better performance than the MV portfolio

Table 8.8 Annualized performance of the S&P 500 index over the entire period and sub-periods: 1990–1999, 2000–2009, and 2010–2014

Performance metrics	Entire period	1990–1999	2000–2009	2010–2014
Annualized return	9.62%	18.21%	−0.95%	15.45%
Annualized std	14.64%	13.43%	16.13%	13.00%
Ratio	0.66	1.36	−0.06	1.19

Table 8.9 Annualized performance (%) of quantile and mean–variance portfolios over three sub-periods: 1990–1999, 2000–2009, and 2010–2014

Performance metrics	MV	Q(0.10)	Q(0.25)	Q(0.50)	Q(0.75)	Q(0.90)
1990–1999						
Annualized ret	14.65	12.22	10.66	13.73	13.43	16.56
Annualized std	14.26	13.64	11.84	11.69	12.72	14.59
Sharpe ratio	1.03	0.90	0.90	1.17	1.06	0.03
2000–2009						
Annualized ret	1.25	3.03	3.57	3.60	2.01	0.56
Annualized std	14.37	14.39	12.74	13.19	13.71	16.26
Sharpe ratio	0.09	0.21	0.28	0.27	0.15	0.03
2010–2014						
Annualized ret	17.40	15.94	15.84	15.35	17.90	19.06
Annualized std	10.49	9.91	9.58	9.73	10.08	11.13
Sharpe ratio	1.66	1.61	1.65	1.58	1.78	1.71

S&P 500 market from 1990 to 2014. In order to study the performance of quantile portfolios across different market conditions, we divide the whole period into three sub-periods: 1990–1999, 2000–2009, and 2010–2014. We are aware that during the sub-period 2000–2009, the US equity market was ranging downward due to recessions caused by the housing market collapse and financial crisis. The index performance metrics are summarized in Table 8.8 for each of the three periods. We see that while annualized returns are approximately 18% and 15% during 1990–2009 and 2010–2014, respectively, it is negative (−0.95%) with larger risk levels during 2000–2009. Indicated by the principle of quantile optimization, the left-tail portfolio would presumably be shining during this period, while the high percentile portfolio would perform relatively well for the other two periods.

The performance metrics of the quantile and MV portfolios for the three sub-periods are listed in Table 8.9. For illustration purposes, we only present the loose constraints case with $\lambda = 0.10$ in the table. We can see that for the bullish periods of 1990–1999 and 2010–2014, the right-tail quantile portfolios, Q(0.75) and Q(0.90), have the best performance, outperforming low quantile portfolios by about 1–4% and MV portfolios by about 2%. For the bearish period of 2000–2009, it is the low-tail portfolios that perform the best: the Q(0.10) portfolio outperforms Q(0.90) by about 2.50% and the MV portfolio by about 2% on an annual basis. In addition, we see again that the median portfolio outperforms the MV portfolio during both the bearish-market period and in most cases during the bullish-market period.

The empirical results show that compared to the one-size-fits-all MV approach, the quantile optimization method offers a more colorful landscape for portfolio construction, thus allowing investors to express different views about the market. Quantile portfolios have characteristics very similar to classical MV portfolios in terms of risk levels, number of holdings, etc. However, they have the ability to deliver much better performance in the context of market conditions and percentile specifications. Specifically, given that the median is a strong competitor to the MV portfolio, it could be used by investors regardless of the market conditions. Another practical benefit of using the quantile approach is that it generates portfolios differently from the mainstream MV method, which makes the trading less concentrated in the market. Hence, the quantile approach could bring tremendous value on executions. More rigorous studies are needed along this line.

8.7 R Commands for Quantile Regression

In this section, we introduce the R package and commands for quantile regression models. The most popular and reliable R package for quantile regression is *quantreg*. The main function for quantile regression estimation is *rq*. We show below the details of quantile regression computation using R with real-world data.

The package *quantreg* was developed by Roger Koenker for estimation and inference of QR models. The package contains many functions to carry out data analysis using linear and nonlinear parametric and nonparametric models for conditional quantiles. The package also includes many commands for plots, which are very convenient for visualization of quantile effects in terms of comparison with classical methods, such as OLS, or between different values of quantiles.

Note that the q̃uantreg package depends on other packages, such as *SparseM*. To load the package into an R session, we need to use the *library* command. You can obtain the details of the package and function *rq* using the *help* command.

QR Package: quantreg

```
> install.packages("quantreg")
> library(quantreg)
> help(package="quantreg")
> help(rq)
> example(rq)
```

In the following, we first focus on estimation, then introduce inference and plots of quantile regression results.

8.7.1 Estimation of QR Models

On QR estimation, we start from a univariate model with one quantile and expanding to a multi-factor model with multiple quantiles. QR is a general methodology, just like OLS. The regression command and formula of *rq* are different from OLS, *lm*, but the commands for summary, coefficient, and residuals are all the same, which we show in examples below. We illustrate these R commands using the gold price model.

Bivariate Model with One Quantile We regress the price of gold on the US unemployment rate. The QR regression command *rq* has the following formula:

$$rq(y \sim x_1 + x_2, data =, tau =),$$

where x_1 and x_2 are factors. For example, we have the price of gold data as below

Price of Gold Data

```
   yyyymm goldPrice Inflation UNEM    USDX   WTI
58 197301     65.14      3.41  5.2  108.19  3.56
59 197302     74.20      3.65  4.9  103.75  3.56
60 197303     84.37      3.87  5.0  100.00  3.56
```

where UNEM is the US unemployment rate, USDX is the UD dollar index, and *goldPrice* is the price of gold with units in dollars per ounce. We would like to estimate the following bivariate model at $\tau = 0.90$:

$$goldPrice = b_0 + b_1 UNEM + \epsilon \tag{8.38}$$

The quantile estimates can be obtained by running the R scripts listed below.

Bi-variate Model with $\tau = 0.90$

```
> fit1=rq(goldPrice ~ UNEM, tau=0.90, data=gold)
> print(fit1)

Coefficients:
```

```
(Intercept)        UNEM
  -113.2719    124.6064
```

If we need values more than the estimates for coefficients, such as confidence bands, we can use the R command *summary* (see below the R scripts). The coefficients and residuals can be obtained by using the *coef* and *resid* commands. The *summary* command produces a 90% confidence band for the estimate.

$$\begin{aligned} \text{summary:} &\quad summary(fit1) \\ \text{coefficients:} &\quad coef(fit1) \\ \text{residual:} &\quad resid(fit1) \\ \text{fitted values:} &\quad fitted(fit1) \end{aligned}$$

Summary of an r Object at $\tau = 0.90$

```
> print(summary(fit1))

Coefficients:
            coefficients lower bd   upper bd
(Intercept) -113.27191   -288.85161  182.81882
UNEM         124.60638     94.31908  145.93423
```

Multi-factor Model with One Quantile Now, we add one more factor, USDX, the US dollar index strength, to the gold price model. We would like to estimate the following two-factor model at $\tau = 0.90$:

$$goldPrice = b_0 + b_1 UNEM + b_2 USDX + \epsilon \tag{8.39}$$

The R scripts below show the quantile estimates for a two-factor model.. Note that in the summary function, we specify the non-iid option for the standard error calculation.

summary(rq(. . .), se="nid")

The parameter "se" pertains to the methodology for standard error computation. There are currently six available methods. The default is "rank." We show the example of "nid" method in the R scripts below.

Multi-factor Model with One Quantile

```
> fit2=rq(goldPrice ~ UNEM+USDX, tau=0.90,data=gold)
> print(fit2)
Coefficients:
(Intercept)        UNEM        USDX
 1131.24542   102.57388   -11.89376

## use non-iid method to calculate standard error
> print(summary(fit2,se="nid"))
Coefficients:
            Value      Std. Error t value   Pr(>|t|)
(Intercept) 1131.24542  109.25881   10.35381   0.00000
UNEM         102.57388   10.44629    9.81916   0.00000
USDX         -11.89376    1.10163  -10.79647   0.00000
```

As more factors are added and sample size increases, it is important to have an efficient computation algorithm. The parameter *method* enables user to choose a proper method for computation solutions.

Multi-factor Model with One Quantile

```
> fit3=rq(goldPrice ~ UNEM+USDX, tau=0.90,data=gold, method="fn")
```

The default method is *br*, which is efficient to solve a problem with a sample size of thousands. If the sample size gets bigger, we suggest readers use the Frisch–Newton interior point method, *fn*.

Multi-factor Model with a Series of Quantiles One of the advantages of QR is that by running a series of taus, QR can deliver the full-distributional view of how factors impact the response variable. The function *rq* accommodates such a feature, with "tau" taking a vector of quantile values, e.g., $tau = c(0.10, 0.90)$. This is demonstrated in the R scripts below.

Multi-factor Model with Two Quantiles

```
> fit4=rq(goldPrice ~ UNEM+USDX, tau=c(0.10, 0.90), data=gold,
   method="fn")
> print(summary(fit4,se="nid"))
```

```
tau: [1] 0.1
Coefficients:
              Value       Std. Error t value    Pr(>|t|)
(Intercept) 1255.30514    35.61965    35.24193    0.00000
UNEM           7.14076     4.24588     1.68181    0.09329
USDX         -11.19456     0.49598   -22.57057    0.00000

tau: [1] 0.9
Coefficients:
              Value       Std. Error t value    Pr(>|t|)
(Intercept) 1131.24542   109.25881    10.35381    0.00000
UNEM         102.57388    10.44629     9.81916    0.00000
USDX         -11.89376     1.10163   -10.79647    0.00000
```

Thus, we see immediately that at different quantiles, US unemployment has very different impacts ($b_1(\tau)$) on the price of gold depending on the level of prices:

$$\hat{b}_1(\tau = 0.10) = 7.14, \quad \hat{b}_1(\tau = 0.90) = 102.57;$$

while for the USDX, the impacts, $b_2(\tau)$, remain about the same

$$\hat{b}_2(\tau = 0.10) = -11.19, \quad \hat{b}_1(\tau = 0.90) = -11.89.$$

Of course, the above judgment can be biased as it is purely based on "visualization." We introduce more formal tests for the significance of a factor, heterogeneity effects, and goodness of fit in the following subsection.

8.7.2 *Inference of QR Estimates*

Continuing with the gold price model, we illustrate how to perform inference studies by R in this subsection.

Local Significance of a Factor: $b(\tau) = 0$ There are two types of local significance: one is for a single quantile, the other is for a series of quantiles. The single quantile case is straightforward, where the test can be done by using the standard error and t-value.

For the case with a series of quantiles, it involves with the process of QR which deals with the whole distribution of quantile effects. The joint and non-joint test for slope parameters can be computed using the method proposed by Khmaladze (1981), which was implemented as *KhmaladzeTest* by Koenker and Xiao (2002).

The R codes below specify two null hypotheses: one is for location effect, and the other is for location-scale shift. For our price of gold model, all effects are significantly different from zero, both jointly and non-jointly.

Local Significance for Quantile Process

```
> ## location effects
> Ktest.location=KhmaladzeTest(goldPrice ~ UNEM+USDX,
  taus = seq(.05,.95,by = .01),nullH="location", data=gold)
> print(Ktest.location)
$nullH
[1] "location"
$Tn
[1] 17.83448
$THn
    UNEM     USDX
6.911201 9.580864

> ### lcoatin-scale effects
>   Ktest.scale   =KhmaladzeTest(goldPrice ~ UNEM+USDX, taus =
    seq(.05,.95,by = .01), nullH="location-scale",data=gold)
>   print(Ktest.scale)
$nullH
[1] "location-scale"
$Tn
[1] 40.19004
$THn
    UNEM     USDX
8.518422 5.506457
```

Heterogeneity Across Quantiles: $b(\tau_1) = b(\tau_2)$] This can be done jointly for all slope parameters or non-jointly for each individual slope parameter. The R command is *anova.rq* with the option "joint." The results from the R codes below show that the p-value for the F-distribution is zero for $b_1(\tau)$ and 0.55 for $b_2(\tau)$. The default test is "Wald," but users can use other tests such as "rank."

UNEM: $H_0 : b_1(0.10) = b_1(0.90),\ H_1 : b_1(0.10) \neq b_1(0.90),\ p - value = 0, reject\ H_0;$

USDX: $H_0 : b_2(0.10) = b_2(0.90),\ H_1 : b_2(0.10) \neq b_2(0.90),\ p - value = 0.55, accept\ H_0.$

Inference: Heterogeneity Test

```
> #anova(object, test = "Wald", joint = TRUE, score = "tau",
> #se = "nid", R = 200, trim = NULL)

>  fit4.tau10=rq(goldPrice ~ UNEM+USDX, tau=0.10, data=gold, method="fn")
>  fit4.tau90=rq(goldPrice ~ UNEM+USDX, tau=0.90, data=gold, method="fn")
> heter=anova(fit4.tau10,fit4.tau90,joint=F)
> print(heter)

Model: goldPrice ~ UNEM + USDX
Tests of Equality of Distinct Slopes: tau in { 0.1 0.9 }

     Df Resid Df F value Pr(>F)
UNEM  1      911 75.1415 <2e-16 ***
USDX  1      911  0.3617 0.5477
```

Goodness of Fit at τ For people who get used to conditional mean methodologies, such as GLS, there is a tendency to have the value of R^2.

> R-squared is evil, that is why there isn't an automated way to compute something similar for quantile regression in the *quantreg* package. – Roger Koenker

Nevertheless, here is an analogue of $R(\tau)$ to R^2, proposed by Koenker:

$$R(\tau) = 1 - \frac{Q_1(\tau)}{Q_0(\tau)},$$

where both $Q_0(\tau)$ and $Q_1(\tau)$ are for the same quantile and $Q_0(\tau)$ is nested within $Q_1(\tau)$. For example:

$$Q_0(\tau = 0.9): \; rq(y \sim 1, tau = 0.9)$$

$$Q_1(\tau = 0.9): \; rq(y \sim x, tau = 0.9)$$

It is expected that $0 \le R^1 \le 1$. For further details, see Koenker and Machado (1999).

In practice, the objective value of a quantile regression can be obtained by using the command "rho." We carry out a quick study below to calculate $R(\tau = 0.90)$ for the two-factor gold price model. For comparison purposes, we also obtain the R^2 value from the OLS estimates:

$$R(\tau = 0.9) = 77\%, \qquad R^2 = 49\%.$$

The local goodness of fit at the 90th percentile of the gold prices is 77% while the global conditional mean fit is 49%. Note that $R(\tau)$ and R^2 use different methods to calculate fitness. Because R^2 is based on the square values, it tends to exaggerate both good and bad fit: make the fitness rosier for a good fit and even worse for a bad fit.

Multi-factor Model with Two Quantiles

```
> fit5=rq(goldPrice ~ UNEM+USDX, tau=0.90,data=gold, method="fn")
> fit6=rq(goldPrice ~ 1, tau=0.90,data=gold, method="fn")

> print(fit5$rho)
[1] 13004.39

> print(fit6$rho)
[1] 57038.21

> ### R(tau)
> Rtau= 1- fit5$rho/fit6$rho
> print(Rtau)
[1] 0.7720057

> ### R-square from OLS
> R2=summary(lm(goldPrice~UNEM+USDX,data=gold))$adj.r.squared
>  cat ("R2 from OLS is: ", R2, "\n")
R2 from OLS is:  0.4879304
```

Just for the sake of comparison, for those who are overly obsessed with R^2, one way to compare QR and OLS apples-to-apples is to use R^2 for quantile regression models. That is, we use the square values of the error term, and the revised formula of rho is

$$\tau \sum_{e_i \geq 0} e_i^2 + (1-\tau) \sum_{e_i < 0} e_i^2,$$

although this distorts the means and ends completely.

8.7.3 Plot of QR Results

The *quantreg* package provides very powerful tools for the visualization of quantile regression results. A widely used R command for quantile regression results is *plot(summary(rq(. . .)))*. For example, we use the following R scripts to generate the plots in Fig. 8.22. Note that the red line is for OLS. The confidence band intervals are set at 90%.

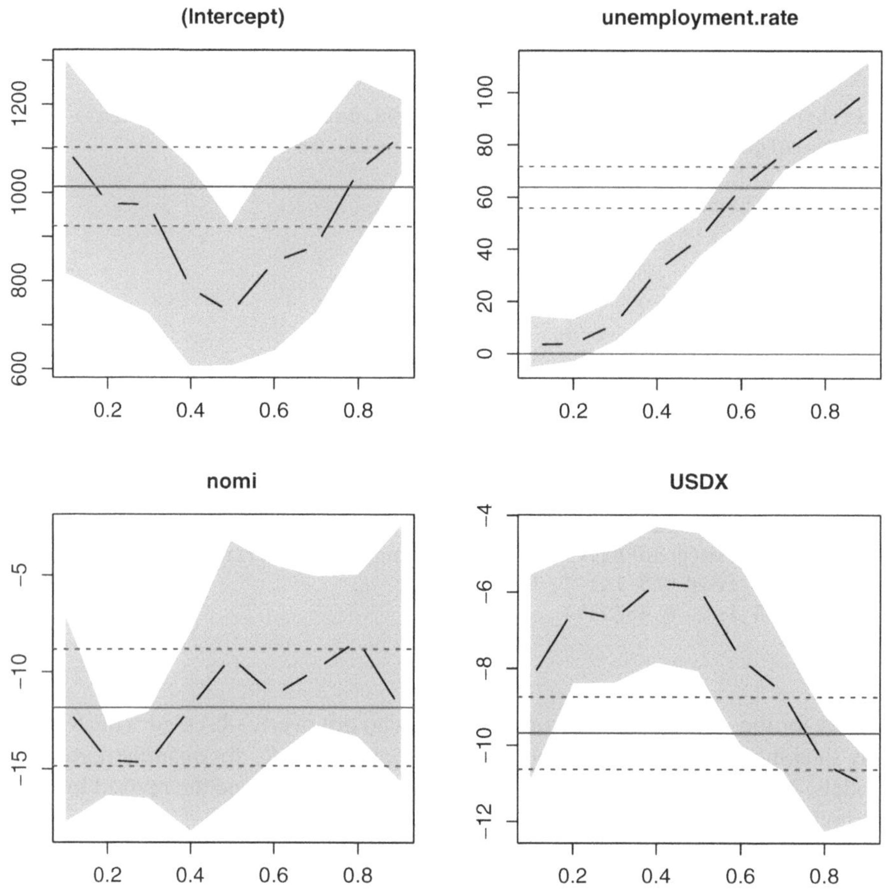

Fig. 8.22 Example of QR plots using the gold price data

QR Plot Example

```
> plot(summary(rq(gold.price~unemployment.rate+nomi+USDX,
    data=gold,tau=c(1:9)/10)))
```

There are other commands and packages for more advanced QR plots, such as nonlinear effects of QR results. See the R section in Koenker (2005) for details.

Keywords, Problems, and Group Project

Part I: Keywords

Quantile regression, linear programming, $\rho_\tau(e_i)$, asymptotics , tail behavior
Conditional value at risk, loss avoidance, quantile optimization
Median portfolio, tail portfolio, gold price forecast
R package quantreg for quantile regression

Part II: Problems

Problem 8.1 Using the gold price data, conduct multi-factor model analyses.

(1) OLS
(2) Quantile regression at $\tau = 0.1, o.25, 0.500.75, 0.90$
(3) Plot the results, using the R command $plot(summary(rq(y\ x, tau, data)))$.

Problem 8.2 Use the data of stocks in Problem 4.5 to build alpha by quantile regression.

$$Q_R(\tau|F) = b_0(\tau) + b_1(\tau)PROF + b_2(\tau)EQ + b_3(\tau)VALUE$$
$$+b_4(\tau)PM + b_5(\tau)MQ$$

(1) Using $\tau = 0.50$, construct median alpha values $ALPHA_{0.50}$.
(2) Change $\tau = 0.10$ and $\tau = 0.90$, construct alpha values at the left tail and right tail, $ALPHA_{0.10}$ and $ALPHA_{0.90}$, respectively.

Problem 8.3 Applying the diagnostics package built in Chap. 5 to the ALPHA at $\tau = 0.50, 0.10, 0.90$, respectively, compare the results with $ALPHA_{ols}$.

Problem 8.4 Using the diagnostics package results, construct a long-only and long-short portfolio, respectively, for ALPHA at $\tau = 0.50, 0.10, 0.90$. Compare performance with $ALPHA_{ols}$ (Table 8.10).

Table 8.10 Comparing performance

	$ALPHA_{ols}$	$ALPHA_{0.50}$	$ALPHA_{0.10}$	$ALPHA_{0.90}$
Long-only				
Long-short				

Part III: Group Project

Problem 8.5 Run quantile optimization with the same constraints and parameters as the mean–variance optimization in Chap. 7 for ALPHA at $\tau = 0.50, 0.10, 0.90$.

$$\max W^T\alpha - \lambda Q(\tau, W)$$

(1) Does the median portfolio with $ALPHA_{0.5}$ outperform mean–variance portfolios with $ALPHA_{ols}$ in terms of return and risk?
(2) For $\tau = 0.10, 0.90$, compare their performance with the median portfolio.

References

Adrian, T., and M. Brunnermeier. 2008. CoVaR, Federal Reserve Bank of New York staff reports.

Bassett, G., R. Koenker, and G. Kordas. 2004. "Pessimistic Portfolio Allocation and Choquet Expected Utility." *Journal of Financial Econometrics* 2(4): 477–492.

Cenesizoglu, T., and A. Timmermann. 2008. "Is the Distribution of Stock Returns Predictable?" Working paper, UCSD-Centre for Economic Policy Research (CEPR).

Engle, R., and S. Manganelli. 2004. "Conditional Autoregressive Value at Risk by Regression Quantiles." *Journal of Business and Economic Statistics* 22(4): 367–381.

Gauss, C.F. 1809. Theoria Motus Corporum Coelestium in Sectionibus Conicis Solum Ambientium (theory of the motion of the heavenly bodies moving about the sun in conic sections), the original work published in 1809 and original translation published, 1857, translated by C.H. Davis, 2004. Mineola: Dover Publications.

Goldberg, L.R., M.Y. Hayes, J. Menchero, and I. Mitra. 2010. "Extreme Risk Analysis." *Journal of Performance Measurement*, Spring, 17–30.

Hendricks, W., and R. Koenker. 1991. "Hierarchical Spline Models for Conditional Quantiles and Demand for Electricity." *Journal of the American Statistical Association* 87: 58–68.

He, X. 2017. "A Conversation with Roger Koenker." *International Statistical Review* 0(0): 1–15.

Khmaladze, E.V. 1981. Martingale approach in the theory of goodness-of-fit tests. *Theory of Probability & Its Applications* 26(2): 240–257.

Koenker, R. 2005. *Quantile Regression*. Cambridge University Press.

Koenker, R., and G. Bassett. 1978. "Regression Quantiles." *Econometrica* 46(1): 33–50.

Koenker, R., and G. Bassett. 1982a. "Robust Tests for Heteroscedasticity Based on Regression Quantiles." *Econometrica* 50: 43–61.

Koenker, R., and G. Bassett. 1982b. "Tests of Linear Hypotheses and l_1 Estimation." *Econometrica* 50: 1577–1584.

Koenker, R., and J.A.F. Machado. 1999. Goodness of fit and related inference processes for Quantile regression. *Journal of the American Statistical Association* 94: 1296–1310.

Koenker, R., and Z. Xiao. 2002. "Inference on the Quantile Regression Process." *Econometrica* 70: 1583–1612.

Ma, L. 2015. "Portfolio Construction by Quantiles: An Alternative to the Mean-Variance Approach." Working paper.

Ma, L., and L. Pohlman. 2004. "Return Forecasting and Portfolio Construction: A Quantile Regression Approach." Working paper.

Ma, L., and R. Koenker. 2006. "Quantile Regression for Recursive Structural Equation Models." *Journal of Econometrics* 134(2): 471–506.
Ma, L., and L. Pohlman. 2008. "Return Forecasting and Portfolio Construction: A Distributional Approach." *European Journal of Finance* 14(5): 409–425.
Ma, L., and L. Pohlman. 2010. "Return Forecasting by Quantile Regression." *Journal of Investing* 19(4): 116–121.
Ma, L., and G. Patterson. 2012. "Is the Price of Gold Overvalued?" *Journal of Investing* 22: 113–127.
Powell, J. 1991. "Estimation of Monotonic Regression Models Under Quantile Restrictions." In *Nonparametric and semiparametric methods in econometrics*, ed. W. Barnett, J. Powell, and G. Tauchen. Cambridge: Cambridge University Press.
Rockafellar, T., and S. Uryasev. 2000. "Optimization of Conditional Value-at-Risk." *Journal of Risk* 2: 21–41.
Siddiqui, M. 1960. "Distribution of Quantiles from a Bivariate Population." *Journal of Research of the National Bureau of Standards* 64: 145–150.
Taleb, N. 2007. *The Black Swan: The Impact of the Highly Improbable*. Random House. ISBN 978–1400063512.
Weitzman, M. 2009. "On Modeling and Interpreting the Economics of Catastrophic Climate Change." *The Review of Economics and Statistics* 91(1): 1–19.

Chapter 9
Quantamental Investment

Abstract In this chapter, we introduce quantamental investment—a frontier investment approach that emerged following the financial crisis of 2008. Quantamental investment combines fundamental and quantitative analysis to try to achieve both depth and breadth. We provide a heuristic definition of quantamental investment, explore how to conduct quantamental investment, and discuss important factors for successful quantamental investments such as team building and corporate culture. In the section on industry insights, we illustrate in detail how to employ a quantamental approach to build a Japanese stock selection strategy. In the R section, we discuss surveying with R.

9.1 Quant and Fundamental

There are many approaches to investment. In the context of how securities are selected, investment approaches generally fall under two categories: fundamental and quantitative. The former picks securities based on fundamental understanding of companies and their prospects, and the latter picks securities based on statistical models incorporating various factors. Needless to say, fundamental investment depends heavily on individual experience and judgment, while quantitative investment depends on historical data and modeling techniques.

Large-scale industry money management started from fundamental investment. Built upon the work of pioneers such as Benjamin Graham, Warren Buffett, and numerous academic researchers, the modern sense of fundamental investment emerged after the Great Depression, though it can be traced back to the period when stocks were born. With the birth of computers and increasing data availability, quantitative investment emerged in the late 1970s, developed during the 1980s, and gained momentum in the late 1990s.

Table 9.1 lists major differences between fundamental and quantitative approaches. In the following, we discuss in detail the fundamental and quantitative approaches with a focus on alpha sources, risk control, portfolio construction, and characteristics.

L. Ma, *Quantitative Investing*, https://doi.org/10.1007/978-3-030-47202-3_9

Table 9.1 Quant versus fundamental

	Fundamental	Quantitative
Alpha source	Company specifics	Factors
Information	Prospective	Historical data
Value	Personal experience	Investment process
Portfolio	Concentrated	Large number of names
Advantage	Depth and proprietary insights	Breadth and systematic

9.1.1 *Fundamental Approach: Achieving Depth with Company Specifics*

A fundamental approach is based on fundamental analysis of companies, industries, and related macro-events. Fundamental analysis is carried out by fundamental portfolio managers and financial analysts. The former are usually responsible for investment decisions and overall portfolio performance, while the latter play more of a supporting role by providing prospective information and analysis. A fundamental portfolio manager usually starts her career as a financial analyst.

Due to limits of time and energy, each analyst can only cover about 20–30 companies within an industry. Typically, there are two analysts covering an industry, with a senior analyst covering large companies and a junior analyst covering mid- and small-size companies.[1]

Company-specific fundamental information is collected through meetings, industry conferences, company visits, and other channels. Both portfolio managers and financial analysts participate in such events. They try to collect any information that will impact the future price of a company. This information includes the following:

- Macro
  - events, policy instruments, sensitivity
- Industry
  - trends, innovations, market shares, commodity prices
- Company
  - core business, other business, products, competition edge, financial conditions, management, M&As

How is this information incorporated into a portfolio? The collected information needs to be analyzed and integrated into investment decisions and portfolio

[1]This also depends on many other factors. For example, a larger industry, a cross-industry investment strategy, or a larger-asset portfolio may have more analysts.

construction. Some typical tools or approaches used by fundamental portfolio managers are:

- Filtering process
  - to filter securities based on some criteria such as profitability ratios, market competitiveness, etc. For example, starting from an investment universe, first remove the stocks with negative earnings for the most recent quarter, then remove the stocks with decreasing market shares, etc.
- Scenario analysis
  - to analyze companies and events based on different scenarios
- Ranking process
  - to rank securities based on some criteria

These are then used as inputs for the portfolio manager to decide on the weights of securities. Some portfolio managers make investment decisions without any process or formula; some employ a process to formulate a portfolio; and some make decisions with the help of an optimizer for risk control. Regardless of methods, personal experience and judgment play significant roles in portfolio construction. Portfolio rebalance relies on similar information and procedures to those presented above.

A typical fundamental portfolio is concentrated, that is, it consists of a few stocks the portfolio manager believes have the potential to deliver outperformance. Risk control is more of an informal mental process. In other words, diversification and scenarios may all be considered implicitly when a fundamental portfolio is constructed, but not with an explicit quantifiable process. After the 2008 financial crisis, more and more fundamental portfolio managers started to adopt an explicit investment process with quantifiable risk metrics. Some even use these factors as inputs for alpha and optimizers for portfolio construction.

The following are some key elements for a fundamental portfolio to achieve better than market and peer performance:

(1) a good understanding of the industry and company
(2) a meaningful way to transform such an understanding into a portfolio
(3) avoiding personal and emotional bias

9.1.2 *Quantitative Approach: Achieving Breadth with Factor Parsimony*

A quantitative approach employs statistical modeling with multiple factors to identify sources and patterns of market inefficiency based on the law of large numbers. Historical data are collected and used to analyze pricing patterns from which alpha and risk models are derived. Risk-controlled alpha is used for portfolio construction, which is typically achieved through optimization. Constraints are usually set explicitly during the investment process, and portfolio performance is monitored quantitatively across factors and other metrics.

A quantitative approach can be realized by an investment process. We have discussed the process and its components in detail in the previous chapters.[2]

A quantitative strategy is carried out by a quantitative investment team, which usually consists of a portfolio management group and a research team. Sometimes, there is an additional risk management or data/IT team to support researchers and portfolio managers. Some quant teams combine research and portfolio management functions in one group such that a quantitative researcher is also a portfolio manager. Since quantitative products are based on modeling and process, research plays a significant role in the overall investment process and final portfolio performance, while portfolio management focuses on daily account management, execution, and trading. Another important responsibility of quantitative portfolio management is to deal with corporate events, which are usually not captured by models.

A quantitative portfolio is derived from alpha, risk, and optimization methods. The alpha is usually delivered by a multi-factor model with each factor/theme capturing an aspect of security pricing. Having a comprehensive understanding of the factors and methods is critical for the performance of a quantitative portfolio. We discussed this in detail in Chaps. 4–8.

Compared with a fundamental portfolio, a typical quantitative portfolio has a large number of securities. Risk management and hence diversification are achieved systematically in the modeling and portfolio construction process. Portfolio rebalancing is usually based on the information decay of alpha or risk models. The investment process can be applied to a new market fairly quickly. However, the success of a new strategy will depend on a good understanding of the market, the factors, and their connection to fundamentals. After all, it is the fundamental information behind the numbers that really matters.

9.2 What Is Quantamental Investment?

Suppose we have a mathematician thinking like a fundamental portfolio manager using common sense. Suppose we have a financial analyst thinking like an economist to build a model. These represent a quantamental mentality. Quant and fundamental approaches are not mutually exclusive; rather, they should be complementary. This is the central idea of quantamental investment. Loosely speaking, any kind of hybrid approach of quant and fundamental methods can be considered quantamental investment. However, strictly speaking, quantamental investment should combine quant and fundamental analysis at every step of the investment process. We define

[2] For example, see Chap. 1 for the overall investment process, Chap. 4 for the alpha process, and Chap. 7 for the portfolio construction process.

quantamental investment as an investment methodology based on the combination of fundamental and quantitative principles in the following aspects:

- Alpha source: company specifics and factor statistics
- Information: both historical and prospective
- Value: customized investment process
- Portfolio: between concentrated and broad coverage
- Advantage: depth and breadth

The information listed above is also presented in the last column of Table 9.2 which compares quantamental with quant and fundamental approaches.

9.2.1 Why Do We Need Quantamental Investment?

Having defined quantamental investment, we now show that quantamental investment—an organic combination of quant and fundamental approaches—is not only possible but will add value. We use public equity markets for illustration purposes. Based on their sources, equity returns can be decomposed into four parts: market, industry, factors, and company specifics (Table 9.3).

$$\text{Total return} = \text{Market} + \text{Industry} + \text{Factor} + \text{Company},$$

Table 9.2 Quantamental investment

	Fundamental	Quantitative	Quantamental
Alpha source	Company specifics	Factors	Company specifics expressed by factors
Information	Prospective	Historical data	Both historical and forward-looking data
Value	Personal experience	Investment process	Customized investment process
Portfolio	Concentrated	Large number of securities	Something in between
Advantage	Depth and proprietary insights	Breadth and systematic	Depth and breadth

Table 9.3 Return decomposition: quant and fundamental

Fundamental
Company specifics: business model, products, innovation, M&A, etc.
Quant
Factor statistics: value, momentum, profitability, earnings quality, etc.
Industry: including seasonality
Market: including currency and commodity prices

where market and industry are common returns, factor returns are the target of quant analysis, and company-specific returns are the target of fundamental analysis. However, factor and company-specific information are not disconnected: Company specifics can be expressed by factors, and factor values serve to characterize company specifics. Thus, fundamental and quantitative analysis are just two different ways to describe the business performance of companies. Therefore, quant and fundamental approaches are not mutually exclusive; they actually complement each other.

However, fundamental and quantitative approaches do focus on different aspects:

(1) Fundamental approaches focus more on prospective information that will impact prices, while quantitative approaches focus more on historical data that have proved to have forecasting power.
(2) Fundamental portfolio managers seek a deep understanding of company-specific information, while quants have a very broad view based on a large number of companies.

These make the combination both necessary and beneficial. Their common features ensure that quant and fundamental analysis can be combined, while their differences ensure that the combination adds value. However, it should be stressed that the combination is not simply A plus B, but the organic integration of A and B.

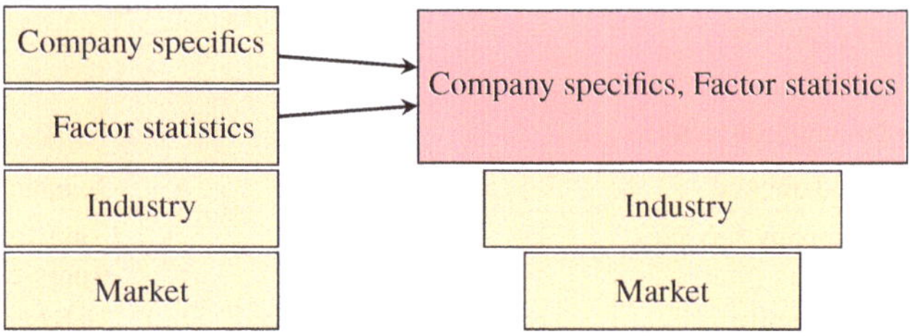

We can also express the analysis above with equations. Suppose R is the total return for a security in financial market M. The security is issued by public company C, which belongs to industry I, with factor values F. The total return can be decomposed into four parts: market, industry, factor, and company. The return from the company specifics is the focus of fundamental investment, while factor parsimony is the focus of quant investment (Table 9.3).

$$\begin{aligned} R &= R_M + R_I + R_F + R_C \\ &= (R_M + R_I) + (R_F + R_C) \\ &= R_\beta + R_\alpha. \end{aligned} \tag{9.1}$$

We can define the alpha from quantamental analysis as

$$\begin{aligned} R_Q &= R_\alpha \\ &= R_F + R_C. \end{aligned} \tag{9.2}$$

Intuitively speaking, quantamental investment simply combines the alpha sources of company-specific and factor characteristics, which are pursued separately by fundamental and quantitative approaches.

Needless to say, it can be very challenging to achieve both depth and breadth, the respective merits of fundamental and quantitative investment. Actually, the combination does not derive from a theoretical design. It emerged in practice from the realistic needs of portfolios. The desire to bring both depth and breadth into a portfolio has had a long history with many trials. In the following section, we describe the emergence and development of quantamental investing.

9.3 Quantamental Investment: Emergence and Development

We defined quantamental investment in the previous section. In this section, we review the emergence and development of quantamental investment with a focus on the rationale for the combined approach.

9.3.1 Emergence

Quantamental investing emerged organically because quant and fundamental approaches complement each other in many respects.

The emergence of the quantamental approach came from the realization of the shortcomings of both approaches—the lack of depth of quant analysis and the lack of breadth of fundamental analysis.

In detail, a typical quantitative approach suffers from:

1. blackbox
   - no one knows what is in the model and how the portfolio is derived except for the machine (computer) and the person who builds it, it looks like a blackbox for an outsider.
2. backlook
   - a quant portfolio is based essentially on historical data, which is a purely retrospective approach with the assumption that history will repeat itself.
3. fundamental insight
   - lack of fundamental intuitions.
4. data mining
   - data can be tortured to confess whatever a quant wants to hear.

5. crowding
   - many quants use similar data sets, employ similar multi-factor models, and use similar optimization methods/tools, ending up with similar portfolio holdings for similar strategies.

These shortcomings did not seem serious when quant emerged and gained momentum. However, as quant became popular, and more and more funds were being managed by quants, the problems became very serious. Many quant strategies employ the same data sources, similar alpha and risk models, and the same optimizers, ending with similar portfolios. This creates two immediate issues: (1) the crowding of portfolio transactions and (2) profit opportunities being explored away fairly quickly. As a result, many quant strategies started to incorporate fundamental inputs, which help differentiate and add depth to a quantitative approach.

Surveying for Prospective Information Many quant shops started to use surveys to collect prospective information. This complements the information from historical data on which quants have traditionally relied.

Company Visits and Meetings Many quant shops send their researchers and PMs to visit companies and attend conferences and meetings where fundamental analysts and PMs gather.

On the other hand, a typical fundamental approach suffers from

1. personal preference
   - the portfolio strongly reflects a portfolio manager's personal preference and mood.
2. concentrated investment
   - given the limited time and energy available to cover companies in depth, only a handful of names can be picked from an investment universe.
3. bias due to past experience
   - a portfolio manager makes decisions based on personal experience.
4. herding effects
   - a fundamental portfolio manager may be more likely to add a "hot" stock to her portfolio.

When a fundamental portfolio achieves outperformance, it can easily be attributed to alpha or stock-picking skills. But when the portfolio underperforms relative to the market or peers, the limitations listed above are realized. Some fundamental investment professionals started to incorporate quant practices such as risk calibration and management, investment process, and factor backtesting, to add rigor and breadth to their portfolios.

Risk Calibration and Management The transition from the fundamental to the quantamental approach started from using risk models, as fundamental portfolio managers need to know the risk exposure of their portfolios. Since many risk vendors use alpha factors in their risk models, fundamental portfolio managers realized that those risk models are not just for risk control, but may actually add value to their portfolio performance.

Screening Process This is the stone-age method of quant analysis. However, a well-thought screening process can be very profound, as the process is usually nonlinear in nature, in the sense of capturing nonlinear factor modeling and portfolio construction. For example, a portfolio manager for a US small-cap value fund can craft her screening process as follows:

(1) Select stocks with certain liquidity criteria.
(2) For those stocks that meet liquidity criteria, screen them for quality, where quality can be defined in many ways such as financial strength.
(3) Among the stocks that pass step 2, select value stocks with potential business growth.
(4) Select the stocks with high momentum, and decide weights.

Backtesting a Factor Analyze historical data to confirm or reject subjective judgments, which are approximated by factors. This enables fundamental investment professionals to observe and experience a large number of securities by adopting factor analysis.

It seemed all natural, necessary, and beneficial to add quant elements to a fundamental approach or vice versa.

9.3.2 Development

Quantamental investment emerged in the late 1990s and early 2000s, when many professionals realized the shortcomings of both quant and fundamental approaches. During this period, these professional investors stressed the importance of incorporating the other approach, and some even added some colors of quantamental approaches here and there. However, not many funds were employing a systematic quantamental approach for investments.

It was during the 2008 financial crisis that people, especially in quantitative investment, realized the importance of the quantamental approach.

Quantamental investment developed rapidly after the financial crisis in the 2010s. Many investment strategies are inked with the name "quantamental." There are many asset management teams named with "quantamental."

9.4 Quantamental Approach: How to Conduct Quantamental Investment

In the previous sections, we defined quantamental investment and reviewed the development of this frontier approach. Next, we focus on how to conduct quantamental investment. We first introduce principles of quantamental investment, then illustrate how to follow these principles in real investments. In practice, not many investors have the chance to start fresh with quantamental investment. Rather, many investment strategies and teams have been either quant or fundamental for many years. If those teams want to adopt a quantamental approach, what is the proper way to do so? We explore this practical question in the second subsection.

9.4.1 Principles of Quantamental Investment

There are many ways to practice quantamental investment. For example, there can be various degrees of combination between quant and fundamental, and there can be many places in the investment process where quant and fundamental are combined. These can all be regarded as quantamental investment. However, they may have very different characteristics. Regardless of their differences, to be quantamental, they all need to follow some key principles:

- Driven by fundamentals
  - ensure depth and avoid data mining
- Proven by data
  - ensure breadth and avoid personal bias
- Sound investment process
  - ensure transparency and consistency

For example, following the three principles above, a quantamental investment strategy can be built with the following practices:

(1) factors and model: built on fundamental intuition, supported by historical data, and reflects prospective information
(2) experience: supported by historical backtesting
(3) portfolio construction: alpha with prospective company-specific information and explicit risk management
(4) strategy: performance achieved through an investment process
A portfolio derived from these practices is a quantamental portfolio. It has both depth and breadth.

Next, we propose an intuitive approach that incorporates fundamental and quantitative analysis to construct a quantamental portfolio. Each step of the investment process is based on a solid foundation. In the first step, the factors are selected based on the driving force or characterization of the soundness of companies' business and their potential performance in the market. Thus, the fundamental analysis contributes greatly during this step. A rigorous factor selection process helps the model avoid data mining issues in the second step, which is often claimed as a serious problem in many quantitative approaches. The next step is learning how to utilize the information contained in the selected factors in the first step to deliver a powerful forecast. While an econometric model will accomplish this objective in a systematic and consistent way, building such a model is not an easy task. One challenge is that the model is nonlinear in factors due to the interaction and nonlinear effects of factors. To capture the true nonlinear relationship between factors, and finally the nonlinear relationship between factors and returns, requires a good understanding of not only the factors but also the statistical model. Another potential issue for model building that is often overlooked by many quantitative modelers is the estimation and interpretation of parameters. We should keep in mind that the model is built on a lot of conditions and assumptions; hence, the validity of estimates and forecasts depends heavily on how such conditions and assumptions hold. Only careful consideration of these issues will ensure the benefits of a hybrid model. Thus, our hybrid approach to investing is not at odds with either the fundamental or quantitative approach. Rather, we take advantage of the strengths of both approaches.

9.4.2 Quantamental Approach: Alpha, Risk, and Investment Process

We now discuss in detail how to proceed with a quantamental approach focusing on alpha exploration, risk management, and the investment process.

Consider alpha exploration first. The alpha sources for quant and fundamental approaches are different. They have low correlations and thus can potentially yield much added value.[3] We summarize alpha sources for fundamental and quantitative approaches in Table 9.4. We show in detail in the next section how to combine the two sets of alpha sources in practice to formulate a quantamental strategy, including

Table 9.4 Low correlation of alpha sources between fundamental and quant approaches

Fundamental	Quantitative
Company specifics	Factors
Forward-looking opinions	Backward filled observations
Many events data with scenarios	Few events data
Survey data, self-collected	Standard data vendor
Discretionary, category	Continuous

[3]Recall from Chap. 5 the IPRAE criteria for evaluating a factor: intuitive, predictive, robust, add value, executable.

how to transform company specifics into a factor, enrich a factor with fundamental specifics, and/or add information from the two approaches directly.

Now consider risk. For a fundamental portfolio, risk is usually linked to the margin of safety in the investor's mind, whereas for a quantitative portfolio, risk is measured by standard deviation or volatility. It is therefore easy to understand that fundamental portfolios in general have higher tracking error, and quantitative portfolios generally do not focus on managing loss aversion. Intuitively, the margin of safety makes more sense from a money management perspective. One quantamental approach is to find a mathematical expression of the margin of safety and incorporate that measurement into portfolio construction. This can be achieved by conditional value at risk, a coherent quantitative risk measurement, which we discussed in detail in the previous chapter.

Regarding investment process, it is a strength of quantitative strategies. Quantamental investments can thus adopt an investment process from quantitative approaches but ensure transparency and incorporate fundamental insights.

9.4.3 Paths from Quant or Fundamental to Quantamental

Many investment teams are either quant or fundamental, and their investment strategies have been in place for a long time. In practice, how can such teams and strategies transition from quant or fundamental to quantamental? We present two paths below corresponding to a quantitative approach and a fundamental approach, respectively.

From Quant to Quantamental Quant with fundamental inputs: prospective, insights, and depth.

- A quant team with a quantitative investment process
- Add and strengthen fundamental inputs
  - Alpha model with fundamental insights; visit companies; add depth
  - Add prospective information: survey; leverage sell-side professional expertise
  - PM, events, meet with CEOs; add connection and depth
  - Hire fundamental analysts or a fundamental PM, but not both

From Fundamental to Quantamental Fundamental with quant inputs: data, process, and breadth.

- A fundamental team with fundamental analysts and PMs
- Add more colors of quant
  - Start from a risk model, calibrate risk exposure, add diversity for risk management
  - Backtest factors and investment methods: add more holdings for alpha breadth
  - Add an investment process, more disciplined
  - Optimization for validation

– Hire a quant analyst and data analyst or a quant PM, but not both

The practical path from either a quant or fundamental to quantamental is to keep the original advantage, add strength from the other approach gradually, and eventually arrive at a true quantamental approach.

9.5 Quantamental Investment: Two Examples

In this section, we show by two examples how to conduct a quantamental investment. Example 1 discusses special items and focuses on how a fundamental portfolio manager can transform company specifics into a factor and thus leverage the merits of quant to add breadth to her portfolio. Example 2 explores the classic challenge—how to evaluate management—but now with a combined approach of both forward-looking fundamental information and historical-data-proven quant factors.

9.5.1 Example 1: Special Items, Transitioning from Fundamental to Quant

The term "special items" is an accounting item that can be used to attribute a company's earnings. A fundamental portfolio manager (FPM) follows companies in her portfolio and knows which companies play creative accounting in their earnings reports. She either avoids such companies in a long-only portfolio or places them in short positions in a long-short portfolio.

The FPM observes that in recent years, more and more companies report negative special items, and she understands very well that some companies do this to dilute negative earnings by

(1) placing a significant portfolio of negative earnings in special items, trying to alleviate negative impacts of business performance through creative accounting.
(2) dividing the negative special items across several fiscal quarters, making the negative amount seem smaller and less noticeable.
(3) allocating more negative special items to the fourth fiscal quarter because the fourth fiscal quarter and fiscal annual results are usually released on the same date, and investors pay more attention to the annual report than to the 4th-quarter report.

She also understands that more often than not, companies with a series of reports of negative special items underperform in stock markets.

However, does the above hold true for other companies in general? Perhaps it would be useful to validate the above understanding and quantify the effects on stock returns by historical data. So the FPM decides to investigate the issue in a systematic way: collect data on special items, analyze (1)–(4) above, connect

Table 9.5 Average percentage of negative, positive, and zero special items by fiscal quarter. Data source: CompuStat

Fiscal quarter	Special items<0	Special items>0	Special items=0
Q1	28%	9%	63%
Q2	32%	10%	58%
Q3	33%	10%	57%
Q4	42%	12%	46%

them with companies' stock prices, identify any patterns, and explore mispricing opportunities to be used in her portfolio.

Quarterly data are available for special items in CompuStat, an industrial data vendor of financial reports. Using monthly data from December 1997 to May 2013 for the combined universe of the S&P/TSX Composite and MSCI USA, the FPM presents the time series data from 2006 to 2013 (Fig. 6.4) which show that indeed there is an increasing trend of special items reporting, especially negative values. Moreover, based on the historical data, the FPM finds that there is clearly a seasonal pattern, as shown in Table 9.5: There are slightly more companies reporting special items during the fourth quarter than any other quarter. This validates the FPM's fundamental understanding: moving towards the earnings announcement date for the fiscal year end, investors pay more attention to annual financial performance results (and possibly new fiscal year guidance) than fourth quarter reports. Therefore, management has a tendency to allocate more negative special items to the fourth fiscal quarter than any other.

Most importantly, the FPM finds that if companies report negative special items during a quarter, they are more likely to do so in the following quarter. This especially holds true for the most negative special items, as the persistence is the most pronounced there. Based on the data available, the FPM constructs a signal for special items with a focus on the left tail:

$$SPI = \frac{1}{8}\sum_{t=1}^{8} \text{sign}(\text{special items}_t)$$

$$SPIleft = SPI \times I(SPI < \text{quantile}(SPI < 0, 0.30)).$$

In other words, the SPI factor is based on the sum of the signs of special items for the past eight quarters. The SPIleft signal is then derived from SPI by using the left tail (30%) of the negative part of the distribution.[4]

The binary definition of SPIleft captures sign persistency of negative special items. To validate this special feature, the FPM divides SPI values into three groups for negative and positive parts, respectively. SPIleft is the most negative third. The

[4]The value of SPI is in the range of $[-1, 1]$.

Table 9.6 The transition probability (%) of SPI across categories between adjacent quarters

Category	−3(SPIleft)	−2	−1	0	1	2	3
−3(SPIleft)	87.52	11.98	0.15	0.35	0	0	0
−2	12.35	70.29	15.96	1.34	0.02	0.03	0
−1	0.13	17.54	68.03	12.69	1.1	0.48	0.04
0	0.37	2.22	11.71	78.65	4.02	2.21	0.83
1	0	0.17	6.47	25.04	50.54	14.19	3.59
2	0	0.1	0.93	14.5	23.35	46.66	14.45
3	0	0.2	0	5.76	3.59	20.88	69.57

transition table (Table 9.6) shows that a company in SPIleft (the value −3) in a given quarter has an 87% chance of remaining in the same bracket the following quarter.

To see how SPIleft is related to future stock returns, the FPM conducts a backtest study. For each firm, SPIleft is constructed using information from the most recent reported quarter. The announcement date is forecasted by adding ninety calendar days to the last fiscal quarter's financial reporting date. For the sake of simplicity, the portfolio is constructed with equal weights for all the companies in the SPIleft bracket. The plots in Fig. 9.1 display the cumulative stock returns of a portfolio based on the present point-in-time SPIleft values. Stock returns are demeaned across countries and sectors. The portfolio is constructed thirty days prior to the forecasted earnings announcement dates and held thirty days after. For the North American region, the average cumulative country/sector demeaned return is −35bp leading up to the announcement date (the top plot). The results for Canada alone are even more striking, yielding nearly 100bp underperformance (the bottom plot). Furthermore, the underperformance continues for about 5–7 days after the earnings announcement date, reaching 150bp in total.

Next, to see how the new factor performs over time, the FPM constructs a time series of returns for a portfolio formulated one month prior to the forecasted announcement date for the companies in SPIleft. Portfolio performance is presented in Fig. 9.2, demonstrating periods of underperformance in North America (the top plot) and more consistent underperformance in Canada (the bottom plot). In North America, the underperformance periods occur after the tech-bubble burst in early 2000 to mid-2002, 2004–2006, and 2008–2009. During the most recent period (2010–2013), returns have been flat for both North America and Canada.

The returns of portfolios and benchmarks are summarized in Table 9.7 for the investment universe in Canada and North America, respectively. The summary statistics include annualized return, risk, and the Sharpe ratio. The FPM presents both total returns and country/sector demeaned returns. The benchmark includes both the equal weighted and market capitalization weighted versions. In the Canadian equity market, a portfolio based on SPIleft yields an annualized return of −1.25%, while the equal weighted benchmark return is 7.88%, which confirms that companies with consistent negative special items will underperform their peers. This is further shown by the demeaned return of −7.20% with a comparable risk level as a benchmark,

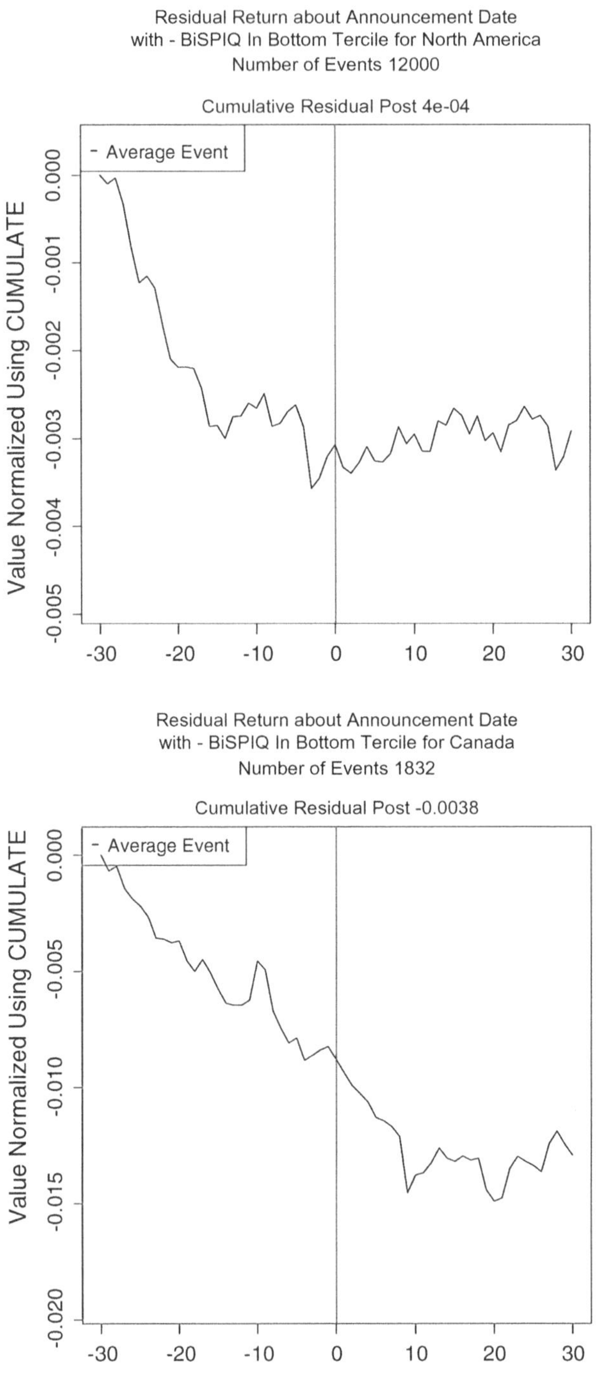

Fig. 9.1 The forecasting power of SPIleft for stock returns around reporting dates: the top plot is for the NA region and the bottom plot is for Canada

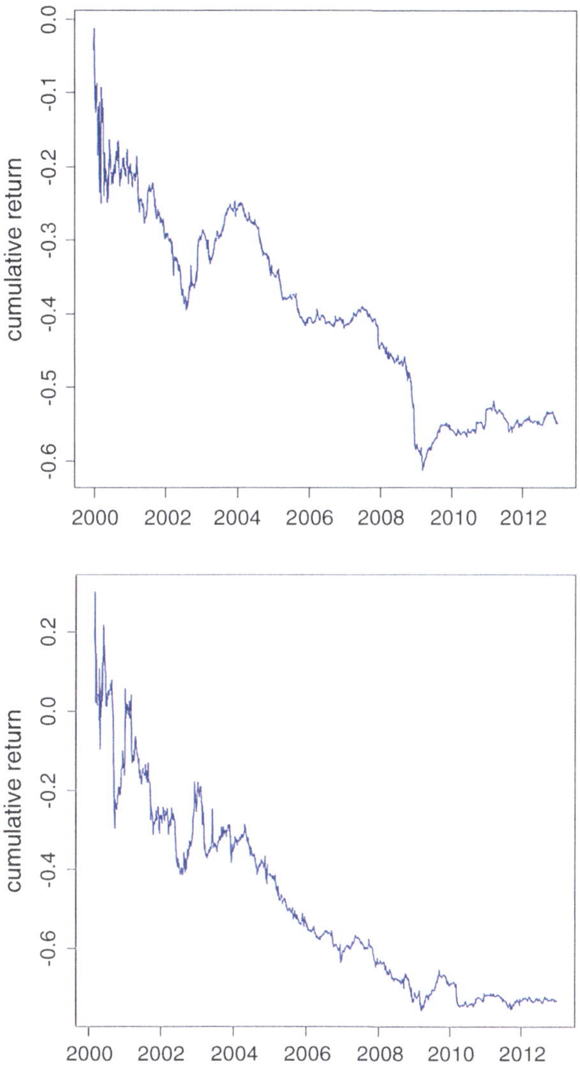

Fig. 9.2 Portfolio performance of SPIleft over time: the top plot is for the NA region and the bottom plot is for Canada

thus indicating a profitable strategy based on the SPIleft signal. A strategy could potentially be formulated by shorting the SPIleft portfolio while hedging out the benchmark performance. In the case of the North American universe (now with the addition of the US equity market), the SPIleft strategy still has the potential for significant profitability but with lower returns and risk levels.

The fundamental portfolio manager has conducted a quantitative analysis to validate her understanding of special items. The new factor SPIleft, based on a deep understanding of fundamentals, can now be applied to hundreds or thousands

Table 9.7 Performance summary for the SPIleft portfolio in Canada (S&PTSX/Composite, top) and NA (MSCI USA and S&PTSX/Composite, bottom)

	Portfolio total	Portfolio demeaned	Benchmark equal weight	Benchmark cap weight
Canada				
Annualized return	−1.25%	−7.20%	7.88%	6.84
Annualized std	25.87%	16.00%	18.20%	16.42
Ratio	−0.05	−0.45	0.43	0.42
NA				
Annualized return	−0.01%	−4.27%	8.86%	4.93
Annualized std	25.51%	7.26%	19.96%	17.39
Ratio	−0.00	−0.59	0.44	0.28

of stocks, thus adding tremendous breadth to her portfolio. Indeed, a quantamental approach is born.

9.5.2 *Example 2: Management Quality Assessment with Both Quant and Fundamental Approaches*

For any public company, its senior management team is critical for its business and stock market performance. A reasonable and fair evaluation of the senior management team is always an important part of portfolio construction. People may think that business performance metrics, such as profits, projects, financial strength, etc., already reflect the management quality. This is correct. However, these are just the results we observe as consequences of past decisions. A senior management team usually has a stable temperament and management style. It is the changes in the future the senior management team will make that will be important for future stock prices. So, from this perspective, evaluating the senior management team is very important. Warren Buffett has an excellent discussion on this:

> When we own portions of outstanding businesses with outstanding managements, our favorite holding period is forever... Once management shows itself insensitive to the interests of owners, shareholders will suffer a long time from the price/value ratio afforded their stock (relative to other stocks), no matter what assurances management gives that the value-diluting action taken was a one-of-a-kind event. – Warren Buffett

Regarding management quality, fundamental and quantitative investments have very different assessment methods (Table 9.8). Fundamental professionals collect information through meetings and due diligence, such as company visits. They also leverage sell-side analysts and other network connections. Fundamental professionals usually try to identify good teams and then overweight the associated companies directly or indirectly in their portfolio. In contrast, quantitative investment tries to identify bad teams. We learned from Chap. 4 that management quality can be

Table 9.8 Fundamental and quantitative approaches to evaluate management quality

	Fundamental	**Quant**
Focus	Identify good CEO	Identify bad CEO
Impacts on portfolio	Overweight	Underweight
Horizon	Short to medium	Long
Methods	Meetings with senior mgmt team	Capital expenditure
	Company visits	External financing
	Sell-side information	Asset growth
	Network connections	Board independence
Data	Description, scoring	Continuous factor values

quantified: how does a senior management team spend money (capital expenditure), and where does it acquire money (external financing)? There are other quant factors along this line such as frequency and magnitude of acquisitions and board dependence. Quantitative factors are built with time series data to measure changes and then compared across companies. Based on a large amount of historical data, the quantitative approach finds that companies with much greater capital expenditure, excessive external financing, and accelerated asset growth are associated with downward stock prices.

Apparently, for management team evaluation, there is little overlap between information from quant and fundamental approaches. Quantamental investments can use both directly.

$$\begin{aligned} MQ_Q &= f(MQ_F, MQ_C) \\ &= MQ_F + MQ_C, \end{aligned} \tag{9.3}$$

where MQ_F is for quant (factor) and MQ_C is for fundamental (company). Equation (9.3) specifies that the combination of quant and fundamental approaches can take the simplest form: A linear addition of quant measurement (MQ_F) and fundamental assessment (MQ_C) will add value by identifying both good and bad teams at the same time. Moreover, the fundamental sources of management evaluation can help portfolio managers make discrete decisions about corporate events when managing live portfolios on a daily basis.

By using the quantamental approach, we can achieve both breadth and depth in the evaluation of management quality.

9.6 Quantamental Investment: Mentality, Team, and Culture

A successful quantamental investment strategy requires many components. Here we focus on four key elements: mentality, approach, team, and culture. We have discussed approach in detail in previous sections. Now we focus on the other three elements:

mentality, team, and culture. Although not related to investment and portfolio directly, they impact quantamental performance both on a daily basis and in a long-term and strategic sense.

Quantamental Mentality It is very important to avoid bias towards either a quant or fundamental approach. In simple terms, we should be able to leverage modern science (econometrics), tools (computer), and data to build an investment strategy we could explain with simple words to our grandmas.

While a quantamental mentality is the part of an iceberg under the ocean, a quantamental approach is the part above the ocean. This is how to realize a quantamental mentality in an investment process. Starting from fundamental insights, we build factors and models and derive a portfolio through an investment process. The quantamental approach produces quantamental products.

Quantamental Team Building a quantamental team is critical because the quality of a team determines the success of quantamental strategies. This is the organizational guarantee for quantamental investment. A quantamental team should have professionals with both fundamental experience and quantitative skills. It is not effective to simply add quantitative professionals and fundamental portfolio managers/analysts together.

First, the head of the team is critical. She herself should be a strong believer in combining quant and fundamental approaches. She is the one who has overall responsibility for quantamental strategies. Moreover, she will implement the combination through the organization chart, team building, and culture cultivation.

Second, the team should be built with the appropriate blend of quant and fundamental expertise. The position level should be based on overall quantamental expertise. There should NOT be a division of quant and fundamental professionals in terms of positions and functions. For example, regarding direct investment professionals, the mid-level positions can include research manager, portfolio manager, trading manager, while senior level positions can include director of research, director of portfolio management, director of trading.

For supporting positions such as data, IT, or product specialists, the basic requirements are described as follows:

IT + data: capable of a quant framework, database, GUI, optimizer settings; deal with dirty and non-traditional data sets;

Marketing/product team: should be able to articulate the major characteristics of quantamental approaches and products.

Quantamental Culture A strong quantamental culture is very important in the long run to reap the benefits of quant and fundamental approaches, not the shortcomings of both. Without a good culture, the team can be weak or dissolve quickly, and portfolio performance can decline without the confidence to improve. It takes a long time to build a quantamental culture, but the benefits can also last for a long time.

The culture should be driven by portfolio performance with a team approach. For example, there should be opportunities for quant and fundamental development, an appropriate level of transparency, teamwork with individual responsibilities, etc. Compensation should be based on performance, with quant and fundamental expertise treated the same.

Together, mentality, approach, team, and culture constitute quantamental investment. Among the four elements, the quantamental approach is the central.

9.7 Industry Insights: A Quantamental Japanese Stock Selection Strategy

To build a stock selection portfolio in Japan, we first build an investment universe close to the MSCI Japan by using rules similar to the MSCI World, an index of developed equity markets. Note that the MSCI Japan is a large cap investment universe representing about 85% of total Japanese stock market capitalization. The plots in Fig. 9.3 show the number of stocks in the universe and the top seven industries by number of companies and weight from Dec 31, 1997 to Mar 31, 2012. There are about 350 stocks in the universe, which is broad enough for quantamental analysis. As also shown in Fig. 9.4, the top five industries in terms of both number of companies and index weight in the universe are capital goods, banking, computers, automobile components, and consumer durables. Together, they represent about 50% of the total weight, which approximately reflects the industry structure of Japan.

From a long historical and global perspective, we present the number of companies and weights of the MSCI Japan in the MSCI World from 1992 to 2012 (Fig. 3.2 in

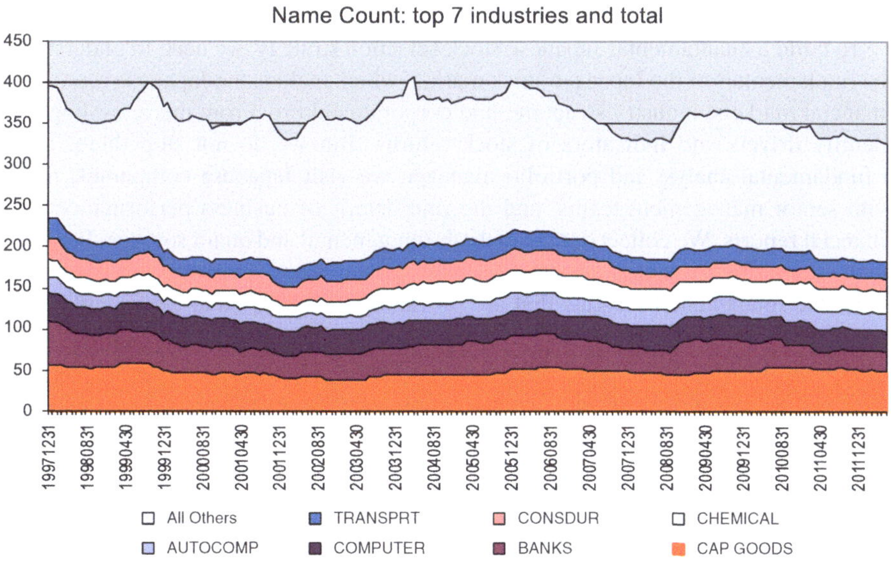

Fig. 9.3 The number of stocks in the large-cap Japanese stock market with in the seven industries from 1997 to 2012

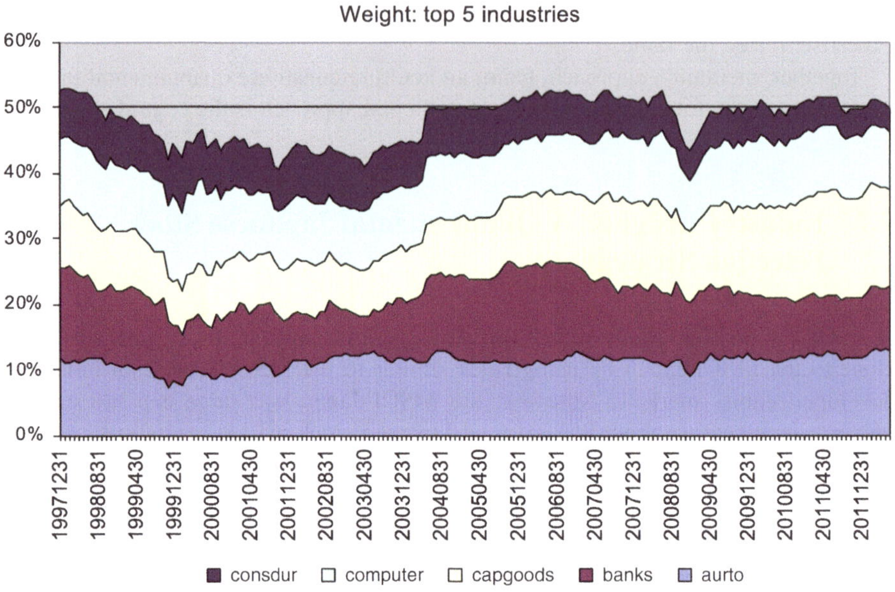

Fig. 9.4 The cap weights of the top five industries in the Japanese stock market from 1997 to 2012

Chap. 3). We see that in merely twenty years, from the 1990s to the 2010s, the number of stocks dropped from 800 to 300, and the weight shrank from 40% to about 8%. This reflects the fact that the Japanese economy and stock market were not growing at the same pace as the rest of the world, especially the USA.

To build a quantamental Japanese stock selection strategy, we need to understand the fundamentals of the Japanese stock market, which include the Japanese economy, financial markets, industry structure, and corporate culture. From there, we hope to identify drivers and indicators of stock returns. But we do not stop there. Like a fundamental analyst and portfolio manager, we visit Japanese companies, meet with senior management teams, and dig into details of business performance and financial reports. We collect data from both fundamental and quant sources. We then apply statistical models to backtest the signals and ideas. Once we are confident about the fundamental insights that are proven by historical data, we formulate an investment process to incorporate both historical and forward-looking information into the portfolio. Thus, we construct a quantamental strategy.

Given the introductory nature of this chapter on quantamental investment, we describe only the framework of fundamental insights for the stock selection strategy in Japan. We discuss each item briefly and then present a quantamental model of a Japanese stock selection strategy.

The Japanese stock market is unique. A good understanding of this market is the first step for any investment. We present below some major attributes in the context of quantamental investment. We discuss each item in detail in the next section.

1. Economic growth and development: stagnation after the 1990s
    - Sluggish economy
    - Inflation and deflation
    - Heavy debt
    - Pension burden due to the aging population
2. Dynamics and chaos in financial markets
    - Equity market
    - Bond market
    - Currency
3. Rationales for a stock selection strategy
    - Macro-level
        - Implications of the current regime for a stock selection strategy
        - Continuity of the current regime in the near future
        - Possibility of the regime change: degree and direction
    - Micro-level
        - Company structure
        - Accounting practices
        - Market participants
4. Quantamental approach for stock selection in the Japanese stock market
    - Fundamental: depth
    - Factor: breadth
    - Investment process: robustness and transparency

9.7.1 Economic Growth and Development: Stagnation After the 1990s

We discuss the economic situation of Japan in terms of GDP growth, inflation, unemployment, debt ratio, and the serious aging issue facing the country. All of these have important impacts on the stock market and companies' performance.

Sluggish Economy Since 1990 After the real estate bubble burst in Japan in the early 1990s, the Japanese economy entered a new era. The GDP growth rate dropped from 5% to about 1%, and unemployment increased from 2% to about 4% at the same time. Among the many factors that contributed to the sluggish economy, the key factor was Japan's loss of competitive edge in the applied technology industries. Of course, other factors such as Japanese currency appreciation and the bad-debt financial system made things even worse. The Japanese economy is heavily export-oriented: the GDP growth rate depends 60% on exportation and 40% on domestic

Table 9.9 Japanese industry trend

Flagship industry	Status	Trend	Competitors
Electronics	Traditional	Down	Korea, China
Automobile	Traditional	Down	Korea, the USA, Germany
Resources	Traditional	Down	China, resource-nationalization countries
Capital goods	Traditional	Stable	
Healthcare	New	Up	Mostly domestic
Energy	New	Up	The USA and others

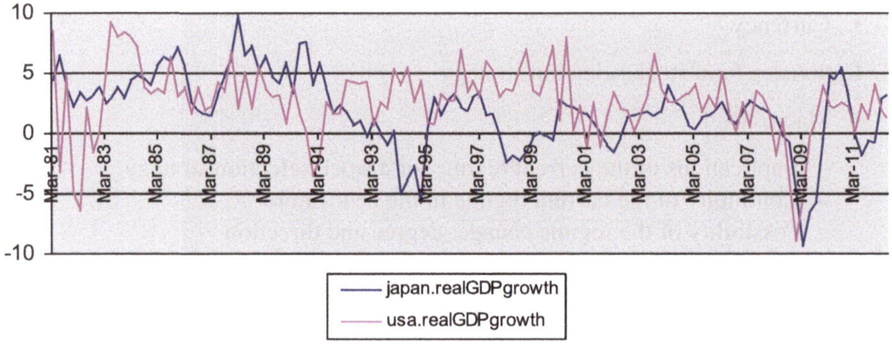

Fig. 9.5 Annual GDP growth rates of Japan and the USA from 1982 to 2012

products. During the last twenty years, on one side, the primary engine of Japanese economic growth—the applied technology industry—has been losing market share, and margins have declined dramatically; on the other hand, a new growth engine has not yet emerged or is only beginning to emerge. Table 9.9 lists Japanese "growth engine" industries and their current trends.

We present the real GDP growth rate for Japan from 1982 to 2012 in Fig. 9.5. For comparison purposes, we also include the US numbers in the plot. We see that the Japanese economy follows the USA on the downside more than the upside after 1991.

We collect data on the Japanese unemployment rate from 1953 to 2012 and plot the annual unemployment rate in Fig. 9.6. We see that the Japanese unemployment rate stayed at about 1–3% for forty years, from 1953 to 1993, then increased steadily to above 5% in 2003, and has remained at the current level of 4–6%. The low unemployment rate before 1993 was largely due to Japanese companies' loyal and lifelong employment practices and the corporate culture during that period, when the economy was very good (GDP growth rate was about 5%). After 1993, the GDP growth rate fell below zero, that is, the economy was in contraction, which made layoffs necessary for companies to reduce costs and survive.

Deflation Unfortunately, the low GDP growth rate has been accompanied by a low inflation rate (Fig. 9.7), the combination of which is called stagnation. The deflation caused huge problems and led to widespread defaults, which made the operation of

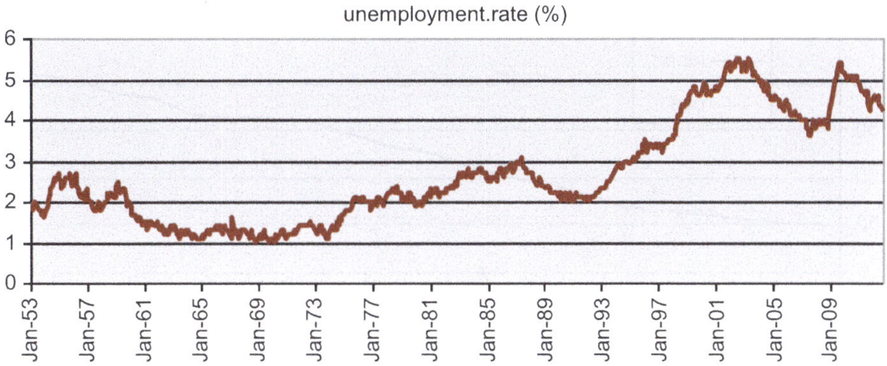

Fig. 9.6 Annual unemployment rate of Japan from 1953 to 2012

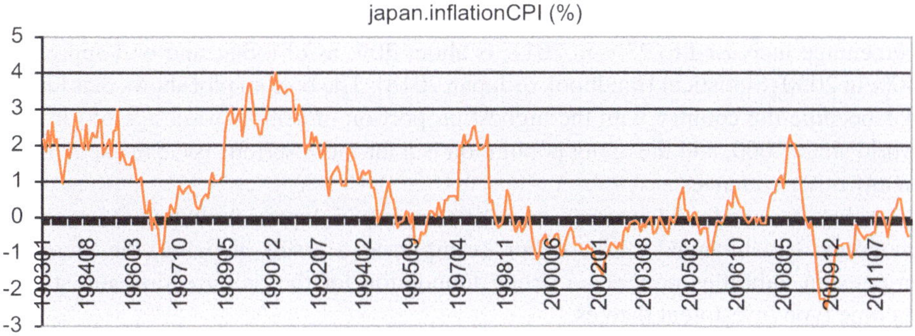

Fig. 9.7 The deflation rate of Japan based on CPI from 1982 to 2012

the banking system very difficult. Fortunately, reforms in the financial sector were successful, which paved a healthy foundation for potential economic growth.

Heavy Debt We use the ratio of debt to GDP to measure debt severity. We present plots for both Japan and the USA in Fig. 9.8. We see that in 1992, the debt ratio was the same, about 50%, for both countries, but 10 years later, it quickly increased to 200% in Japan and only 100% in the USA. Heavy debt was an issue for the Japanese economic structure and hindered its economic growth, but it was not as serious as in some other countries, such as Greece and Italy. This is because Japan has large holdings of overseas assets. Note that the debt level is also related to the political and social systems.

Pension Burden The aging population is a unique feature of Japan and imposes a heavy burden on the pension system to support seniors (aged over 65). We present two plots in Fig. 9.9: the top plot shows the aging pattern in Japan in 1950, 2011, and 2050 (projection); the bottom plot displays the proportion of seniors in the population by country from 1950 to 2050. We see from the top plot that the Japanese age structure looked like a pyramid in 1950 and then changed gradually to the shape

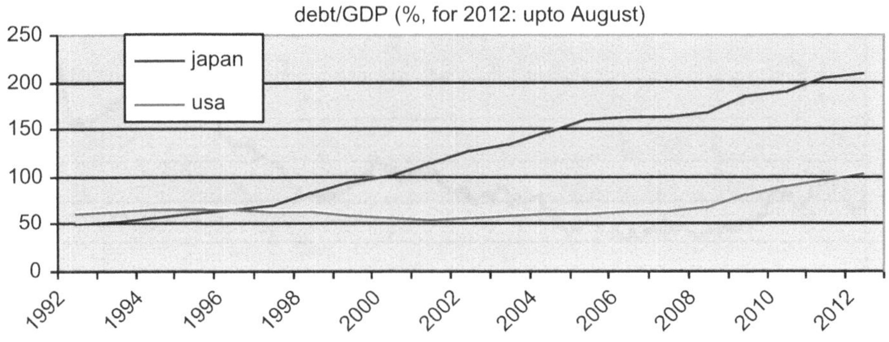

Fig. 9.8 Ratio of debt to GDP of Japan and the USA from 1992 to 2012

of an ice cream. In 1950, only about 5% of the Japanese population was over 65. That percentage increased to 23% in 2011, is about 30% as of today, and will approach 40% in 2050 (Statistical Handbook of Japan 2018). The bottom plot shows that Japan has become the country with the highest proportion of seniors (over age 65) in the world since 2000, and the aging population is a far more serious issue in Japan than in any other countries.

Aging has profound impacts on every aspect of Japan, including its society, economy, and financial markets. For example, as a result of aging, the increase in pension liabilities provides a strong foundation for the risk aversion and stable income type investment themes.

9.7.2 *Dynamics and Chaos in Financial Markets*

Japan's economic situation over the last twenty years is clearly reflected in its financial markets. During this period, Japanese financial markets can be characterized by a chaotic equity market, low-yield bond market, and appreciating currency market. We present the price movements for each market in Fig. 9.10. Note that depending on data availability, we use different periods for the three markets. The top plot is for the equity market, the middle plot is for the bond market (10-year government bond yield), and the bottom plot is for currency movements against the USD. We discuss each market briefly.

Equity Market From 1997 to 2012, the Japanese stock market was very volatile, with significantly low returns compared with the US stock market. We calculate annualized return and risk with two kinds of weighting schemes: equal and market cap. For comparison purposes, we include the same metrics for the large-cap stocks of the US market. We present a summary of Japanese and US stock market performance in Table 9.10. We see that during the period from Dec 31, 1997 to June 30, 2012, the annualized return of the Japanese stock market was −1.60% for the cap-weighted

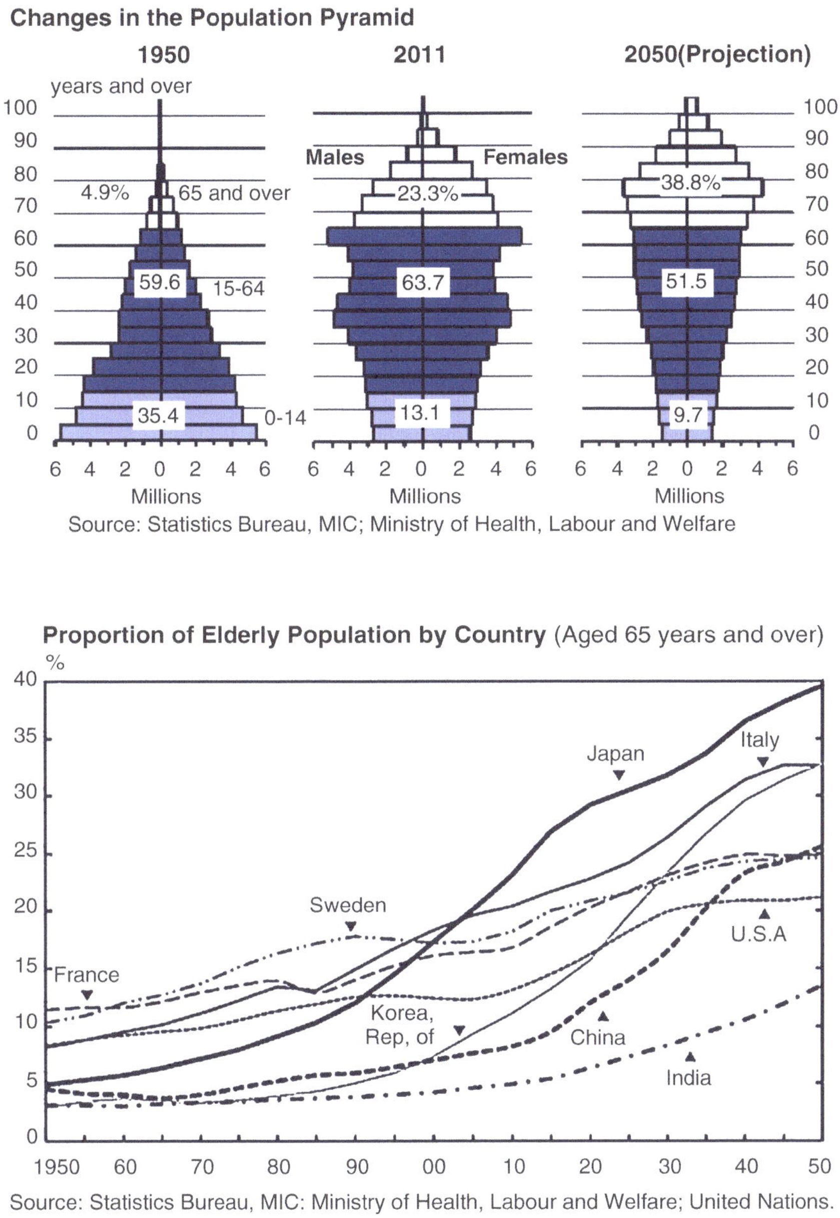

Fig. 9.9 Population age structure (top plot) and trend by country (bottom plot) from 1950 to 2050. Data source: Statistics Bureau, MIC; Ministry of Health, Labour and Welfare

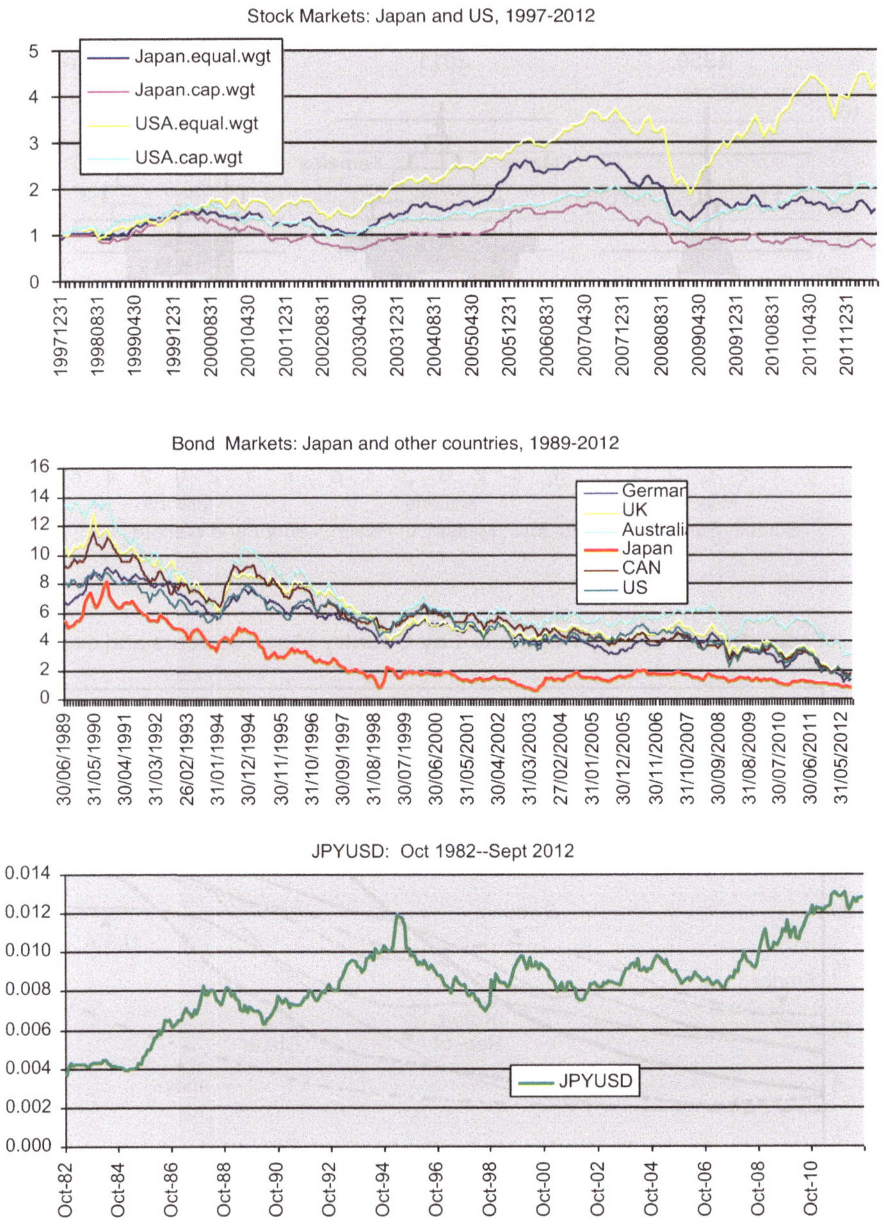

Fig. 9.10 The Japanese stock market (top plot), bond market (middle plot), and currency movements (bottom plot)

Table 9.10 Annualized performance of the Japanese and US stock markets, 1997–2012. The words "equal" and "cap" indicate the performance is equal or market cap weighted

	Japan equal	Japan cap	US equal	US cap
Annualized return	3.12	−1.60	10.52	5.07
Annualized risk	17.69	18.51	18.47	16.53
Sharpe ratio	0.18	−0.09	0.57	0.31

Table 9.11 Annualized performance of JPY/USD, 1982–2012

	Currency JPY/USD, 1982–2012
Annualized return	4.32
Annualized risk	11.44
Sharpe ratio	0.38

version and 3.12% for the equal-weighted version. The values are 5.07% and 10.52% for the US stock market, indicating a very chaotic stock market in Japan during this period. In both markets, small stocks performed better than large stocks, and the Japanese stock market had a somewhat higher volatility among large-cap stocks. In terms of Sharpe ratios—risk-adjusted returns—the Japanese stock market had a tough period and underperformed dramatically relative to the US stock market. This is a reflection of Japanese economic stagnation during this period.

Bond Market Regarding bond markets, the Japanese ten-year government bond had the lowest yield compared to other developed countries from 1982 to 2012. It increased from 4–5% in the 1980s to 6–8% in the early 1990s and then dropped continuously thereafter for twenty years to less than 1% in 2012. Japan's low bond yield is related to several factors, such as low economic growth, low returns in equity markets, and heavy debts. Considering the latter, if yield increased by 1%, it would impose a heavy burden on the economy. On the other hand, the low yields of bond products do not imply high returns of stocks in general. In Japan, many seniors live on the income from bond yields. The bond yield and inflation both had to be kept low.

Currency The currency movements between JPY and USD depend on monetary and other government policies, such as international trade. In general, currency depreciation encourages exports. Currency returns can be very important for foreign investors.

$$R_{foreign} = (1 + R_{local})(1 + R_{currency}) - 1,$$

where R_{local} is the local return and $R_{currency}$ is the currency return.

The Japanese currency appreciated significantly between 1982 and 2012, about 4.32% every year (Table 9.11). One USD was worth about 250 JPY in the 1980s but only about 80 JPY in the 2010s. The Japanese Yen's appreciation had very negative impacts on the international trade of Japan with the USA: it discouraged exports and encouraged imports. In terms of investing in the Japanese market, investing with US currency without any currency hedging would add 4.32% return each year, just because of the change in the exchange rates between USD and JPY!

9.7.3 Rationale for a Stock Selection Strategy in Japan

To construct a quantamental portfolio in Japan, we conduct quantamental analysis at both the macro- and micro-levels. At the macro-level, we analyze various aspects of the current economic regime, which is significantly related to the stock market, and explore their implications for stock performance. We also discuss potential regime changes and their impacts on the financial markets. At the micro-level, we summarize information on special features of Japanese company structure, typical accounting practices, and market participants.

Macro-Level

1. What are the implications of the current economic regime for a stock selection model? We summarize major features of the current regime, current trends, and a forecast of the future regime for the next 1–2 years in Fig. 9.11. We then explore their implications for the Japanese stock market, particularly from the stock selection perspective.
2. Will the current regime continue in the near future? For a short period (the next 1–2 years), the Japanese economy and monetary/fiscal instruments will likely continue along the same track (please see the column "next 1–2 years" in Fig. 9.11). However, in the long run, if the US economy is up, the Japanese economy will be in better shape given the influence of the US economy on Japan's export-oriented economic structure. However, as the Japanese economy depends more and more on the AP region, the improvement of the US economy will have diminishing impacts on Japan. The main issue with the Japanese economy is the lack of an engine for long-term growth coupled with short-term political turmoil and an aging problem. The transition from the old industry structure to a new one will most likely take a long time.
3. What if the regime changes dramatically? The financial reform was carried out successfully, which made the bad-debt less of an issue, although the reform process was very slow as is typical in the Japanese culture. Given that the aging issue has become more and more serious in Japan, any current or future reform will take an even longer time. If the economic regime does change dramatically, the financial markets will respond accordingly, and we need to adapt our investment strategies to the new regime. The following factors should be considered if we are to construct an adaptive model or macro-guided stock selection strategies for the Japanese stock market:

   i) political system
   ii) connection with the USA
   iii) connection with the AP region, especially China
   iv) fiscal policy
   v) monetary policy
   vi) banking system

	Current level/trend	Next 1-2 years	Impacts on equity market	Implications for stock seleciton alpha
Unemployment	[4, 5]%, highest since 1950s	stay the same level	negative	value, mean reversion
GDP growth	2012Q3, -0.9%	low, increase	depress the stock market	value in general, but some short-term momentum
Inflation	[-2, +2]%	low, stay in the track	deflation increases defaults	long-term growth
Currency	highest levels, 1USD=80JPY	stay the same till US is strong	JPY appreciation has negative impacts	currency overlay
Bond yield	low, <2% for a long time	stay low	neutral	investor seek dividend
Bad debt (structure)	less an issue now	better	foreign investment, bankig in better shape	firms are cash rich
Export orientaiton	shift from US to China and AP	volatile	sensitive	foreign sales growth less attractive
Aging	25% populaiton > 65 years old	increase	risk aversion, liability senstive	dividend, stability, value

Fig. 9.11 Features of the current Japanese economic regime and their implications for stock selection strategies

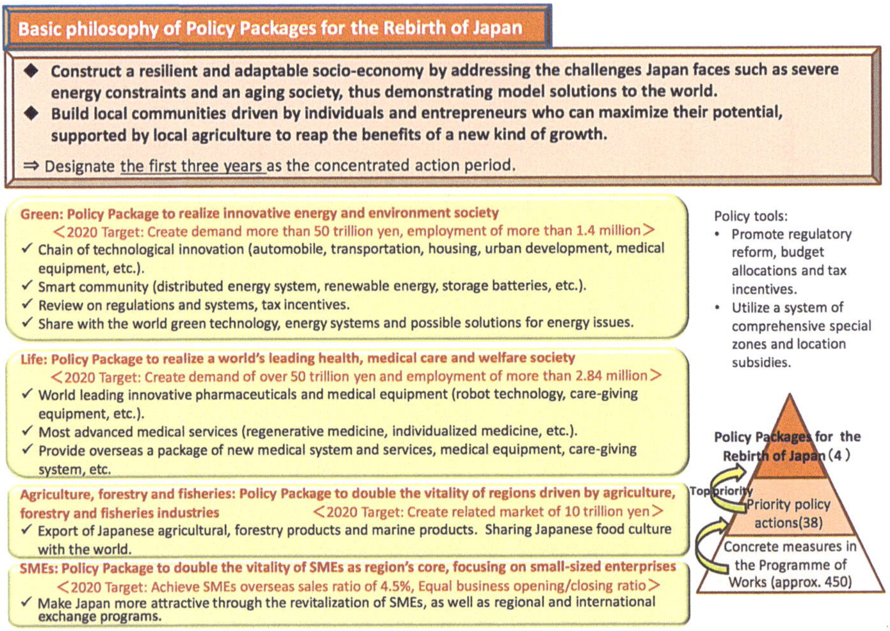

Fig. 9.12 The long-term targets and development plan set up by the National Policy Unit of Japan Cabinet. Data source: *Cabinet Decision: Rebirth of Japan* published on July 31, 2012

The following are the long-term targets and development plans from the National Policy Unit of Japan Cabinet (Cabinet Decision: Rebirth of Japan, July 31, 2012). The overall target is "overcoming deflation and medium- to long-term economic and fiscal management." The target economic growth rates and inflation rate are set:

- Nominal growth rate: 3%
- Real growth rate: 2%
- Inflation rate: 1%

Figure 9.12 lists plans and development targets across the Japanese social, economic, and political system (set up by the National Policy Unit of Japan Cabinet) Cabinet Decision: Rebirth of Japan (2012).

Micro-Level We now discuss microstructural aspects of the Japanese stock market. These special features at the micro-level have significant implications for the construction of a quantamental portfolio in Japan. Here, we present a summary based on our study of a quantamental portfolio. We leave the details for the readers to explore further, as each item will have different implications for different investment strategies.

1. Company structure

- Ownership: has developed from family-owned to family-controlled and now to modern corporate organization
- Governance: strong family/clan impacts, weak corporate governance
- Organization: a parent company with many subsidiaries
- Shareholders: inter-linked equity holdings among companies and subsidiaries
- Operation: focus on long-term business growth
- Finance: fostered special relationship with a lender bank

2. Business practices

- Accounting: follow Japan GAAP, which is similar to US GAAP
- Fiscal term: most firms have the fiscal year end on March 31
- Culture: conservative

  a) focus on strategic plans
  b) less creative accounting to inflate net income

- Dividend: many firms pay dividends but with low dividend payout ratios
- Value: geared towards long-term business growth, not shareholders' interests

3. Equity market participants

- equity stakes held by many other companies, interholding
- domestic institutional investors: conservative and risk-averse
- foreign investors: many are opportunistic and short term

For a quantamental portfolio in Japan, the depth comes from our deep understanding of the Japanese economy, industry, and companies. The breadth comes from parsimonious statistical modeling for a large number of stocks. The link between the two is the factors: Depth is realized through the contents of factors, while breadth is achieved by applying the factors to a large number of companies.

9.7.4 *Quantamental Approach: A Japanese Stock Selection Portfolio*

Based on the quantamental analysis above, we summarize the unique features of the Japanese stock market and hence the differences from the US stock market in Table 9.12. This not only explains how the Japanese stock market is a different animal but also identifies the unique drivers and indicators for stock returns in the Japanese stock market.

To build a quantamental portfolio, we follow the quantamental process specified in the previous section: Collect both historical data and forward-looking information; validate fundamental insights using data; process information through an investment process; and build a portfolio with reasonable constraints and robust risk management.

Table 9.12 Quantamental analysis of the Japanese stock market

Classical for the USA	Why not in Japan?	Structural reasons	What works in Japan
Cash is king	Rich in cash	Tradition	Firms without much cash
Price momentum	Sluggish market	Economic stagnation	Price reversal
Accruals	Less earnings manipulation	Family controlled	Operating results
Management quality	Loose governance	Connections	Long-term business growth
Profitability	Total sales not important	Competition edge matters	Profit margin
Analysts' estimates	Follow the crowd	Conservative culture	Extreme values

- Deep understanding
  - Macro-level: country, culture, industry, etc.
  - Micro-level: company, market microstructure, etc.
- Information
  - Prospective
  - Historical
- Factors
  - Realization of depth
  - Data proven
- Investment process
  - Alpha model
  - Risk management
  - Portfolio construction

In the following subsections, we briefly discuss each step towards building a quantamental stock selection portfolio in Japan.

9.7.4.1 Information: Historical and Prospective

The information collection and processing is done at different levels—the macro, industry, and company levels. Historical information is usually provided by industry data vendors. For example, for companies' financial statements, we can get historical data from CompuStat for the US companies and Worldscope for international companies. The prospective information is collected from various sources, such as professional forecasts for macroeconomic indicators, sell-side analysts, industry evaluations, company visits, and meetings at the industry and company levels.

Of course, we need to analyze both historical and prospective information and establish a consistent way to use it. For example, if the prospects for economic growth are not bright, we would not expect the overall stock market to grow as much, which would imply that necessary industries (such as pharmaceuticals and utilities) will outperform, high-beta stocks should be avoided, and value factors will outperform momentum factors. The same analysis can be applied at the industry and company levels, and these will all be incorporated into the portfolio.

9.7.4.2 Factors: Data Proven the Depth with Breadth

While a deep understanding of the market, industry, and companies adds depth to an investment, data availability makes large-scale statistical analysis possible. It is the factor building stage that we realize both depth and breadth—using data to validate deep understanding and achieve breadth.

Japan is a very different market from the USA: for a quantamental strategy to be successful, we need to understand how they are different as well as the reasons behind the differences. However, the potential causes and effects we identify should be validated by data. We consider the following signals for a quantamental stock selection strategy in the large cap universe of Japan:

- Cash: asymmetric; the less one is penalized but the more one is not rewarded
- EM: local analysts; survey, culture is conservative, only extremes are used
- MQ: board and family control are interconnected
- Margin: more important, reflects competitive edge
- Events sensitivity: earthquake, utilities, aging, etc.

We propose a linear alpha model as specified in Fig. 9.13. Note that many signals represent both historical and prospective information.

9.7.4.3 Investment Process

We have an investment process for a quantamental stock selection strategy, including information collection and usage, alpha building, risk management, and portfolio construction. A meaningful process adds robustness and transparency to the portfolio.

Regarding live portfolio management, immediate follow-up on events and reasonable investment judgment are crucial. But judgments should be based on both fundamental and quantitative analysis. For example, for management of domestic

Fig. 9.13 A linear quantamental Japanese stock selection model

theme name	signal name	signal wgt(%)	theme wgt(%)
VALUE			35
	B2P	70	
	DIV2P	20	
	CFO2P	10	
PVM			20
	PM1m	75	
	IPM3m	25	
PROF			15
	FCF2EV	60	
	EBITDA2EV	20	
	FCF2NOA	20	
EQ			5
	accrualsCF	60	
	CFg.left	20	
	EPSg.left	20	
MQ			10
	Tag	30	
	xfinBS	50	
	foreignASSETSg.left	10	
	foreignSALESg.left	10	
MS			15
	EPSDiff	70	
	DPSDiff	20	
	SHI	10	

events, we need to conduct sensitivity analysis such as how different utility companies respond to earthquakes, and then incorporate this into the portfolio both as a part of the investment process and the portfolio manager's quick decisions when similar events occur. On the other hand, for the management of foreign events, particularly US spillover effects, the same quantamental principles should be applied but with a different focus as Japan usually follows the US's actions closely, such as monetary and fiscal policy; therefore, these can be prepared early given the lag between US policy changes and Japan's response.

9.7.5 Quantamental Alpha: Model Efficacy

We build a linear alpha model using the weights specified in Fig. 9.13. The weights are decided based on comprehensive consideration of each theme's character, such as the forecasting power, information decay, relationship among themes, and mandates of a strategy. For a long-only stock selection strategy with a medium- to long-term investment horizon, we specify the weights as follows:

$$ALPHA = 35\%VALUE + 20\%PVM + 15\%MS + 15\%PROF + 10\%MQ + 5\%EQ.$$

Alpha Performance by Quintile We first investigate the model's efficacy using simple quintile performance. There are no risk model and no constraints involved. All performance metrics are based solely on the alphas. For comparison purposes, we also add the quantitative alpha: the alpha that was built without any fundamental components but with the same quantitative information as used in the quantamental

Table 9.13 Annualized active performance of quantamental alpha and quantitative alpha based on monthly data Jan 1998–May 2012. LS refers to Q5–Q1, the "Universe" column refers to equal-weighted returns in the investment universe

	Q1	Q2	Q3	Q4	Q5	LS	Universe
Quantitative Alpha							
Annualized ret (%)	−1.73	−0.36	−1.70	−0.78	4.51	6.24	1.81
Annualized std (%)	6.41	3.94	3.86	4.32	6.24	11.32	17.57
IR	−0.27	−0.09	−0.44	−0.18	0.72	0.55	0.10
Quantamental Alpha							
Annualized ret (%)	−8.11	−5.78	−1.35	5.12	9.98	18.09	1.81
Annualized std (%)	6.62	3.85	3.28	3.85	6.12	11.63	17.57
IR	−1.23	−1.50	−0.41	1.33	1.63	1.56	0.10

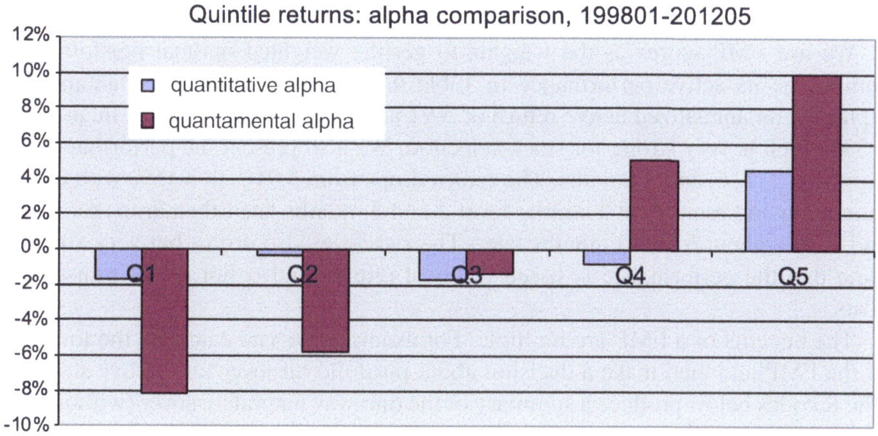

Fig. 9.14 Annualized active returns by quintile for both quantamental alphas and quantitative alphas, Jan 31, 1998 to May 31, 2012

alpha. We summarize the annualized performance across quintiles in Table 9.13, where Q5 and Q1 correspond to the highest and lowest alpha scores. The returns are in local currency (JPY) and equal weighted (demeaned across the universe), and the data is monthly from Jan 1998 to May 2012.

Note that the quintile return within each quintile is the average of returns for each period for stocks in that quintile minus the equal weighted universe return for the same period. Therefore, it is the active or excess return. The annualized standard deviation is for the active returns. The ratio is thus the information ratio. This ratio is meant to check active performance across quintiles, which is typical for a long-only stock selection portfolio. We also plot the returns in Fig. 9.14, where we see clearly that quantamental alpha outperforms quantitative alpha with both higher returns and lower risks. Moreover, the quantamental alphas have a much higher monotonicity than quantitative alphas, which will be reflected in a better performing portfolio, all else being equal.

Quantamental Alpha with Risk Added We now add the risk model to make the portfolio closer to the live one. We use a risk model for Japan and employ the mean–variance method to obtain an optimized portfolio.[5] We generate the factor mimic portfolio (FMP) based on the alpha, risk, and mean–variance optimization.

To see the monotonicity and forecasting power of risk-adjusted alpha, we use the principles of quintile analysis to obtain equal-weighted active returns within each quintile based on the risk-adjusted scores and plot the cumulative active returns for the five portfolios over time. The term "active" indicates the total return minus the equal-weighted universe return (benchmark). The finger chart is used to check the monotonicity of the risk-adjusted alpha over time. The results are presented in Fig. 9.15. Note that the values are displayed with log 10 and quintiles are labeled from zero to four. The finger chart shows very strong monotonicity of active returns associated with risk-adjusted alphas indicating that risk model scores do not distort the original efficacy of the model alpha.[6]

We use FMP scores as the weights to get the weighted optimal portfolio and summarize its active performance in Table 9.14. The risk-adjusted quantamental alpha has an annualized active return of 5.91 and risk of 2.54, resulting in an IR of 2.33, which is very strong for stock selection. We also present the performance for lags of 1, 2, 3, 6, and 9 months. The return drops from 5.91% to 3.15% with a one-month lag and remains at a similar level 2 and 3 months later, then drops to 2.58% and 2.26% about 6 and 9 months later. The risk level also drops, but very slightly. Note that the performance is based on local returns and is not net of transaction costs.

The benefits of a FMP are multiple. For example, we can calculate the turnover of the FMP and then make a decision about portfolio turnover for the live strategy. The R codes below produce a summary of the one-way natural turnover (without any constraints) of the risk-adjusted quantamental alphas. The average turnover is about 33%, meaning that 16% of the portfolio is sold and bought each month on average. In other words, we would have a completely new portfolio in about half a year. This is in line with the medium investment horizon we initially set as a target. We present more detailed information—time series data on monthly turnover—in Fig. 9.16.

Turnover: Risk-Adjusted Quantamental Alpha

```
> japan.alphaFMPturnover$OnewayFMPturnover
   Min. 1st Qu.  Median    Mean 3rd Qu.    Max.
  24.16   30.34   32.28   33.28   36.14   49.35
>
```

[5]We use GEM2S, an MSCI Barra risk model used widely by industry for international stock markets.

[6]The plot of cumulative quintile returns is informally called a finger chart in the industry.

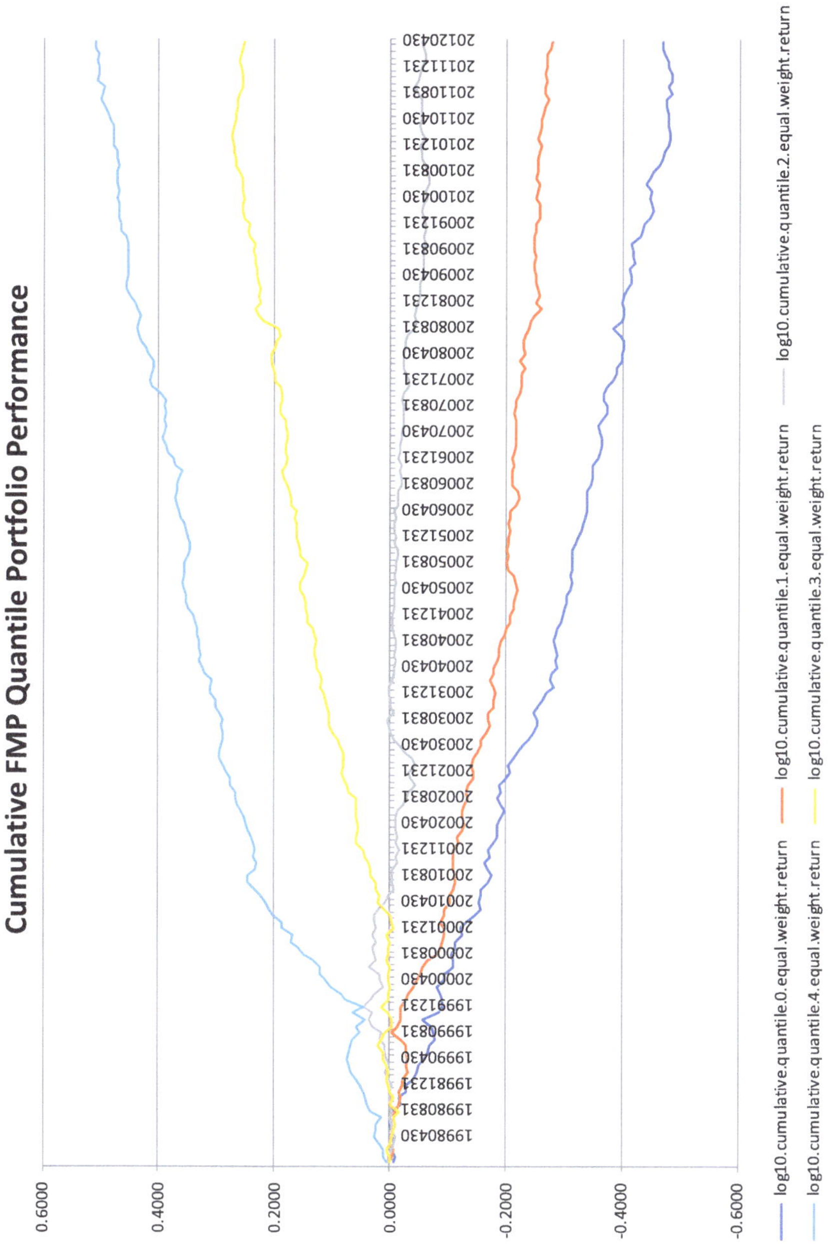

Fig. 9.15 Cumulative active returns by quintile for risk-adjusted quantamental alphas, Jan 31, 1998 to May 31, 2012

Table 9.14 Annualized active performance of FMP quantamental alpha based on monthly data, Jan 1998–May 2012. L is the lag in months

	FMP Portfolio	L=1	L=2	L=3	L=6	L=9
Annualized ret (%)	5.91	3.15	2.88	3.16	2.58	2.26
Annualized return (%)	2.54	2.36	2.30	2.12	2.10	2.12
IR	2.33	1.34	1.25	1.49	1.23	1.07

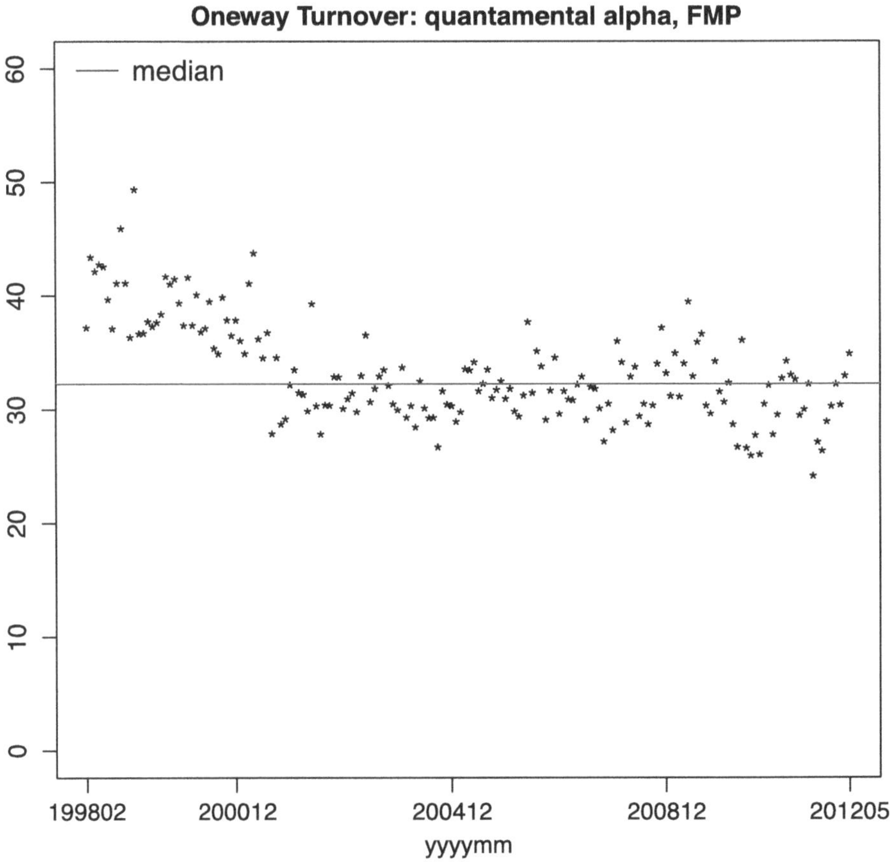

Fig. 9.16 The monthly one-way turnover of the optimized portfolio based on risk-adjusted quantamental alphas, Feb 1998 to May 31, 2012. The red line is the median

We can also explore the risk characteristics of the FMP portfolio and provide bases for the actual constraints, such as holdings of names, exposure to classical risk factors, and Japanese market special exposures such as earthquakes. We can also use the FMP portfolio to conduct sensitivity analysis for certain special events, such as a change of prime minister in Japan or a US monetary policy change.

9.7.6 Quantamental Portfolio Performance with Actual Constraints

We carry out a simulation study by using practical constraints and portfolio construction parameters. We conduct two types of backtests: one is for the total-return long-only portfolio, and the other is for the active-return long-only portfolio. Backtests are from Jan 1998 to May 2012 with monthly rebalancing at the end of each previous month. All returns are in USD, and the portfolio returns are after the transaction cost, which is assumed to be 1% for a two-way 100% turnover.

Note that to follow industry practice, the constraints are set at both a soft and a hard level. A soft constraint should not be violated during the optimization process if possible, but it can be violated if necessary. A hard constraint simply cannot be violated or the optimization will fail.

1. Total-return portfolio For the total-return strategy, we focus on maximizing total return for a given risk level, that is, the goal is to maximize the Sharpe ratio. We set parameters and constraints for portfolio construction as follows. Note that the constraints are at the total level. At the security level, the portfolio weight of each security will not exceed 2% in general and absolutely no more than 2.5%. At the industry level, total weights of all securities in an industry will not exceed 20% in general and absolutely no more than 22%. The minimum holding is 5bps, which is to limit small positions.

   Parameters: lambda = 500, two-way turnover = 10%

   Constraints: assets, soft/hard = 2/2.5%;

   industry, soft/hard = 20/22%;

   min.position = 5bps

2. Active-return portfolio For the active-return strategy, we focus on the active return for a given risk level, that is, the goal is to maximize the information ratio. We set the portfolio construction parameters and constraints as follows. The active-return strategy relies more on alpha, so we decrease the risk-aversion parameter from 500 to 100 and increase the turnover. Note that the constraints are at the active level. For example, the active overweight and underweight for a security relative to the benchmark is 2.5% (in general 2%), and the active weight for an industry is no more or less than 5% (4.5% in general) of the benchmark weight of that industry in the universe. The minimum position for a security in the portfolio is 5bps.

   Parameters: lambda = 100, two-way turnover = 20%

   Constraints: assets, soft/hard = +/- 2/2.5%;

   industry, soft/hard = +/- 4.50/5%;

		Quantitative Alpha		Quantamental Alpha	
Total Return Strategy	BM	portfolio	active	portfolio	active
annu return	1.22%	3.83%	2.61%	5.53%	4.31%
annu risk	18.98%	13.19%	13.58%	13.42%	13.27%
ratio	0.06	0.29	0.19	0.41	0.33
Active ReturnStrategy	BM	portfolio	active	portfolio	active
annu return	1.22%	3.94%	2.72%	7.91%	6.69%
annu risk	18.98%	18.48%	5.48%	20.85%	7.29%
ratio	0.06	0.21	0.50	0.38	0.92

Fig. 9.17 The annualized performance of backtest portfolios with quantamental and quantitative alphas, February 1998 to May 31, 2012

$$\text{min.position} = 5\text{bps}$$

We present the simulation results for the portfolio based on quantamental alphas in Fig. 9.17 for both total-return and active-return strategies. For comparison purposes, we also have the portfolio performance derived from quantitative alphas. We see that though both quantitative and quantamental portfolios outperform the benchmark, the quantamental portfolio performs far better than the quantitative portfolio, and the latter could barely work as a real strategy given its weak performance relative to the index. In contrast, the quantamental portfolio has strong performance: the total-return portfolio has an annualized return of 5.53% and risk of 13.42%, compared with 1.22% and 18.92% of the benchmark; the active-return portfolio has an active return of 6.69% and active risk of 7.29%, resulting in a strong information ratio of 0.92.

Thus, we have shown that the quantamental approach indeed adds significant value to portfolio performance. The value added arises from the combination of both depth and breadth, the respective strengths of fundamental and quantitative approaches.

9.8 Surveying with R

In this section, we discuss how to design a survey to collect useful information, process survey information, and employ it in quantamental strategies. The purpose of surveying is to get useful information from credible sources. This information will complement historical data from public sources or vendors. Such information can be hard to evaluate with an individual quantitative signal, for instance, accounting principles and asset quality. Such information can also be hard to quantify, such as opinions about potential events, e.g., a dividend increase or capital raise. The information collected through a survey will become proprietary data for investors.

We specify below some basic rules we need to adhere to when conducting a survey and using survey information for quantamental analysis.

- Sources of survey
  - credible
  - deep understanding of the companies and industry
  - sustainable relationship
  - multiple sources
  - regular surveying
- Questions in the survey
  - In compliance with laws and policies
  - Not captured by public reports
  - Prospective events
  - Replies are simple and score/category based
- Information processing and usage
  - storage in database
  - making signals
  - alpha or risk component
  - portfolio construction parameters or constraints
  - live portfolio management

9.8.1 Designing a Survey to Collect Quantamental Information

As we have mentioned in the previous chapters, a classical factor or alpha model usually has a correlation with forward returns of less than 10% and an R^2 of the OLS model of less than 5%. This implies that 90% of information about returns is not captured by traditional quant factors and alpha models. This is so for many reasons. For example, the model is not based on prospective information, or the historical information is already reflected in the prices; or sometimes, it is very challenging to quantify the information, this is particularly true for events.

On the other hand, for fundamental portfolio managers, much information is collected randomly and not stored in a database or tested systematically. Rather, information is used based on personal experience and judgment.

In quantamental analysis, we can combine fundamental and quant approaches to complement each other. The survey approach is an example of a quantamental approach in the sense that prospective information is collected regularly, processed by systematic statistical analysis, integrated into a robust investment process, and eventually reflected in the portfolio and hence performance.

- Complement quant
  - forward-looking and hard to quant information
- Complement fundamental

Category	Survey Question	Answer Sheet				
MQ	Please rank the management quality of each company. 1 being low quality and 5 being high quality.	1	2	3	4	5
AQ	Please rank the overall mine quality of each company. 1 being low quality and 5 being high quality.	1	2	3	4	5
DIV	What, if any, is the expected material dividend change?	No change	decrease	increase		
CAP	Will the company need to raise capital	No	Yes	Not sure		
	Would it be difficult to raise capital, if necessary?	1	2	3	4	5
EVENT	Are there any major event risks in the next 3-6 months?	No	Yes	Not sure		
	Would these positively or negatively impact the stock price?	Positive	Negative	Not sure		
MA	Will there be M&A in 6 months?	No	Yes	Not sure		
	If Yes, this company is a: Take out target	No	Yes			
	If Yes, this company is a: Acquirer	No	Yes			
	If Acquirer, which targets make cost sense?					
COST	Cost guidance for the following fiscal year 1 is lower 5 is higher	1	2	3	4	5
PROD	Production guidance for the following fiscal year 1 is lower and 5 is higher	1	2	3	4	5
GEORISK	Geopolitical risk for the mines the companies owned 1 is low risk and 5 is high risk	1	2	3	4	5
	Please specify the major risks, if any.					

Fig. 9.18 Sample survey questions for gold mining companies

– a systematic method of information collection, sourcing, storage, and usage

We present a general framework for survey questions below. Subjects include quality of management and assets, products, operation, capital structure, M&A, and events. Note that some questions are only relevant to specific industries.

- Management quality
- Asset quality
- Operation
- Products
- Capital
- Events
- M&A

We illustrate this process using an example survey for gold mining companies. For gold mining companies, asset or mine quality is very important because it determines cost and risk. However, mine quality is hard to quantify using only a few metrics, such as grade or cash cost, since mine quality is also related to location, regime, political stability, safety, etc. Another factor is that there are many M&A activities in the mining industry, which have large impacts on stock prices. We present a sample questionnaire for gold mining companies in Fig. 9.18.

9.8.2 Using R to Process Information

Once a survey is designed, R can be used to help collect, store, and process information. We summarize the procedure in the list below and then briefly discuss each item.

1. Prepare survey questions and files.
2. Conduct survey and collect information.
3. Store and access information.
4. Process survey results and incorporate them into the investment process.
5. Evaluate survey sources.

1. *Prepare survey questions and files* We first need to create IDs. There are two sets of IDs: one is the company and the other is the survey source that covers that company. There may be multiple sources covering the same company, so each survey source should have a unique ID. The survey source ID should be affiliated with many features of that source, such as language, location, currency, etc. Each survey can be identified by the survey source ID and the stamp date, the date when the survey information is received. These ID files can be created and maintained using R scripts.

Survey ids

```
company,  survey source, survey question,
Boeing,     ABC,           Asset Quality,

survey answer, surveying date, filling date;
4,               20191120,     20191123
```

2. *Conduct survey and collect information* A survey should be conducted in a user-friendly way for the survey sources. An efficient way to conduct a survey is a web-based survey. The survey source can open the link, answer the survey questions whenever it is convenient, save the results, and submit them.[7] Also, the survey source should be allowed to change answers and resubmit their answers, but the changes should be recorded with time stamps.
   Survey tools can be connected with R so that the web-based survey will be transformed into data collected by R.
3. *Store and access information* Web-based survey information can be collected and combined by R and then added to a specific database. The company, survey source, and date stamp comprise a unique identification for a record. R has

[7]For example, LimeSurvey is widely used in the investment industry.

powerful packages to work with databases. Please refer to Chapter 6 for details. It should be noted that once the original data is recorded and stored in a database, it should be read-only so that no one can modify the original survey information.

4. *Process survey results* The survey information stored in the database should be easily accessible. R can be used to retrieve and process the original data. For quantamental analysis, the survey questions are designed specifically for some purpose. They can be used to build a new signal or strengthen existing signals; as part of an investment process as risk factors; as parameters for portfolio construction; or as part of the considerations for live portfolio management. For example, if we believe in the survey information about a company being the acquired target in an M&A deal, we need to incorporate such information during the portfolio construction process, say, by having a list of such names with recommended actions, like the one below.

Long only:	no underweighting for the names on the list
Long short:	no short positions for the names on the list

We should pay special attention to these names during live management.

5. *Evaluate survey sources* For quality control, the survey sources should be evaluated periodically for their timeliness and accuracy. This information should then inform the usage of collected information. For example, if we know that survey source ABC has more credible information on dividend changes for companies in industry XYZ than other survey sources, and if we formulate a signal for dividend changes, source ABC should be given more weight:

$$DIVchg = \sum_{i=1}^{K} w_i DIVchg_i,$$

where w_k for survey source ABC should be higher than other survey sources.

To summarize, surveying is an effective way to gather information for quantamental investing. Of course, there are many other approaches that may enhance the performance of quantamental portfolios.

It should be addressed here that quantamental investing only emerged in the early 2000s, and it still has a long way to go. There are potentially many avenues for improvement. Many practitioners have accumulated excellent experience in this field. We would like to conclude this chapter and also the book with a Chinese saying, "throw out a brick to attract jade," based on a story that took place in China during the 900s.

> In the Tang dynasty, there was a famous poet named Zhao Gu, who was admired by many local scholars including Chang Jian. One day, Chang heard that Zhao was going to visit a local temple, LingYinSi. Early in the morning that day, Chang arrived at the temple and wrote two sentences, the first half of a poem, on the wall, and then hid in an office. A few hours later, Zhao came to visit the temple. Impressed by the two sentences on the wall, Zhao wrote the other two sentences to complete the poem and then went home. Chang came out

> after Zhao left, and to no surprise, he happily read the completed poem as the second half was much better than his own first half.—story of Pao Zhuan Yin Yu, from *Jing De Chuan Deng Lu*, vol.10, written in 1004–1007 by Shi Daoyuan.

Similarly, I hope the introductory contents of this book can be helpful and serve as a foundation for further thoughts and applications in quantitative investing.

Keywords, Problems, and Group Project

Part I: Keywords

Quantamental, depth, breadth, prospective information, data proven
Survey, stagnation, aging population, Japanese stock selection strategy
R scripts for survey

Part II: Problems

Problem 9.1 Discuss challenges of quantamental investments.

Problem 9.2 Nonlinear effects are closer to the real world than linear, but it is easy to fall into the trap of data mining. How can a quantamental approach help to address this issue?

Part III: Group Project

Problem 9.3 Recall that in Problem 2.4, you did a project to help on a family asset. Now, redo the project with a quantamental approach and compare with your previous report.

References

Cabinet Decision: Rebirth of Japan. 2012. A Report by National Policy Unit of Japan Cabinet, July 31, 2012.

Statistical Handbook of Japan. 2018. A Report by Statistics Bureau, Ministry of Internal Affairs and Communications, Japan, ISSN 0081–4792, Fall, 2018.

Index

L. Ma, *Quantitative Investing*, https://doi.org/10.1007/978-3-030-47202-3

The manufacturer's authorised representative in the EU is Springer Nature Customer Service Centre GmbH, Europaplatz 3, 69115 Heidelberg, Germany. If you have any concerns regarding our products, please contact ProductSafety@springernature.com

Printed and bound by CPI Group (UK) Ltd, Croydon, CR0 4YY
05/10/2026
02228124-0018